HANDBOOK OF
SIMPLIFIED
TELEVISION SERVICE

HANDBOOK OF

SIMPLIFIED

TELEVISION SERVICE

JOHN D. LENK

Consulting Technical Writer

PRENTICE-HALL, INC., *Englewood Cliffs, New Jersey 07632*

Library of Congress Cataloging in Publication Data

LENK, JOHN D
  Handbook of simplified television service.

  Includes index.
  1. Television—Repairing. I. Title.
TK6642.L456   621.388'87   76-43046
ISBN 0-13-381780-6

This Reward paperback edition
published 1979 by Prentice-Hall, Inc.

Printed in the United States of America

10 9 8 7 6 5 4 3

PRENTICE-HALL INTERNATIONAL, INC., *London*
PRENTICE-HALL OF AUSTRALIA PTY. LIMITED, *Sydney*
PRENTICE-HALL OF CANADA, LTD., *Toronto*
PRENTICE-HALL OF INDIA PRIVATE LIMITED, *New Delhi*
PRENTICE-HALL OF JAPAN, INC., *Tokyo*
PRENTICE-HALL OF SOUTHEAST ASIA PTE. LTD., *Singapore*
WHITEHALL BOOKS LIMITED, *Wellington, New Zealand*

To Irene, Karen, Mark, Brandon, and Lambie

Contents

Preface

The purpose of this book is to provide a simplified system of television service for many types and models of television receivers in use. Originally, all receivers used vacuum-tube circuits, and provided only black and white pictures. Later, color receivers were added, followed by partial solid-state receivers (hybrid receivers), and all-solid-state receivers. Today, many television receivers include a number of auxiliary devices and features, such as remote control, automatic color control (ACC), automatic fine tuning (AFT), automatic frequency and phase control (AFPC), plug-in modules for simplified service, and many others.

It is virtually impossible, in one book, to cover service for all such receivers and their related circuits. Likewise, it is impractical to attempt such comprehensive coverage, although such attempts have been made, since rapid technological advances soon made those books obsolete.

To overcome this problem, we shall concentrate on a *basic approach to television service;* an approach that can be applied to any receiver, those now being manufactured, those to be manufactured in the future, and receivers now in use. Keep in mind that older black and white vacuum-tube receivers, even though no longer being manufactured, still require service (perhaps more so than the latest models).

The service approach described here is based on the techniques found in the author's highly successful troubleshooting works: *HANDBOOK OF PRACTICAL SOLID STATE TROUBLESHOOTING and HANDBOOK OF BASIC ELECTRONIC TROUBLESHOOTING.*

As is described in Chapter 1 of this book, the author's techniques follow a *basic troubleshooting sequence,* which includes failure *symptom analysis, localizing* trouble to a module in the receiver (plug-in, circuit board, or on the chassis), *isolating* trouble to a circuit within the module, and *locating* the specific trouble within the circuit.

Chapter 2 of the book is devoted to test equipment used in television service. This is particularly important since television receiver faults are often best located by analyzing test results (oscilloscope waveforms, voltage and resistance measurements, response to input signals, etc.). Chapter 2 covers black and white, as well as color test equipment, and includes basic operating principles and characteristics.

Chapter 3 provides basic service procedures for a black and white television receiver, and for some color circuits, using the test equipment described in Chapter 2. Emphasis is placed on "universal" service procedures that apply to all receivers now in existence, and to those that may be found in the future.

Chapter 4 provides service for color television receivers, using the test equipment described in Chapter 2. This includes color set-up procedures such as purity, convergence, and linearity adjustments, as well as testing color sync, chroma demodulator, and matrix circuit characteristics with oscilloscopes, generators, and vectorscopes. Again, the procedures are "universal" in nature so that they apply to the widest number of television receivers and circuits.

Chapter 5 describes service hints and tips. These include how to get the most out of service literature, what to expect in the way of service literature, how to test antenna systems, and how to analyze television broadcast test patterns.

Chapter 6 is devoted entirely to solid-state television, and is concerned with practical troubleshooting techniques. A consistent format is used throughout the chapter. First, typical solid-state television circuit diagrams are given, along with the basic theory of operation. Next, the recommended troubleshooting approach for that particular type of circuit is discussed. Then a group of typical troubles is described, along with the most likely causes of such trouble.

The author has received much help from many organizations and individuals prominent in the field of television service. He wishes to thank them all, particularly B&K-Precision, Dynascan Corporation and RCA Commercial Engineering. The author also wishes to express his appreciation to Mr. Joseph A. Labok of Los Angeles Valley College for his help and encouragement.

JOHN D. LENK

HANDBOOK OF
SIMPLIFIED
TELEVISION SERVICE

1

Introduction To Television Service

This chapter is devoted to the basics of TV service. On the assumption that you may not know how television works or that you need a refresher, we shall start with descriptions of TV broadcast and reception for black and white as well as color programs. These descriptions are kept to the block-diagram level since you are interested in service (not theory) and so that the descriptions will apply to the greatest number of TV sets (new and old, vacuum-tube, solid-state, etc.). It is not intended that this introduction provide a complete course in TV broadcast and reception, but a summary.

With TV basics established, we shall describe the author's universal troubleshooting approach in detail. This is followed by discussions of how the universal troubleshooting approach can be applied to the specifics of TV service. Throughout these discussions, references are made to the remaining chapters where you will find detailed information on troubleshooting and service procedures.

1-1. BASIC BLACK AND WHITE TELEVISION BROADCAST SYSTEM

As shown in Fig. 1-1, the basic TB broadcast system consists of a transmitter capable of producing both AM and FM signals, a television camera, and a microphone. TV program sound (or audio) is transmitted through the microphone, which modulates the FM portion of the transmitter. The picture (or video) portion of the program is broadcast by means of the AM transmitter.

1

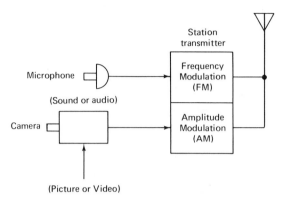

Fig. 1-1. Basic television broadcast system

The TV broadcast channels are approximately 6 MHz wide, with the sound (FM) carrier at a frequency 4.5 MHz higher than the picture (AM) carrier. For example, on VHF channel 8, the picture (AM) is transmitted at a frequency of 181.25 MHz, whereas the sound (FM) is transmitted at 185.75 MHz. Channel 8 occupies the band of frequencies between 180 and 186 MHz. A listing of all VHF and UHF channel frequencies is given in Chapter 5.

The sound portion of the TV signal uses conventional FM broadcast principles, which need not be described here. However, the picture portion of the signal is unique to TV in that both picture information and synchronizing pulses must be transmitted on the AM carrier.

From a troubleshooting standpoint, operation of the TV picture channel can be understood by reference to Fig. 1-2, which shows the relationship between the picture tube in the TV receiver and the station camera picture tube. Both tubes have an electron beam, which is emitted by the tube cathode and strikes the tube surface. Both tubes have horizontal and vertical sweep systems that deflect the beam so as to produce a rectangular screen (or "raster") on the tube surface. The vertical sweep is at a rate of 60 Hz, whereas the horizontal sweep is at 15,750 Hz.

The electron beam of the station camera picture tube is modulated by the amount of light that strikes the camera tube screen. In turn, the transmitted AM signal is modulated by the electron beam. Thus, the amplitude of the transmitted AM signal is determined by the intensity of the light at any given instant. The position of the electron beam at a given instant is set by the horizontal and vertical sweep circuits, which, in turn, are triggered by pulses in the camera. These same pulses are transmitted on the AM carrier and act as synchronizing pulses ("sync pulses") for the TV receiver horizontal and vertical sweep circuits.

The electron beam of the TV receiver is modulated by the transmitted

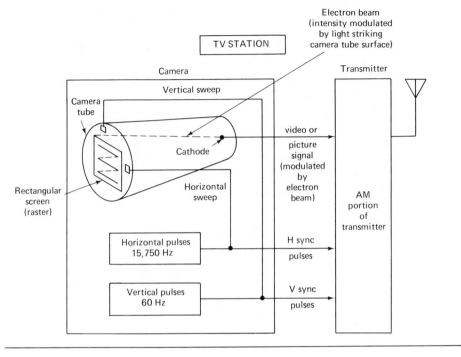

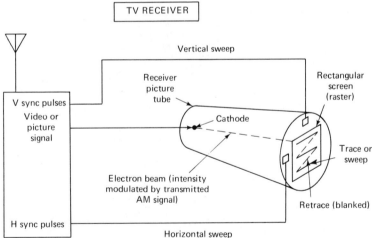

Fig. 1-2. Relationship between camera and receiver picture tubes

AM signals so as to "paint" a picture on the tube screen. The amplitude of the AM signal determines the intensity of the light produced on the TV screen at any instant. For example, if the camera sees an increase in light, the electron beams in both tubes (camera and receiver) are increased, and the TV receiver tube shows an increase in light. The TV receiver tube horizontal and vertical deflection systems are triggered by the horizontal and vertical synchronizing pulses transmitted on the AM portion of the TV broadcast signal. Thus, the electron beam of the TV receiver follows the beam in the camera tube. Both beams are at the same corresponding spot, at the same instant.

Assume that the camera is focused on a white card with a black numeral 3 at the center, as shown in Fig. 1-3. As the camera tube electron beam is swept across the surface, light is reflected from the card onto the camera tube surface. When the beam passes across the white card background portion of the reflected light, the beam intensity is maximum. Beam

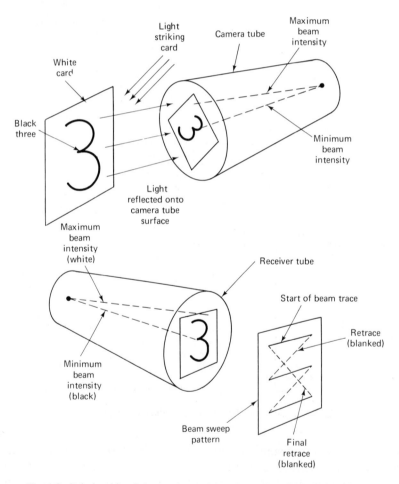

Fig. 1-3. Relationship of electron beams in camera and receiver picture tubes

intensity drops to minimum when the beam is at any portion of the light reflected from the numeral 3. This varying electron beam modulates the transmitted AM carrier.

The electron beam starts at the top of the camera tube screen, sweeps across to one side, and is blanked while it returns to the other side (retraces), until the beam finally reaches the bottom of the screen. The beam is then blanked and returned to the top of the screen. Thus, if the camera is focused on the numeral 3 pattern, the camera translates the entire pattern, line by line, into a picture signal (a voltage that varies in amplitude with intensity of the reflected light).

At the TV receiver, the picture tube electron beam follows the camera beam in both position and intensity. In the example of Fig. 1-3, both electron beams increase when they are swept across any of the white background. Likewise, both beams decrease when they are swept across any portion of the numeral 3. Thus, the TV receiver picture tube reproduces the numeral 3 in black and white (black, or minimum beam intensity on the numeral portion of the sweep; white, or maximum intensity on the card background portion).

1-2. BASIC BLACK AND WHITE RECEIVER CIRCUITS

Figure 1-4 is the block diagram of a black and white TV receiver. The diagram shown is a *composite of several types of TV receivers* and is presented as a point of reference for troubleshooting. In the following paragraphs we shall describe the basic principles of all TV receivers. The *detailed circuit descriptions* for a similar receiver are provided in Chapter 6.

1-2.1   Low-voltage Power Supply

The basic function of the low-voltage power supply is to provide direct current to all circuits of the receiver, except the high voltages required by the picture tube. The high voltages are supplied by the *flyback circuit* (more properly known as the horizontal output and high-voltage circuit), as described in Sec. 1-2.2. The low-voltage power supply consists essentially of a rectifier (typically a full-wave bridge), followed by a regulator (typically a Zener diode and transistor regulator for a solid-state receiver).

With vacuum-tube receivers, the outputs from the low-voltage power supply are typically in the 200- to 300-V range. The low-voltage power supplies in most solid-state sets produce about 10 to 12 V, although some provide 25 to 30 V. In many solid-state TV receivers, the set can be operated with self-contained rechargeable batteries; thus, the low-voltage power supply also provides for recharging the batteries.

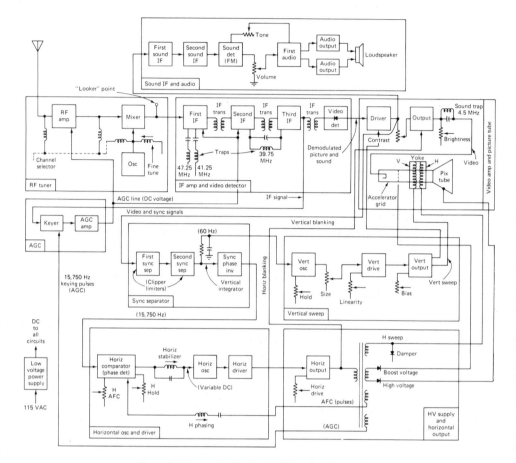

Fig. 1-4. Composite black and white television receiver

1-2.2   High-voltage Supply and Horizontal Output

Although these circuits have more than one function in all TV receivers, they do not always have the same functions in all sets. The main functions of the circuit are (1) to provide a high voltage for the picture tube, and (2) to provide a horizontal deflection voltage (horizontal sweep) to the picture tube deflection yoke.

In most receivers, the circuits also supply a boost voltage for the picture tube focus and accelerating grids. In solid-state sets, the circuits may also supply a voltage for the video output transistor (Sec. 1-2.8) since the output transistor usually requires about 40 to 70 V (instead of the 12 V available from the low-voltage power supply). Often, the circuits also supply an AFC (automatic frequency control) signal to control the frequency of the horizontal oscillator (Sec. 1-2.3) and an AGC (automatic gain control) signal to AGC circuits (Sec. 1-2.9).

The circuits generally have only one input, no matter what combination of outputs are provided. The circuit receives pulses from the horizontal driver (Sec. 1-2.3). These pulses are at a frequency of 15,750 Hz and are synchronized with the TV broadcast picture transmission.

It is important that you note and remember inputs and outputs in any TV set being serviced. The troubleshooting technique discussed throughout this book is based on a comparison or study of inputs and outputs at each circuit. For example, in the simplest of terms, assume that the input to the high-voltage supply and horizontal output circuit is normal (15,750) Hz synchronizing pulses of correct amplitude) but that one or more of the outputs (high voltage, horizontal sweep, boost voltage, AGC, AFC, etc.) is absent or abnormal. This pinpoints trouble to some part of the high-voltage supply and horizontal output circuit.

The circuits generally have only one adjustment control, the *horizontal drive* control. This control sets the amplitude of the horizontal sweep. In some receivers, the horizontal drive also sets the amount of high voltage available to the picture tube. The amount of high voltage is determined by the size and type of picture tube. Typically, a 23-in. picture tube requires between 25 and 30 kV.

It is also important that you note and remember adjustment controls for all circuits of the TV set. Proper adjustment of controls is a vital part of any troubleshooting technique.

1-2.3   Horizontal Oscillator and Driver

The horizontal oscillator and driver circuits provide the drive signal to the horizontal output and high-voltage supply (Sec. 1-2.2). These signals are at a frequency of 15,750 Hz and are synchronized with the picture transmission by means of sync pulses from the sync separator (Sec. 1-2.5). Thus, the horizontal oscillator and driver circuits have one imput and one output.

In addition to the basic input and output, most horizontal oscillator and driver circuits have a feedback signal that is taken from the output and fed back to the input. The purpose of the feedback signal is to provide an AFC system for the horizontal sweep circuits. (Such AFC systems are found on many vacuum-tube receivers and on almost all solid-state TV receivers.)

The AFC system is used to ensure that the horizontal sweep signals are synchronized (both for frequency and phase) with picutre transmission, despite changes in line voltage and temperature or minor variations in circuit values. The AFC action is accomplished by comparing the sync pulses with the horizontal sweep signals (for frequency and phase) in the phase detector portion of the circuits. Deviations of the horizontal sweep signals from the sync pulses cause the horizontal oscillator to shift in frequency or phase as necessary to offset the initial (undesired) deviation.

For example, if the horizontal sweep decreases in phase from the sync pulses, the horizontal oscillator is increased in phase by action of the phase detector.

Although there are many types of AFC circuits used for control of the horizontal oscillator (typically a blocking oscillator), in no case is the oscillator triggered directly by the sync pulses. Instead, the sync pulses and comparison pulses (feedback from the horizontal sweep) produce a variable dc control voltage that is applied to the horizontal oscillator. Any change in this control voltage shifts the horizontal oscillator frequency and phase.

In addition to automatic changes in the control voltage, the voltage can be manually set by the horizontal adjustment controls. These adjustment controls are not standard on all receivers. As a minimum, there will be a *horizontal hold* control, which permits manual adjustment of the control voltage applied to the horizontal oscillator. Some sets include an AFC *control,* which also sets the oscillator control voltage. When both controls are used, the AFC is generally a back-panel adjustment (not readily accessible to the user), whereas the horizontal hold is a readily accessible control for the user.

The circuits may also include a *horizontal stabilizer* control (sometimes called the *horizontal frequency* control), which is a parallel-tuned circuit between the phase detector and horizontal oscillator. This circuit sets phasing of the oscillator. Another adjustment found on some receivers is the *horizontal phase* (or phasing) control, which is a series-tuned circuit in the feedback line between the output and the phase detector. This phasing circuit is series-tuned to adjust the phase of the comparison pulses (from the output) to the phase detector. Note that the horizontal stabilizer and phase controls *do not normally adjustment,* even during troubleshooting, unless parts have been replaced.

1-2.4   Vertical Sweep Circuits

The vertical sweep circuits provide a vertical deflection voltage (vertical sweep) to the picture tube deflection yoke. The vertical circuits also supply a blanking pulse to the picture tube [usually through the video amplifier (Sec. 1-2.8)]. This pulse blanks the picture tube during retrace of the vertical sweep. The vertical signals are at a frequency of 60 Hz and are synchronized with the picture transmission by means of vertical sync pulses from the sync separator (Sec. 1-2.5). Thus, the vertical sweep circuits have one input and two outputs.

Figure 1-4 shows that the vertical sweep circuits include the vertical oscillator, driver, and output. (In most receivers, the vertical deflection yoke is also considered as part of the vertical sweep circuits, just as the horizontal yoke is usually a part of the horizontal output circuits.) In some solid-state receivers, the functions of vertical driver and vertical output are combined into one transistor.

The number and type of controls are not standard for the vertical sweep circuits of all receivers. However, as a minimum, there is a *vertical hold,* a *vertical size,* and a *vertical linearity* control. On many receivers, there is also a *vertical bias* control.

The vertical hold controls set the frequency of the vertical oscillator (60 Hz) and thus synchronizes the oscillator (generally a blocking oscillator) with the input sync pulses. The vertical size control is essentially a gain control used to set the amplitude of the vertical oscillator signal (a sawtooth sweep) applied to the driver and output. The vertical linearity control determines the linearity of the sawtooth sweep (or determines the portion of the sweep used) applied to the output and vertical deflection yoke. If the sawtooth sweep is not linear (or a nonlinear portion of the sweep is used), the picture will not be linear.

The vertical bias control (when used) sets bias on the vertical output transistor. Thus, adjustment of the bias control can affect both vertical size and linearity to some extent. However, the vertical bias control does not normally require adjustment, even during troubleshooting, unless the vertical output transistor (or its related parts) have been replaced. The primary function of the vertical bias control is to compensate for variations in replacement vertical output transistors.

1-2.5   Sync Separator

The sync separator circuits function to remove the vertical and horizontal sync pulses from the video circuits (Sec. 1-2.6) and apply the pulses to the vertical (Sec. 1-2.4) and horizontal (Sec. 1-2.3) sweep circuits, respectively. Thus, the sync separator circuits have one input and two outputs. The sync separator circuits also function as clippers and/or limiters to remove the video (picture) signal and any noise. Thus, the horizontal and vertical sweep circuits receive sync pulses only and are free of noise and video signal (in a properly functioning receiver).

Sync separator circuits are not standardized, particularly in solid-state TV. That is, there are several methods used to obtain the desired clipping and limiting action. No matter what system is used, the vertical sync pulses are applied through a capacitor/resistor low-pass filter (often called the *vertical integrator*) to the input of the vertical sweep circuit. The horizontal sync pulses are applied to the AFC section (phase detector) of the horizontal sweep circuits. Here, the incoming sync pulses are compared with the horizontal sweep as to frequency and phase (Sec. 1-2.3).

Generally, there are no adjustment controls in the sync separator circuits. An exception in some circuits is a control that sets the bias on one stage. This control makes it possible to set the level of clipping/limiting action.

1-2.6   RF Tuner

The functions of an RF tuner in a TV receiver are essentially the same as the RF sections in other AM and FM receivers. The typical tuner in a TV receiver consists of an RF amplifier, mixer, and oscillator. Each TV channel consists of two carriers: the FM sound carrier and the AM picture carrier. Both carriers are amplified by the RF amplifier. Both carriers are mixed with signals from the oscillator to produce two intermediate-frequency (IF) signals in the mixer. These two signals are amplified by the IF amplifiers (Sec. 1-2.7). Thus, the RF tuner has one input and one output, even though there are two carriers.

The RF tuner also receives automatic gain control (AGC) signals from the AGC circuits (Sec. 1-2.9). These AGC signals are generally applied only to the RF amplifier (not the mixer and oscillator). The AGC signals function to control gain of the RF amplifier in the presence of carrier signal variations. For example, if the carrier signal strength increases, the AGC signals decrease the RF amplifier gain and thus offset the initial increase in carrier strength.

The VHF portion of the RF tuner has provisions for tuning each channel. Usually, tuning is accomplished by slug-tuned coils, with one set of coils for each channel. Typically, there will be three coils for each channel (one for RF amplifier, one for mixer, and one for oscillator). However, the coil arrangement varies with the type of RF tuner circuit. There is usually a *fine tune* control on most RF tuners. The fine tune is a front-panel (user) adjustment control. Typically, the fine tune control is a slug-tuned coil in parallel with the oscillator coil. Some VHF tuners will also have other adjustments, such as traps and filters at the input and output. However, there is no standardization of such adjustments.

In most sets, the UHF tuner is physically separate from the VHF tuner. (In pre-1962 receivers, you may not find any UHF tuner.) The functions of the UHF tuner are essentially the same as for the VHF tuner, although there are circuit differences. For example, most UHF tuners do not have RF amplifiers, and the mixer uses a diode (rather than a transistor or vacuum tube). The UHF circuits are tuned by means of resonant cavities, rather than tuning coils and capacitors. AGC is generally not used in UHF tuners. Also, most UHF tuners have continuous tuning, similar to an AM broadcast radio, rather than fixed, channel-by-channel tuning (in steps).

There are two basic types of channel selection for VHF tuners (also used in a very few modern UHF tuners of the non-continuous-tune type). Most RF tuners are of the turret type, where a separate set of drum-mounted coils is used for each channel, and the entire drum rotates when the channel is selected. A few tuners are of the switch type, which have series-connected coils mounted on wafer switches.

1-2.7   IF and Video Detector

The IF and video detector circuits amplify both picture and sound signals from the RF tuner mixer, demodulate both signals for application to the video amplifier (Sec. 1-2.8) and sound IF amplifier (Sec. 1-2.10), provide signals to the AGC circuits (Sec. 1-2.9), and trap (or reject) signals from adjacent channels.

In most receivers, both picture and sound outputs from the video detector are applied to the video amplifier circuits where both signals are amplified by at least one stage of the video amplifier. At this point, the sound signals are applied to the input of the sound IF amplifier, and the picture signals are applied to the remaining stages in the video amplifier (and to the sync separator). In a few receivers, the sound signals from the video detector are applied directly to the sound IF amplifiers.

In addition to the signal input and output, most stages of the IF amplifier receive an AGC signal from the AGC circuits. These signals, which are in the form of a varying dc bias voltage, are the same as those applied to the RF tuner (Sec. 1-2.6).

The IF and video detector circuits do not usually contain any operating or adjustment controls, as such. However, the circuits do contain a number of IF transformers which must be aligned to provide the proper IF bandwidth. Generally, there is one transformer at the input and output of the IF amplifier as well as one transformer between each IF stage. Thus, in a three-stage IF amplifier there will be four transformers. The transformers are usually slug-tuned, although some may be tuned by means of variable capacitors.

As is the case with most FM receivers (and some wide-band AM receivers), the IF stages are stagger-tuned. That is, each IF transformer is tuned to a different frequency (one tuned to the high end, one to the low end, and the remainder tuned to various points in between). This gives the IF amplifiers suficient bandwidth to pass both the picture and sound signals.

In addition to the IF transformers, there are a number of traps in most IF amplifier circuits. These traps are tuned circuits (usually a fixed capacitor and slug-tuned coil) used to reject and remove signals from adjacent TV channels. Traps are necessary since TV signals may be broadcast simultaneously on adjacent channels, and the combination of wide-band broadcast channels together with wide-band IF circuits may result in overlap of signals.

1-2.8   Video Amplifier and Picture Tube Circuits

The video amplifier circuits have several functions, and there are many circuit configurations in use. In a "typical" receiver, the video amplifier circuit has three inputs and three outputs.

The primary input is the demodulated picture and sound signal from the video detector output (Sec. 1-2.7). Secondary inputs are blanking pulses from the vertical and horizontal sweep circuits. As discussed, some solid-state receivers do not have horizontal blanking pulses. Instead, the brilliance of the picture tube is kept low enough so that the electron beam horizontal retrace cannot be seen. In the "typical" circuit of Fig. 1-4, the three inputs (picture information, vertical blanking pulses, and horizontal blanking pulses) are applied to the picture tube (usually at the cathode). These outputs control the picture tube electron beam intensity.

The pulses blank the picture tube (cut off the electron beam) during the retrace period of the horizontal and vertical sweeps. During the trace periods of both sweeps, the picture signal varies the intensity of the electron beam so as to "paint" a picture on the picture tube screen. As discussed, the electron beam of the picture tube "follows" the broadcast camera tube beam, in both intensity and position, so as to reproduce the picture focused in the camera lens.

The sound signal is prevented from being applied to the picture tube by means of a trap between the video amplifier output stage and the picture tube input. (In most receivers, the picture tube input is at the cathode. In a very few receivers, the picture tube control grid is used as the input.) The trap (often called the sound trap or the 4.5-MHz trap) is usually a parallel fixed capacitor and slug-tuned coil, tuned to 4.5 MHz.

The sound signal is applied to the input of the sound IF and audio circuits (Sec. 1-2.10) after amplification by the video amplifier first stage (often called the driver or video driver). The picture and sync pulse information does not pass to the audio circuits (in a properly functioning receiver).

The picture signal and sync pulses are applied to the input of the sync separator circuits (Sec. 1-2.5) after amplification by the video driver. The sound signal is also present at this point. However, since the sound information is FM and is at a frequency far removed from either the horizontal or vertical sweep frequencies (4.5 MHz compared to 60 Hz and 15,750 Hz), the sound information does not affect the sync pulses. As discussed in Sec. 1-2.5, the picture signal is removed in the sync separator by clipping and limiting action. Thus, only the sync pulses are applied to the sweep circuits.

In virtually all TV receivers, the video amplifier and picture tube circuits have at least two controls. These are the *contrast* control and the *brilliance* (or *brightness*) control. There are many configurations for these two controls. Generally, the brilliance or brightness control determines the amount of fixed bias applied to the picture tube cathode. This sets the intensity of the electron beam (with or without a signal present). The contrast control determines the amount of drive signal applied to the picture tube. An increase in drive signal produces an increase in contrast. In the circuit of Fig. 1-4, the contrast control is, in effect, a volume or gain control between the video driver and video output stages.

1-2.9   AGC Circuits

There are many types of AGC circuits. Most solid-state receivers use keyed, saturation-type AGC. The RF tuner and IF stage transistors connected to the AGC line are forward-biased at all times. On strong signals, the AGC circuits increase the forward bias, driving the transistors into saturation, thus reducing gain. Under no-signal conditions, the forward bias remains fixed.

Keyed AGC circuits have two inputs. One input is from the IF circuit (Sec. 1-2.7) output at the IF center frequency. The other input is from the horizontal output circuit (Sec. 1-2.2) at a frequency of 15,750 Hz. The AGC output is a varying dc bias voltage (controlled by bursts of IF signals). A portion of the IF signal is taken from the IF amplifiers and is pulsed, or keyed, at the horizontal sweep frequency. The resultant keyed bursts of signal control the amount of dc voltage produced on the AGC line. The AGC output is applied to both the IF amplifiers and the RF tuner.

A few AGC circuits are provided with an AGC control that sets the level of AGC action. However, the trend is away from adjustable AGC circuits.

1-2.10   Sound IF and Audio

The sound IF and audio circuits of a TV receiver are essentially the same as those of an FM broadcast receiver. The circuits amplify the 4.5-MHz sound carrier, demodulate the FM sound (remove the audio signals), and amplify the audio signals to a level suitable for reproduction on the loudspeaker. The sound carrier is at a frequency of 4.5 MHz since the picture and sound broadcast carriers are 4.5 MHz apart on all channels. After the AM carrier is demodulated by the video detector (Sec. 1-2.7), only the FM carrier remains.

In a typical TV receiver, the FM carrier is demodulated or detected by means of a ratio detector, which follows two stages of IF amplification. However, some receivers use a discriminator-type sound detector after the IF amplifiers. The IF amplifier stages and FM detector (ratio or discriminator) require alignment. The alignment procedures are similar to those of an FM broadcast receiver.

The audio portion of the circuits is also similar to the audio circuits of AM or FM broadcast receivers. Generally, the audio circuits include a *volume* control and a *tone* control. These are front-panel controls, readily available to the user.

1-3. BASIC COLOR TELEVISION BROADCAST SYSTEM

As shown in Fig. 1-5, the basic color TV broadcast system consists of a transmitter, a color camera, a signal matrix, a brightness amplifier, a color amplifier, and a 3.58-MHz oscillator. As in the case of black and white, the TV program sound (or audio) is transmitted through an FM transmitter. The picture (or video) portion of the program is broadcast by means of amplitude modulation. The same channels and frequencies are used for color as well as black and white.

In addition to the horizontal and vertical sweeps, the color signal is made up of two components: brightness and color information. The brightness portion contains all the information pertaining to the details of the picture and is commonly referred to as the Y or *luminance* signal. The color portion is called the *chrominance* signal and contains the information pertaining to the hue and saturation of the picture.

The camera in Fig. 1-5 contains three separate image pickup tubes, one tube for each of three colors: red, green, and blue. The camera tubes divide the scene being scanned into three colors (red, yellow, and blue). The relative intensity of the scene is also contained in this signal. The output signal from the camera contains the luminance and chrominance information.

The signal is separated in the signal matrix, and the individual components are amplified by the brightness and color amplifiers. The color information (chrominance) is impressed on a 3.58-MHz subcarrier (in the form of phase shift) and is then transmitted together with the Y signal

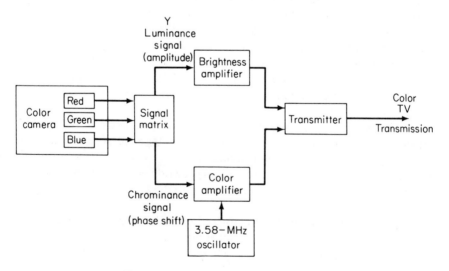

Fig. 1-5. Basic color TV transmitting system

(which is in the form of instantaneous amplitude shift or level). This method of color broadcast provides a compatible signal that can be reproduced on black and white as well as color receivers. The black and white receivers require only the luminance or Y signal (which is the equivalent of the picture information transmitted on the AM carrier of a black and white broadcast).

A compatible color signal must fit within the standard 6-MHz television channel and have horizontal and vertical scanning rates of 15,750 and 60 Hz, plus a video bandwith not in excess of 4.25 MHz. The color information is interleaved with the video information and transmitted within the 6-MHz television channel on the 3.58-MHz subcarrier, as shown in Fig. 1-6. The color subcarrier is actually 3.579545 MHz above the video carrier, since this does not interfere with reception on black and white receivers.

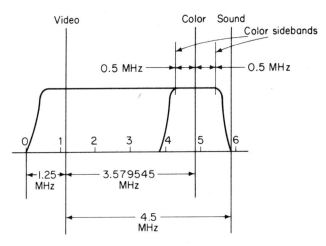

Fig. 1-6. Location of color subcarrier in 6-MHz TV channel

The color signals are synchronized by transmitting approximately eight cycles of the 3.58-MHz subcarrier oscillator signal (of the correct phase) along with the color signal. This signal is called the *burst* sync signal (or may also be known as the *color burst*) and is added to the "back porch" of the horizontal sync pulse, as shown in Fig. 1-7. The location of the burst signal will not interfere with horizontal sync on black and white receivers, since the burst signal occurs during the blanking or retrace portion of the sync pulse.

1-3.1 Black and White Versus Color Signals

A black and white presentation requires only that the amplitude or luminance (or Y) signal be transmitted. A color broadcast also requires that phase-shift information (representing colors) be transmitted. In color

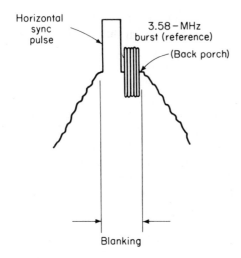

Fig. 1-7. Location of 3.58-MHz reference burst on horizontal blanking signal

television, the hue of color is determined by the instantaneous amplitude of the signal.

Figure 1–8 shows the phase and amplitude required to produce each of six basic colors. For example, at any given instant, if the amplitude is at 0.64 (with an amplitude of 0 for black and 1.0 for white) and the phase of the 3.58-MHz signal is 76.6° (with the burst signal considered to be at 0°), the receiver picture tube will produce pure red. If the amplitude is changed to 0.45 and the phase to 192.8°, a pure blue will be produced. If the amplitude is changed to 0.59 and the phase to 119.2°, the picture tube will produce magenta (which is a combination of red and blue). If the phase of the 3.58-MHz signal is swept from 0 to 360°, a complete rainbow of colors will result.

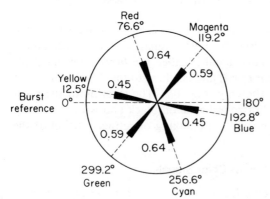

Fig. 1-8. Phase and amplitude relationships of NTSC color signals

1-4. BASIC COLOR RECEIVER CIRCUITS

Figure 1–9 is the block diagram of a color TV receiver. Again, the diagram shown is a composite of several types of color TV receivers and is presented as a point of reference for troubleshooting. In the following paragraphs we shall describe the basic principles of all color TV receivers. The detailed circuit service procedures for color are provided in Chapter 4. As shown in Fig. 1-9, the circuits of a color receiver are very similar to those of a black and white receiver, with two exceptions: the luminance or Y channel and the chrominance channel.

1-4.1 Luminance Channel Operation

The luminance or Y channel consists of the detector, first video amplifier, delay line, and second video amplifier. (Note that a separate sound detector is used in the signal path prior to the video detector in most color receivers. This provides good separation of the sound and video signals.)

The composite video signal is detected and amplified in the first stage of the luminance channel. A portion of this signal is fed to the sync and color stages. The rest of the signal continues on to the delay line. This delay is required so that the luminance information (amplitude) arrives at the picture tube at the same time as the color information (phase relationship). The narrow bandwidth of the color circuit causes the chrominance signal to take longer to reach the picture tube. Additional amplification is given to the luminance signal in the second video amplifier before it is applied to the cathodes of the picture tube.

1-4.2 Chrominance Channel Operation

The composite video signal is fed to the band-pass amplifier (sometimes called the color IF) from the first video amplifier. The chrominance signal is separated from the composite color signal, is amplified, and is fed to the inputs of the demodulators. The burst amplifier removes the burst signal from the chrominance signal and applies the burst signal to the color phase detector.

A color killer stage is used to prevent signals at frequencies near 3.58 MHz from passing through the color circuit during the reception of black and white broadcasts. In the absence of a color broadcast (no color burst signals being transmitted), the color killer biases the band-pass amplifier off, thus disabling all of the chrominance channels. In most circuits, the color killer also receives pulses from the horizontal sweep circuits (usually from the flyback transformer). The horizontal pulses operate the color

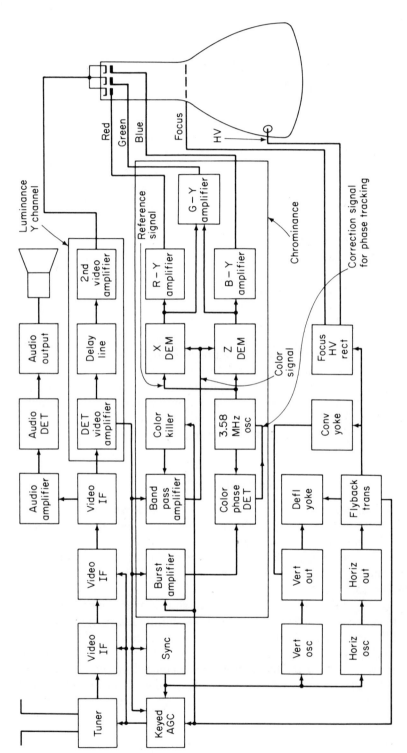

Fig. 1-9. Typical color receiver circuit

killer so as to bias the band-pass amplifier off during the horizontal sweep retrace (so that no color information will pass during the retrace period).

The 3.58-MHz oscillator provides a locally generated reference signal for demodulation of the composite color signal. The color phase detector compares the 3.58-MHz oscillator signal with the burst signal and develops a correction voltage to keep the oscillator locked in phase with the burst signal. In most circuits, the burst amplifier (or its control circuit) receives pulses from the horizontal sweep. These pulses gate the burst amplifier on only during the burst signal (immediately after the horizontal sync pulses, during the retrace blanking period, as shown in Fig. 1-7). Thus, the 3.58-MHz oscillator is locked in phase with the burst at the beginning of each horizontal sweep.

As shown in Fig. 1-9, the X and Z demodulators detect the amplitude and phase variations of the chrominance signal to recover color information. At any given instant, the X and Z demodulators receive two signals: a 3.58-MHz signal from the reference oscillator (locked in phase to the reference burst) and a 3.58-MHz signal from the band-pass amplifier (at a phase and amplitude representing the color at that instant).

The R-Y (red luminance), B-Y (blue luminance), and G-Y (green luminance) amplifiers amplify the color signals and apply them to the picture tube input (grids in this case).

1-4.3   Makeup of Colors

Color picture tubes produce most colors by mixing three colors (red, blue, and green). (Operation of the color picture tube is described in Sec. 1-4.4.) Any color can be created by the proper blending of these three colors; that is, any color can be created if the three colors are present, each at a given level of intensity or brightness. This is due to a characteristic of the human eye. When the eye sees two different colors of the same brightness level, the colors appear to be of different brightness levels. For example, the eye is more sensitive to green and yellow than to red or blue.

The combination of any two primary colors will produce a third color. In color television, the process of mixing colors is *additive* and is dependent on the self-illuminating properties of the picture tube screen (not on the surrounding light source, as found with the subtractive color-mixing process familiar to paints and printing).

If two primary colors, such as red and green, are combined, the primary colors will produce a secondary color of yellow. The combination of blue and red produces magenta. Also, a secondary color added to its complementary primary will produce white. For example, yellow + blue = white, cyan + red = white, and magenta + green = white. Thus, any color can be produced if the three colors (red, blue, and green) are mixed in the proper proportions (that is, if the amplitudes of the three signals at the picture tube inputs are at the proper levels).

For the purposes of our discussion, it is sufficient to understand that the three picture tube inputs will receive signals of the correct amplitude proportions (identical to the proportions existing at the color camera for any given instant) provided that (1) the receiver 3.58-MHz oscillator is locked in phase to the burst signal and (2) the demodulators receive a 3.58-MHz color signal that is of a *phase and amplitude corresponding to the color proportions* existing at the color camera.

1-4.4   Color Picture Tubes

Figure 1-10 illustrates a color picture tube and its related components. Keep in mind that there are many types of color tubes. However, they all have certain points in common (from a troubleshooting or service standpoint). Figure 1-10 illustrates the basic elements of features common to all color picture tubes.

The color picture tube is a cathode-ray tube, similar to that used for black and white receivers. Like the black and white tube, the color picture tube has a magnetic deflection yoke which deflects the electron beam. The yoke contains two sets of coils, one for vertical deflection and one for horizontal. The vertical and horizontal sweep signals are applied to the corresponding set of coils so that the electron beam traces out a rectangular raster on the screen in the normal manner. Here, the similarity to a black and white picture tube ends.

The color picture tube has *three* complete electron guns. Thus, three electron beams are produced, one for each gun. The color picture tube also contains a shadow mask and a three-color phosphor dot screen within the vacuum glass envelope.

The *screen* consists of several million phosphor dots of the primary television colors (red, green, and blue). These dots are arranged in trios and

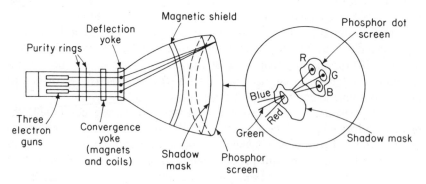

Fig. 1-10. Color picture tube

produce colored light when struck by electrons. Even though the dots are placed next to each other, they appear (to the human eye) to be superimposed so that the colors are (apparently) blended.

The *shadow mask* is a thin plate of metal that has a number of small holes, each centered over a phosphor dot trio. This mask is positioned in front of the screen and controls the landing of the electron beams on the phosphor dots.

The *three electron guns* (one for each color) are positioned approximately 120° apart and are precisely aimed so that the individual electron beams will pass through the shadow mask at different angles to strike the three corresponding phosphor dots. The control grid controls the amount of light leaving the gun and thus regulates the amount of colored light emitted from the phosphor screen.

1-4.5   Color Purity

For the color picture tube to reproduce color pictures correctly, the individual electron beams must strike phosphors of only one color. Each electron beam is then capable of producing a pure field of either red, blue, or green. *Purity* refers to the uniformity of the hue and brightness over the entire area of each color field and over the entire area of the combination of all three. For example, red is pure when the red display is a uniform red with no contamination from blue or green.

Color purity is determined by where the three electron beams strike the screen and can be adjusted by means of the purity magnet and deflection yoke. Typical color purity adjustment procedures are described in Chapter 4.

1-4.6   Convergence

Convergence is the registry of the three beams at the same point in the shadow mask across the entire screen. The individual aiming of the beams must be corrected so that each passes through the same holes in the mask and strikes the same phosphor dot trio.

Static convergence is controlled by convergence magnets which correct the travel of the individual electron beams. The static magnets converge the beams in the center of the screen but cannot compensate for the curvature of the screen to converge the beam at the edges. Such compensation (known as *dynamic convergence*) is accomplished by electromagnetic coils mounted on the convergence yoke. Sawtooth voltages from the vertical and horizontal circuits are fed to the coils, which then alter the static magnetic field in unison with the scanning. Typical color convergence adjustments are described in Chapter 4.

1-5. THE BASIC TROUBLESHOOTING FUNCTIONS

Troubleshooting can be considered as a step-by-step logical approach to locate and correct any fault in the operation of equipment. In the case of TV receiver troubleshooting, there are seven basic functions required.

First, you must study the receiver using service literature, schematic diagrams, etc., to find out how each circuit works when *operating normally.* In this way, you will know in detail how a given receiver should work. If you do not take the time to learn what is normal, you will never be able to distinguish what is abnormal. For example, some receivers simply have a better picture than other receivers, even in the presence of poor signals. You could waste hours of precious time (money) trying to make the inferior receiver perform like the quality set if you do not know what is "normal" operation.

Second, you must know the function of, and how to manipulate, all receiver controls and adjustments. It is difficult, if not impossible, to check out a receiver without knowing how to set the controls. Besides, it will make a bad impression on the customer if you cannot find the channel selector, especially on the second service call. Also, as a receiver ages, readjustment and realignment of critical circuits are often required.

Third, you must know how to interpret service literature and how to use test equipment. Along with good test equipment that you know how to use, well-written service literature is your best friend.

Fourth, you must be able to apply a systematic, logical procedure to locate troubles. Of course, a "logical procedure" for one type of receiver is quite illogical for another. As an example, it is quite illogical to check operation of an automatic fine tune (AFT) circuit on a receiver not so equipped. However, it is quite logical to check automatic gain control (AGC) on all receivers. For that reason, we shall discuss logical trouble-shooting procedures for various types of TV receivers, in addition to basic troubleshooting procedures.

Fifth, you must be able to analyze logically the information of an improperly operating receiver. The information to be analyzed may be in the form of performance, such as the appearance of the picture, or may be indications taken from test equipment, such as voltage readings. Either way, it is your analysis of the information that makes for logical, efficient troubleshooting.

Sixth, you must be able to perform complete checkout procedures on a receiver that has been repaired. Such checkout may be only simple operation, such as switching through all channels and checking the picture. At the other extreme, the checkout can involve complete realignment of the receiver, including IF and RF stages, etc. Either way, checkout is always required after troubleshooting.

One reason for the checkout is that there may be *more than one trouble*. For example, an aging part may cause high current to flow through a resistor, resulting in burnout of the resistor. Logical troubleshooting may lead you quickly to the burned-out resistor. Replacement of the resistor will restore operation. However, only a thorough checkout will reveal the original high current condition that caused the burnout.

Another reason for after-service checkout is that the repair may have produced a condition that requires readjustment. A classic example of this is where replacement of a part changes circuit characteristics, such as a new transistor in an IF stage requiring complete realignment of all IF stages.

Seventh, you must be able to use the proper tools to repair the trouble. As a minimum for television repair, you must be able to use soldering tools, wire cutters, longnose pliers, screwdrivers, and socket wrenches. If you are still at the stage where any of the above tools seem unfamiliar, you are not ready for television service, even simplified service.

In summary, before starting any troubleshooting job, ask yourself these questions: Have I studied all available service literature to find out how the receiver works (including any special circuits such as automatic color control, or AGC)? Can I operate the receiver controls properly? Do I really understand the service literature and can I use all required test equipment properly? Using the service literature, and/or previous experience on similar receivers, can I plan out a logical troubleshooting procedure? Can I analyze logically the results of operating checks, as well as checkout procedures involving test equipment? Using the service literature and/or experience, can I perform complete checkout procedures on the receiver, including realignment, adjustment, etc., if necessary? Once I have found the trouble, can I use common hand tools to make the repairs? If the answer is "no" to any of these questions, you are simply not ready to start troubleshooting. Start studying instead?

1-6. THE UNIVERSAL TROUBLESHOOTING APPROACH

The troubleshooting function discussed in Sec. 1-5 can be summarized into a universal approach with four major steps. These steps include

Determine the trouble symptom;
Localize the trouble to a functional unit;
Isolate the trouble to a circuit;
Locate the specific trouble, probably to a specific part.

Let us examine what is being accomplished by each step.

1-6.1   Determining Trouble Symptoms

Determining symptoms means that you must know what the equipment is supposed to do when it is operating normally and, more important, that you must be able to recognize when the normal job is not being done. Everyone knows what a TV receiver is supposed to do, but no one knows how well each set is to perform (and has performed in the past) under all operating conditions (with a given antenna, lead-in, location, etc.).

All receivers have operating controls and two built-in aids for evaluating performance (the loudspeaker and the picture tube). Using the normal and abnormal symptoms produced by the loudspeaker and picture tube, you must analyze the symptoms to ask the questions, How well is this set performing, and where in the receiver could there be trouble that will produce these symptoms?

The "determining symptoms" step does not mean that you charge into the receiver with screwdriver and soldering tool, nor does it mean that test equipment should be used extensively. Instead, it means that you make a visual check, noting both normal and abnormal performance indications. It also means that you operate the controls to gain further information.

At the end of the "determining symptoms" step, you definitely know that something is wrong and have a fair idea of what is wrong, but you probably do not know just what area of the equipment is faulty. This is established in the next step of troubleshooting.

1-6.2   Localizing Trouble to a Functioning Unit

Most electronic equipment can be subdivided into units or areas which have a definite purpose or function. The term "function" is used in television troubleshooting to denote an operation performed in a specific area of the receiver. For example, in a basic black and white receiver, the functions can be divided into RF, IF, audio, video, picture tube, and power supply.

To localize the trouble systematically and logically, you must have a knowledge of the functional units of the receiver and must correlate all the symptoms previously determined. Thus, the first consideration in localizing the trouble to a functional unit is a valid estimate of the area in which the trouble might be in order to cause the indicated symptoms. Initially, several technically accurate possibilities may be considered as the probable trouble area.

As a classic (oversimplified) example, if both picture and sound are poor, the trouble might be in either the RF or IF stages, since these functional areas are common to both picture and sound reproduction. On the other hand, if the picture is good but the sound is poor, the trouble is probably in the audio stages, since these functional areas apply only to sound.

Use of diagrams. Television troubleshooting involves (or should involve) the extensive use of diagrams. Such diagrams may include a *functional block diagram* and almost always includes *schematic diagrams.* (*Practical wiring diagrams,* such as found in military-type service literature, are almost never available for television service.)

The block diagram (such as shown in Figs. 1-4 and 1-9) illustrates the functional relationship of all major sections or units in the receiver. The block diagram is thus the most logical source of information when localizing trouble to a functioning unit or section. Unfortunately, not all television service literature is provided with a block diagram. It may be necessary to use schematic diagrams.

The schematic diagram (such as shown in Chapters 4 and 6) shows the functional relationship of all parts in the receiver. Such parts include all transistors, capacitors, transformers, diodes, etc. Generally, the schematic presents too much information (not directly related to the specific symptoms noted) to be of maximum value during the localizing step. The decisions being made regarding the probable trouble area may become lost among all the details. However, the schematic is very useful in later stages of the total troubleshooting effort and when a block diagram is not available.

In comparing the block diagram and the schematic during the localizing step, note that each transistor shown on the schematic is usually represented as a block on the block diagram. This relationship is typical in most service literature.

The physical relationship of parts is usually given on *component location diagrams* (also called parts placement diagrams). These location or placement diagrams rarely show all parts. Instead, they concentrate on major parts such as transistors, transformers, diodes, and adjustment controls. For this reason, location diagrams are the least useful when localizing trouble. Instead, the location diagrams are most useful when locating specific parts during other phases of troubleshooting.

To sum up, it is logical to use a block diagram instead of a schematic or location diagram when you want to make a valid estimate as to the probable trouble areas (to the faulty functioning unit). The use of a block diagram also permits you to use a troubleshooting technique known as *bracketing* (or good input/bad output). If the block diagram includes major test points, as it may in some well-prepared service literature, the block will also permit you to use test equipment as aids for narrowing down the probable trouble cause. However, test equipment is used more extensively during the isolation step of troubleshooting.

1-6.3   Isolating Trouble to a Circuit

After the trouble is localized to a single functional area, the next step is to isolate the trouble to a circuit in the faulty area. To do this, you concentrate on circuits in the area that could cause the trouble and ignore the remaining circuits.

The isolating step involves the use of test equipment such as meters, oscilloscopes, and signal generators for *signal tracing* and *signal substitution* in the suspected faulty area. By making valid educated estimates and properly using the applicable diagrams, bracketing techniques, signal tracing, and signal substitution, you can systematically and logically isolate the trouble to a single defective circuit.

Repair techniques or tools to make necessary repairs to the equipment are not used until after the specific trouble is located and verified. That is, you still do not charge into the equipment with solder tools and screwdriver at this point. Instead, you are now trying to isolate the trouble to a specific defective circuit so that, once the trouble is located, it can be repaired.

1-6.4 Locating the Specific Trouble

Although this troubleshooting step mentions only "locate the specific trouble," it also includes a final analysis, or review, of the complete procedure, along with using repair techniques to remedy the trouble once it has been located. This final analysis will permit you to determine whether some other malfunction caused the part to be faulty or whether the part located is the actual cause of the equipment trouble.

When trying to locate the trouble, inspection using the senses—sight, smell, hearing, and touch—is very important. This inspection is usually performed first, in order to gather information that may more quickly lead to the defective part. Among the things to look for during the inspection using the senses are burned, charred, or overheated parts, arcing in the high-voltage circuits, and burned-out parts.

In receivers where it is relatively easy to gain access to the circuitry, a rapid visual inspection should be performed first. Then the active device— vacuum tube or transistor—can be checked. A visual inspection is always recommended as the first step in all solid-state receivers and in most vacuum-tube television. A possible exception is where access to circuit parts is difficult but where vacuum tubes can be easily removed and tested (or substituted).

Next in line for locating the specific trouble is the use of an oscilloscope to *observe waveforms* and a meter to *measure voltages*. The last test usually is the use of a meter to make *resistance* and *continuity* checks to pinpoint the defective part. After the trouble is located, you make a final analysis of the complete troubleshooting procedure to verify the trouble. Then you repair the trouble and check out the receiver for proper operation.

Note that in the service literature for most TV receivers the waveforms, voltages, and resistances are given on the schematic. In a few cases, the information is given in chart form (following the military style). Either way, you must be able to use test equipment to observe the waveforms and make the measurements. For that reason, the function and use of test equipment during troubleshooting is discussed frequently throughout the remainder of this book.

1-6.5 Developing a Systematic, Logical Troubleshooting Procedure

The development of a systematic and logical troubleshooting procedure requires

A logical approach to the problem,

Knowledge of the equipment,

Interpretation of test information,

The use of information gained in each step.

Some television technicians feel that a knowledge of the equipment involves remembering past failures as well as remembering such things as the location of all test points, all adjustment procedures, etc. This approach may be valid when you troubleshoot only one type of receiver, but it has little value in developing a basic troubleshooting procedure.

It is true that recalling past receiver failures may be helpful, but you should not rely upon the possibility that the same trouble will be the cause of a given symptom in every case. In any television receiver, there are many possible troubles which can give approximately the same symptom indications.

Likewise, you should never rely entirely upon your memory of adjustment procedures, test point locations, etc., in approaching any troubleshooting problem. This is one of the reasons for having service literature which contains diagrams and information about receivers. Later in this book (particularly in Chapter 5) we shall discuss more about the specific use and types of information presented in service literature. It is important that you learn to be a systematic, logical troubleshooter, and not a memory expert.

1-6.6 Relationship between Troubleshooting Steps

Thus far, we have established the overall troubleshooting approach. Now let us make sure that you understand how each troubleshooting step fits together with the others by analyzing the relationships.

The first step, *determining the symptoms,* requires the use of senses, observation of the receiver performance, previous knowledge of receiver operation, the manipulation of the operating controls, and the possible recording of notes. Determining the symptoms presupposes the ability to recognize improper indications, to properly operate the controls, and to note the effect that the controls have on trouble symptoms.

The second step, *localizing the trouble to a functional area,* depends on the information observed in the first step, plus the use of a functional block diagram (or possibly the schematic) and reasoning. During the second step, ask yourself the question, What functional area could cause the indicated symptoms? Then bracketing, or narrowing down the probable defect to a single functioning area, is used along with the test equipment to actually pinpoint the faulty function. The observations in this step refer to noting the indications of the testing devices used to localize the trouble.

The third step, *isolating the trouble to a circuit,* uses all of the information gathered up to this time. The main difference between this step and the second step is that now schematic diagrams are used instead of block diagrams, and test equipment is used extensively.

The fourth step, *locating the trouble,* is where all the findings are reviewed and verified to ensure that the suspected part is the cause of the failure. The final step also includes the necessary repair (replacement of defective parts, etc.) as well as a final receiver checkout.

Example of relationship between troubleshooting steps. To make sure that you understand the relationship between troubleshooting steps, let us consider an example. Assume that you are troubleshooting a receiver, you are well into the "locate" step of testing a suspected circuit, and you find *nothing wrong with that circuit.* That is, all waveforms, voltage measurements, and resistance measurements within the circuit are normal. What is your next step?

You could assume that nothing is wrong—that the problem is "customer trouble." This is absolutely *incorrect.* First, there must be something wrong in the receiver since some abnormal symptoms were recognized before you got to the locate step. Never assume anything when troubleshooting; either the receiver is working properly, or it is not working properly; either observations and measurements are made, or they are not made. You must draw valid conclusions from the observations, measurements, and other factual evidence, or you must repeat the troubleshooting procedure.

Repeating the troubleshooting procedures. Some technicians new to service work assume that repeating the troubleshooting procedure means starting all over from the first step. In fact, some service literature recommends this action. The recommendation is based on the fact that it is possible for anyone, even an experienced technician, to make mistakes. When performed logically and systematically, the troubleshooting procedure will keep mistakes to a minimum. However, voltage and resistance measurements can be erroneously interpreted, waveform observations or bracketing can be incorrectly performed, or numerous other mistakes can be caused by simple oversight.

In spite of the recommendation by some other service literature writers, the author contends that "repeat the troubleshooting procedure" means *retrace* your steps, one at a time, until you find the place where you went wrong. Perhaps it was a previous voltage or resistance measurement erroneously interpreted in the locate step, or perhaps a waveform observation or bracketing step was incorrectly performed in the isolate step. Whatever the cause, it must be found logically and systematically by taking a *return path* to determine where you went astray.

1-7. APPLYING THE TROUBLESHOOTING APPROACH TO TV SERVICE

Now that we have reviewed the basic operating principles of television and have established a basic troubleshooting approach, let us discuss how the approach can be applied to the specifics of television service. The remainder of this chapter is devoted to examples of how the approach can be used to troubleshoot TV receivers. These examples are generalized. Detailed descriptions of TV troubleshooting examples are given in Chapters 3 through 6.

1-8. TROUBLE SYMPTOMS

It is not practical to list all symptoms of troubles that may occur in all television receivers. However, the list in Fig. 1-11 covers most troubles for a black and white television receiver. That is, most troubles have been grouped into functional areas (or circuits) of the receiver. These areas or circuit groups correspond to those of the block diagram (Fig. 1-4) and the discussions of Sec. 1-2.

Some of the symptoms listed point to only one area of the receiver as a *probable cause* of trouble. Other symptoms could be caused by more than one area of the circuit. For example, if there is no vertical sweep (the picture tube screen shows only a horizontal line) but other receiver functions are normal (good sound), the trouble is *most likely* in the vertical sweep circuits. On the other hand, if both sound and picture are weak or poor, the trouble *could be* in the RF tuner, in the IF and video detector, or possibly in the driver of the video amplifier and picture tube circuits.

In the discussions of the localize, isolate, and locate steps, we shall give examples of how the symptoms can be used as the first step in pinpointing trouble to an area, to a circuit within the area, and finally to a part within the circuit. Before going into these steps, here are some notes regarding symptoms of TV receiver troubleshooting.

LOW VOLTAGE POWER SUPPLY
   No sound and no picture raster
   No sound, no picture raster, transformer buzz
   Distorted sound and no raster
   Picture pulling and excessive vertical height
HIGH VOLTAGE POWER SUPPLY AND HORIZONTAL OUTPUT
   Dark screen
   Picture overscan
   Narrow picture
   Foldback or foldover
   Nonlinear horizontal display
HORIZONTAL OSCILLATOR
   Dark screen
   Narrow picture
   Horizontal pulling (loss of sync)
   Horizontal distortion
VERTICAL SWEEP
   No vertical sweep
   Insufficient height
   Vertical sync problems
   Vertical distortion (nonlinearity)
   Line splitting
SYNC SEPARATOR
   No sync (horizontal or vertical)
   No vertical sync
   No horizontal sync
   Picture pulling
RF TUNER
   No picture, no sound
   Poor picture, poor sound
   Hum bars or hum distortion
   Picture smearing and sound separated from picture
   Ghosts
   Picture pulling
   Intermittent problems
   UHF tuner problems
IF AMPLIFIER AND VIDEO DETECTOR
   No picture, no sound
   Poor picture, poor sound
   Hum bars or hum distortion
   Picture smearing, pulling or overloading
   Intermittent problems
VIDEO AMP AND PICTURE TUBE
   No raster (dark screen)
   No picture, no sound
   No picture, sound normal
   Contrast problems
   Sound in picture
   Poor picture quality
   Retrace lines in picture
   Intermittent problems
AGC
   No picture, no sound
   Poor picture
SOUND IF AND AUDIO
   No sound
   Poor or weak sound

Fig. 1-11. Typical trouble symptoms

1-8.1 Determining Trouble Symptoms

It is not difficult to realize that there definitely is trouble when electronic equipment will not operate. For example, there obviously is trouble when a TV receiver is plugged in and turned on and there is no picture, no sound, and no pilot light. A different problem arises when the equipment is still operating but is not properly doing its job. Using the same TV receiver, assume that the picture and sound are present but that the picture is weak and that there is a buzz in the sound.

Another problem in determining trouble symptoms is improper use of the equipment by the operator. In complex electronic equipment, operators are usually trained and checked out on the equipment. The opposite is true of home entertainment equipment used by the general public. However, no matter what equipment is involved, it is always possible for an operator (or customer) to report a "trouble" that is actually a result of improper operation. For these reasons, *you* must first determine the signs of failure, regardless of the extent of malfunction, caused by either the TV receiver or the customer. This means that you must know how the equipment operates normally, and how to operate the equipment controls.

1-8.2 Recognizing Trouble Symptoms

Symptom recognition is the art of identifying the *normal* and *abnormal* signs of operation in electronic equipment. A trouble symptom is an undesirable *change* in equipment performance or a deviation from the standard. For example, the normal television picture is a clear, properly contrasted representation of an actual scene. The picture should be centered within the vertical and horizontal boundaries of the screen. If the picture should suddenly begin to roll vertically, you should recognize this as a trouble symptom because it does not correspond to the normal performance which is expected.

Now assume that the picture is weak, say due to a poor broadcast signal in the area or a defective antenna. If the RF and IF stages of the receiver do not have sufficient gain to produce a good picture under these conditions, you could mistake this for a trouble symptom, unless you really knew the equipment. A poor picture (for this particular model of receiver operating under these conditions) is not abnormal operation, nor is it an undesired change. Thus, it is not a true trouble symptom and should be so recognized.

1-8.3 Equipment Failure Versus Degraded Performance

Equipment failure means that either the entire equipment or some functional part of the equipment is not operating properly. For example, the total absence of a picture on the screen of a TV receiver when all controls are properly set is a form of equipment failure, even though there may be sound from the loudspeaker. Degraded performance occurs whenever the equipment is working but is not presenting normal performance. For example, the presence of hum in the receiver loudspeaker is degraded performance, since the receiver has not yet failed but the performance is abnormal.

1-8.4   Evaluation of Symptoms

Symptom evaluation is the process of obtaining more detailed descriptions of the trouble symptoms. The recognition of the original trouble symptom does not in itself provide enough information to decide on the probable cause or causes of the trouble, because many faults produce similar trouble symptoms.

To evaluate a trouble symptom, you generally have to operate controls associated with the symptom and apply your knowledge of electronic circuits, supplemented with information gained from the service literature. Of course, the mere adjustment of operating controls is not the complete story of symptom evaluation. However, the discovery of an incorrect setting can be considered a part of the overall symptom evaluation process.

1-8.5   Example of Evaluating Symptoms

When the screen of a TV receiver is not on (no raster), there obviously is trouble. The trouble could be caused by the brightness control being turned down (assuming that the power cord is plugged in and that the on-off switch is set to on). However, the same symptom can be produced by a burned-out picture tube or a failure of the high-voltage power supply, among many other possible causes. Think of all the time you may save if you check the operating controls first, before you charge into the receiver with tools and test equipment.

To do a truly first-rate job of determining trouble symptoms, you must have a complete and thorough knowledge of the normal operating characteristics of the receiver. Your knowledge helps you to decide whether the receiver is doing the job for which it was designed. In most service literature, this is more properly classified as "knowing your equipment."

In addition to knowing the receiver, you must be able to operate properly all the controls in order to determine the symptom—to decide on *normal or abnormal* performance. If the trouble is cleared up by manipulating the controls, the trouble analysis may or may not stop at this point. By using your knowledge of the receiver, you should find the reason the specific control adjustment removed the apparent trouble.

1-9. LOCALIZING TROUBLE

Localizing trouble means that you must determine which of the major functional areas in a receiver is actually at fault. This is done by systematically checking each area selected until the actual faulty one is found. If none of the functional areas on your list show improper performance, you must take a return path and recheck the symptom information (and observe more information, if possible). There may be several circuits which could be causing the trouble. The localize step will narrow the list to those in one functional area, as indicated by a particular block of the block diagram.

The problem of trouble localization is simplified when a block diagram and a list of trouble symptoms (such as shown in Figs. 1-4 and 1-11) are available for the TV receiver being serviced. Keep in mind that these illustrations apply to a "typical" or composite receiver. However, the general arrangement shown in Figs. 1-4 and 1-11 can be applied to any black and white receiver. Thus, the illustrations serve as a universal starting point for trouble localization.

1-9.1 Bracketing Technique

The basic bracketing technique makes use of a block diagram or schematic to localize the trouble to a functional area. Bracketing (sometimes known as the *good input/bad output* technique) provides a means of narrowing the trouble down to a circuit group and then to a faulty circuit. Symptom analysis and/or signal-tracing tests are used in conjunction with, or are a part of, bracketing.

Bracketing starts by placing brackets (at the good input and the bad output) on the block diagram or schematic. Bracketing can be mental, or it can be physically marked with a pencil, whichever is most effective for you. No matter what system is used, with the brackets properly positioned, you know that the trouble exists somewhere *between* the two brackets.

The technique is to move the brackets, one at a time (either the good input or the bad output), and then make tests to find if the trouble is within the new bracketed area. The process continues until the brackets localize a circuit group.

The most important factor in bracketing is to find where the brackets should be moved in the elimination process. This is determined from your deductions based on your knowledge of the equipment and on the symptoms. All moves of the brackets should be aimed at localizing the trouble with a minimum of tests.

1-9.2 Examples of Bracketing

Bracketing can be used with or without actual measurement of voltages or signals. That is, sometimes localization can be made on the basis of symptom evaluation alone. In practical TV service, both symptom evaluation and tests are usually required, often simultaneously. The following examples show how the technique is used in both cases.

Assume that you are servicing a TV receiver and that you find a "no-sound and no-picture raster" symptom. That is, the power cord is plugged in, and the power switch is on, as is the pilot light. The low-voltage power supply is a logical suspect as the faulty circuit group. You place a good input bracket at the 115-Vac input, and a bad output bracket at the dc output, as shown in Fig. 1-12. To confirm the symptom, you measure both the dc output voltage (or voltages) and the 115-Vac input. If the input is normal but one or more of the output voltages is absent or abnormal, you have localized the trouble to the low-voltage power supply circuits. The next step is to isolate the trouble to a specific circuit in the power supply, as discussed in Sec. 1-10.

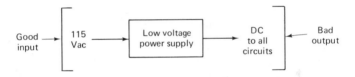

Fig. 1-12. Example of bracketing for "no sound and no picture raster" symptom

From a practical troubleshooting standpoint, it is possible that the low-voltage power supply output voltages are normal but you still have a "no-sound and no-picture" symptom. For example, the lines carrying the dc voltages to other circuit groups could be open. This can be checked by measuring the voltages at the circuit end of all lines, as well as at the power supply end.

Also, it is possible that a failure in another circuit can cause the power supply output voltage to be abnormal. For example, if there is a short in one of the circuits on the dc supply line, the dc output voltage will be low. Of course, this will show up as an abnormal measurement and will be tracked down during the "isolate" step of troubleshooting.

As another example of bracketing, assume that there is no vertical sync. That is, all receiver functions are normal but the picture rolls vertically. You could start by placing a good input bracket at the input to the vertical sweep circuits and a bad output bracket at the vertical coils of the deflection yoke, as shown in Fig. 1-13. However, from a practical standpoint, your first move should be adjustment of the vertical hold control.

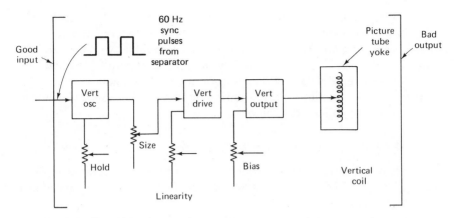

Fig. 1-13. Example of bracketing for "no vertical sync" symptom

If the trouble is not cleared by adjustment of the vertical hold control, then confirm the good input bracket by measuring the input sync pulses. Make this measurement at the input to the vertical oscillator, rather than at the sync separator output, as shown in Fig. 1-13. It is possible that the line between the sync separator and vertical oscillator is open or partially shorted. (A completely shorted line would probably cause failure of the sync separator circuits and would affect the horizontal sweep circuits as well.)

If the vertical sync pulses are normal at the input of the vertical oscillator but there is no vertical sync (even with adjustment of vertical hold), you have localized the trouble to the vertical sweep circuits. The next step is to isolate the trouble to a specific circuit in the vertical sweep group, as discussed in Sec. 1-10.

1-9.3 Localization with Plug-in Modules

The localization procedure can be modified when the circuits of a TV receiver are located on plug-in modules. (The trend in present-day receivers is toward the use of plug-in modules.) With such receivers, it is possible to replace each module, in turn, until the trouble is cleared. For example, if replacement of the vertical sweep module restores normal operation, the defect is in the vertical sweep module. This can be confirmed by plugging the suspected defective module back into the equipment. Although this confirmation process is not a part of theoretical troubleshooting, it is wise to make the check from a practical standpoint. Often, a trouble symptom of this sort can be caused by the plug-in module making poor contact with the chassis connector or receptacle.

Some service literature recommends that tests be made before arbitrarily replacing all plug-in modules. This is usually based on the fact that the plug-in modules are not necessarily arranged on a functional area basis. Thus, there is no direct relationship between the trouble symptoms and the modules. In such cases, always follow the service literature recommendation. Of course, if plug-in modules are not readily available in the field, you must make tests to localize the trouble to a module (so that you can order the right module, for example).

1-9.4   Which Circuit Groups to Test First

When you have localized trouble to more than one circuit group you must decide which group to test first. Several factors should be considered in making this decision.

As a general rule, if you *make a test that eliminates several circuits,* or circuit groups, make that test first, before making a test that eliminates only one circuit. This requires an examination of the diagrams (block and/or schematic) and knowledge of how the receiver operates. The decision also requires that you apply logic in making the decision.

Test point accessibility is the next factor to consider. A test point can be a special jack located at an accessible spot (say at the rear of the chassis). The jack (or possibly a terminal) is electrically connected (directly or by a switch) to some important operating voltage or signal path. At the other extreme, a test point can be *any point* where wires join or where parts are connected together.

Another factor (although definitely not the most important) is past experience and history of *repeated receiver failures.* Past experience with identical or similar receivers and related trouble symptoms, as well as the probability of failure based on records or repeated failures, should have some bearing on the choice of a first test point. However, all circuit groups related to the trouble symptom should be tested, no matter how much experience you may have on the receiver. Of course, the experience factor can help you decide which group to test first.

Anyone who has any practical experience in troubleshooting knows that all of the steps in a localization sequence can rarely proceed in textbook fashion. Just as true is the fact that troubles listed in receiver service literature very often never occur in the receiver you are servicing. These troubles are included in the literature as a guide, not as hard and fast rules. In some cases it may be necessary to modify your troubleshooting procedure as far as localizing the trouble is concerned. The physical arrangement of the receiver may pose special troubleshooting problems. Likewise, experience with similar equipment may provide special knowledge which can simplify localizing the trouble.

1-9.5 Universal Trouble Localization

In the following paragraphs we shall describe a universal trouble localization for a typical black and white receiver. The procedures are based on the assumption that the receiver circuit arrangement is as shown in Fig. 1-4. Thus, it is possible to group the troubles as shown in Fig. 1-11. The details of these localization procedures are discussed further in the remaining sections of this chapter and in Chapters 3 and 6.

If there is a "no-sound and no-raster" symptom, start localization by checking the input and output of the low-voltage power supply.

If the raster is present but there is no sound and no picture on the raster, start localization by checking the signal at the output of the IF and video detector and at the output of the driver in the video amplifier and picture tube circuits. Start at the same point if you have a poor picture, poor sound trouble symptom (or any symptom that points to circuits common to both sound and picture).

If the driver output is abnormal but the IF output is good, you have localized the trouble to the video driver. If the IF output is abnormal, you have localized the trouble to the IF and video detector, RF tuner, or AGC circuits.

To eliminate the AGC circuits as a trouble suspect, apply a fixed dc voltage to the AGC line, and check operation of the receiver. (This is known as *clamping* the AGC line. Clamping is discussed throughout this chapter and other chapters.) If operation is normal with the AGC line clamped but not when the clamp is removed, you have localized trouble to the AGC circuits.

It should be noted that AGC circuit problems are often difficult to localize, because a keyed AGC circuit as shown in Fig. 1-4 (and as in millions of receivers) requires two inputs (pulses from the horizontal section and IF signals from the IF section).

For example, if the IF amplifiers are defective, the AGC circuits will not receive proper IF signals and will not produce a proper dc voltage to the AGC line. In turn, the lack of proper AGC voltage can cause the IF amplifiers to operate improperly. Conversely, if the AGC circuits are defective, the IF amplifiers will not receive proper AGC voltages and will not deliver a proper IF signal to the AGC circuits.

As a simple guide, if clamping the AGC line eliminates the problem, the trouble is most likely localized to the AGC circuits. In that event, check both inputs (pulses and IF) to the AGC circuits.

If the AGC line is clamped and the IF and video detector output is not normal, check the signal at the IF section input. (This is at the input of the first IF amplifier, which is also the RF tuner output. Many receivers have a readily accessible test point at this input/output junction, often known as the "looker point").

If the signal at the looker point is abnormal, the trouble is localized to the RF tuner. If the looker point signal is normal but the video detector output is abnormal, you have localized trouble to the IF and video detector section.

1-9.6 Obvious Symptoms

Some trouble symptoms will lead you to an obvious localization process. For example, if there is a problem (in picture or sound, or both) on only one channel, the RF tuner is the likely suspect. If the problem is one of sound only, with a good picture, the sound IF and audio circuit group is suspect. If there is good sound, and a good picture raster, but a poor picture, start localization at the video amplifier and picture tube circuit group.

If horizontal sync is good, along with good sound, but there is a vertical problem (lack of sync, insufficient height, poor linearity, etc.), look for problems in the vertical sweep circuits. If there is good horizontal and vertical sync but the picture is narrow or there is obvious distortion, start with the high-voltage supply and horizontal output circuit group (probably with the horizontal output stage). If both horizontal and vertical sync appear to be abnormal, start by checking the input and both outputs of the sync separator.

1-9.7 Ambiguous Symptoms

The localization process is not so obvious if troubles occur in other sections. The horizontal circuits are a particular problem. For example, if the horizontal oscillator or horizontal driver fails completely, there will be no pulses to the high-voltage supply and horizontal output circuits. Thus, there will be no high voltage to the picture tube (and no picture raster). This same trouble symptom can be caused by failure of the high-voltage transformer, the high-voltage output stage, or the picture tube itself. (In some receivers, the high-voltage and output circuit group also supplies voltages to the picture tube accelerator grids. If these voltages are absent, there will be no raster.) Also, in some receivers, a failure in the video amplifier and picture tube circuits can cut the picture tube off.

If you are faced with a "no-raster" or "dark screen" symptom but you have sound, there are two logical courses of localization. As a first choice, if convenient, check the high voltage to the picture tube as well as the voltages to all picture tube elements (accelerator grids, cathode, heater, etc.). As a second logical choice, if it is not convenient to check the picture tube voltages, check the input to the high-voltage and horizontal output group (which is also the output of the horizontal oscillator and driver circuit group). This second choice of localization must also be followed if you find the picture tube voltages normal. If the signal at the horizontal oscillator and driver output is normal, the high-voltage and horizontal output group is suspect. If the signal is abnormal, the horizontal oscillator and driver is the likely problem area.

Failure in the horizontal circuits can also affect the AGC circuit. If there are no horizontal pulses to the keyer, the dc output from the AGC circuits will be absent. This is a particular problem since failure of the AGC can affect both the RF tuner and IF circuits, which, as discussed, affect overall performance of the receiver.

When the raster is present, and you have good sound, but there is abnormal horizontal sync (or distortion and/or picture pulling), start with the horizontal oscillator and driver group. Check *both* the input pulses from the sync separator and the feedback pulses from the horizontal output.

1-10. ISOLATING TROUBLE TO A CIRCUIT

The first two steps (symptoms and localization) of the troubleshooting procedure give you the initial symptom information about the trouble and describe the method of localizing the trouble to a *probable faulty* circuit group. Both steps involve a minimum of testing. In the isolate step, you will do extensive testing to track the trouble to a specific faulty circuit.

1-10.1   Isolating Trouble in IC and Plug-in Equipment

It is common to use ICs in many solid-state receivers. For example, in some receivers, the entire IF and video detector circuit group is replaced by a single IC. All parts of the circuit group, except the IF transformers, are included in the IC. In such cases, the trouble can be isolated to IC input and output but not to the circuits (or individual parts) within the IC. No further isolation is necessary since parts within the IC cannot be replaced on an individual basis.

This same condition is true of some solid-state receivers where groups of circuits are mounted on *sealed,* replaceable boards or cards. Note that not all plug-in modules are sealed. Some have replaceable parts.

1-10.2   Using Diagrams in the Isolation Process

No matter what physical arrangement is used, the isolation process follows the same reasoning you have used previously: the continuous narrowing down of the trouble area by making logical decisions and performing logical tests. Such a process reduces the number of tests which must be performed, thus saving time and reducing the possibility of error.

A block diagram is a very convenient tool for the isolation process, since it shows circuits already arranged in circuit groups. Unfortunately, as discussed, you may or may not have a block diagram supplied with your service literature. You must work with a schematic diagram.

With either diagram, if you can *recognize circuit* groups as well as *individual circuits,* the isolation process will be much easier. For example, if you can subdivide (mentally or otherwise) the schematic diagram of the receiver you are servicing into circuit groups, rather than individual circuits, you can isolate the group (or remove the group from possible fault) by a *single test* at the input or output for the group. The block diagram of Fig. 1-4 has been subdivided into circuits and logical functional groups so that you can use the diagram as a pattern for subdividing any TV receiver schematic.

No matter what diagram you use during isolation, you are looking for three major bits of information: the *signal path* (or paths), the *waveforms* along the signal paths, and the *operating/adjustment controls* in the various circuits along the signal paths. In Fig. 1-4, the signal paths are indicated by solid lines between blocks. Arrows on the lines indicate the direction of signal flow. The operating/adjustment controls are connected to the block representing the circuit affected by the controls. Waveforms are given on TV schematic diagrams at several points, usually at the input and output of each circuit group, and possibly at each circuit. Compare this arrangement with the information shown on the typical schematic of Chapter 6.

1-10.3 Comparison of Waveforms and Signals

In the simplest form, the isolation step is done by comparing the actual waveforms produced along the signal paths of the receiver circuits against the waveforms shown in the service literature diagrams. This is known as *signal tracing.* The isolation step may also involve *injection* or *substitution* of signals normally found along the signal paths. Signal tracing and injection are discussed in later paragraphs of this chapter and throughout the book. With either technique, you check and compare inputs and outputs of circuit groups and circuits in the signal paths.

In vacuum-tube receivers, the input signal is injected at the grid, and the output signal is traced at the plate (or possibly the cathode). In solid-state receivers, the input signal is injected at the base, and the output signal is traced at the collector (or possibly the emitter). These input/output relationships are shown in Fig. 1-14.

For a circuit group, the input is at the *first* base (or grid) in the signal path, whereas the output is at the *last* collector (or plate) in the *same* path. With any circuit group, the input signal to the group is injected at one point,

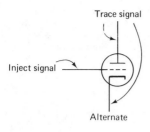

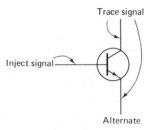

Fig. 1-14. Input/output relationships of vacuum tubes and transistors (for signal tracing and injection)

and then the output signal is obtained at a point several stages further along the same signal path. To determine the signal-injection and output points of a circuit group, you must find the *first* circuit of the group in the signal path and the *final* circuit of the group in the *same* path.

Signal paths are discussed further in Sec. 1-10.4. For now, consider the following example. In the block diagram of Fig. 1-4, the input for the audio amplifier circuit group is at the volume control, whereas the output is at the loudspeaker. Note that the volume control is (simultaneously) at the input of the audio amplifier group, and at the output of the sound IF group. Since the volume control of most receivers is readily identifiable (and accessible) it forms a good input/output test point for universal troubleshooting.

Before going into a study of the signal paths, keep the following points in mind. The symptoms and related information obtained in the previous steps (symptoms and localization) should not be discarded now or at any time during the troubleshooting procedure. From this information you can identify those circuit groups that are probable trouble sources. Likewise, note that the physical location of the circuit groups within the receiver has no relation to their representation on the diagrams (block or schematic). You must consult part placement diagrams to find physical location.

1-10.4   Signal Paths in Television Receivers

There are six basic types of signal paths, no matter what circuit group or circuit arrangement is used. These types are shown in Fig. 1-15 and are summarized as follows:

A *linear* path is a series of circuits arranged so that the output of one

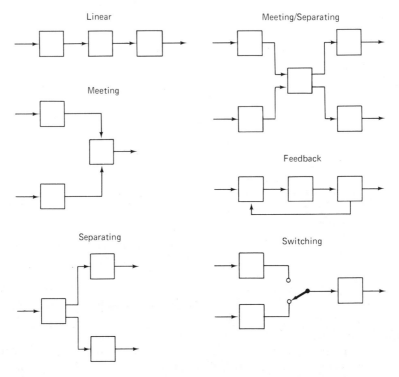

Fig. 1-15. Types of signal paths in a television receiver

circuit feeds the input of the following circuit. Thus, the signal proceeds straight through the circuit group without any return or branch paths. As shown in the block diagram of Fig. 1-4, the main path through the IF amplifier and video detector as well as the sound IF and audio circuits are examples of linear paths.

A *meeting path* is one in which two or more signal paths enter a circuit. The paths to the mixer of the RF tuner and to the keyer of the AGC circuit group are both meeting paths.

A *separating path* is one in which two or more signal paths leave a circuit. The paths from the vertical sweep, horizontal output, and sync separator circuit groups are separating paths.

A *meeting/separating path* is one in which a single stage has multiple inputs and multiple outputs. The paths to and from the driver of the video amplifier and picture tube circuit group are meeting/separating paths.

A *feedback path* is a signal path from one circuit to a point or circuit preceding it in the signal flow sequence. The AGC circuit group and the AFC circuit of the horizontal oscillator and driver are examples of feedback paths.

A *switching path* contains a selector of some sort (usually a switch) that provides a different signal path for each switch position. The paths in the RF tuner through the channel selector are switching paths.

1-10.5   Signal Tracing Versus Signal Substitution

Both signal-tracing and signal substitution (or signal-injection) techniques are used frequently in troubleshooting all types of television receiver circuits. The choice between tracing and substitution depends on the test equipment used. As discussed in Chapter 2, some signal generators designed for television service have outputs that simulate signals found in all major signal paths of a receiver (RF, IF, sound IF, audio, video pulses, etc.). If you have such a generator, signal injection is the logical choice, since you can test all of the circuit groups on an individual basis (independent of other circuit groups). However, it is possible to troubleshoot receivers with signal tracing alone (and this technique is recommended by many technicians).

Signal tracing is done by examining the signals at test points with a monitoring device such as an oscilloscope, multimeter, or loudspeaker. In signal tracing, the input probe of the indicating or monitoring device used to trace the signal is moved from point to point, with a signal applied at a fixed point. The applied signal can be generated from an external device, or the normal signal associated with the equipment can be used (such as using the regular broadcast signal to trace through a television receiver).

Signal substitution is done by injecting an artificial signal (from a signal generator, sweep generator, etc.) into a circuit or circuit group (or to the complete television receiver) to check performance. In signal injection, the injected signal is moved from point to point, with an indicating or monitoring device remaining fixed at one point. The monitoring can be done with external test equipment, or the picture tube or loudspeaker.

Both signal tracing and substitution are often used simultaneously in troubleshooting television receivers. For example, when troubleshooting the RF tuner and IF sections it is common practice to inject a sweep signal at the input and monitor the output with an oscilloscope.

1-10.6   Half-split Technique

The half-split technique is based on the idea of simultaneous elimination of the maximum number of circuit groups or circuits with each test. This will save both time and effort. The half-split technique is used primarily when isolating trouble in a linear signal path but can be used with other types of signal paths. In using the half-split system, brackets are placed at good input and bad output points in the normal manner, and the symptoms are studied. Unless the symptoms point definitely to one circuit group or circuit which might be the trouble source, the most logical place to make the first test is at a convenient test point *halfway between* the brackets.

1-10.7   Isolating Trouble to a Circuit Within a Circuit Group

Once trouble is definitely isolated to a faulty circuit group, the next step is to isolate the trouble to the faulty circuit within the group. Bracketing, half-splitting, signal tracing, signal injection, and *knowledge of the signal path* in the circuit group are all important to this step and are used in essentially the same way as for isolating trouble to the circuit group.

1-11. LOCATING A SPECIFIC TROUBLE

The ability to recognize symptoms and to verify them with test equipment will help you make logical decisions regarding the selection and localization of the faulty circuit group. This ability will also help you to isolate trouble to a faulty circuit. The final step of troubleshooting—locating the *specific* trouble—requires testing of the various branches of the faulty circuit to find the defective part.

The proper performance of the locate step will enable you to find the cause of trouble, repair it, and return the receiver to normal operation. A follow-up to this step is to record the trouble so that, from the history of the receiver, future troubles may be easier to locate. Also, such a history may point out consistent failures which could be caused by a design error.

1-11.1   Locating Troubles in Plug-in Modules

Because the trend in modern television receivers (and most other electronic equipment) is toward IC and sealed-module design, technicians often assume that it is not necessary to locate specific troubles to individual parts. That is, they assume all troubles can be repaired by replacement of sealed modules. Some technicians are even trained that way. The assumption is not true.

While the use of replaceable modules often minimizes the number of steps required in troubleshooting, it is still necessary to check circuit branches to parts outside the module. Front-panel operating controls are a good example of this, since such controls are not located in the sealed units. Instead, the controls are connected to the terminal of an IC, circuit board, or plug-in module.

1-11.2   Inspection Using the Senses

After the trouble is isolated to a circuit, the first step in locating the trouble is to perform a preliminary inspection using the senses. For example, burned or charred resistors can often be spotted by visual observation or by smell. The same holds true for oil-filled or wax-filled parts, such as some capacitors, coils, and transformers.

Overheated parts, such as hot transistor cases, can be located quickly by touch. The sense of hearing can be used to listen for high-voltage arcing between wires or wires and the chassis, for "cooking" or overloaded or overheated transformers, or for hum or lack of hum, whichever the case may be. Although all of the senses are used, the procedure is referred to more frequently as a *visual* inspection.

1-11.3   Testing to Locate a Faulty Part

Testing vacuum tubes. Vacuum tubes are relatively easy to replace (compared to transistors or IC units). For that reason, two common practices have been developed over the years when troubleshooting vacuum-tube receivers. One practice is to test all vacuum tubes, by substitution, as a first step in troubleshooting. The other practice is to remove and test all tubes on a tube tester, as a first troubleshooting step. Neither practice is valid.

With tube substitution, the usual procedure is to replace tubes one at a time until the receiver again works normally, then the *last* tube replaced is discarded, and all the other original tubes are reinserted in their respective sockets. There are several problems with this approach.

Some oscillator or high-frequency circuits may operate with one new tube and not with another because of the differences in interelectrode capacitance between the tube elements (a good tube may react like a bad tube). When removing or inserting the tubes, rocking or rotating them may result in bent pins or broken weld wires where the pins enter the envelope. If there is more than one bad tube in the receiver, substituting good tubes one at a time and reinserting the original tube before substituting the next tube will not locate the defective tube. Finally, if the replacement tube becomes defective immediately after substitution, there definitely is circuit trouble, and further troubleshooting is required anyway.

Testing all tubes on a tube tester, as the first troubleshooting step, is also not recommended. Because this procedure has been followed so religiously in the past, the practice has led to the misconception that defective tubes are the cause of all or most receiver failures. Even if defective tubes cause 50% (or more) of all receiver failures, the process of removing tubes, checking them on a sometimes marginal tube tester, and replacing them with new tubes, as a *first step without further circuit checking,* is a waste of time (and is poor troubleshooting practice).

This does not mean that the tube should never be checked first, after the trouble is isolated to a circuit. For example, the power can be turned on and the tube filaments checked first for proper warm-up. If the tube envelopes are glass, a visual inspection will show whether the filament is burned out. For metal-envelope tubes, you can feel the envelope to find out whether the filament is lit.

This type of test may speed the troubleshooting effort by quickly locating a tube having a burned-out filament. If the tube does not warm up properly, remove the tube and check it on a tube tester or substitute a new (known good) tube, whichever is most convenient. In either case, the complete circuit should be checked to determine if the tube burned out naturally from long use or from some trouble in the circuit. Simply replacing the tube without checking the rest of the circuit does not complete the location of trouble. You still must verify whether the burned-out tube is the cause or the effect of trouble.

The procedure just described works well for checking tubes when the tube filaments are all connected in *parallel*. With parallel filaments, when the filament of one tube burns out, that tube (and only that tube) will show a bad (unlit) filament. The filaments of tubes connected in *series* present more of a problem. When one filament burns out, all the filaments in the string will be unlit. This condition is shown in Fig. 1-16 and is typical for millions of television receivers. With such circuits it is more difficult to determine which filament in the string is the bad one.

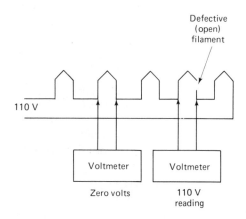

Fig. 1-16. Method of locating defective (open) filament in vacuum tube television receivers (where filaments are connected in series)

Removing the tubes one at a time and checking their filaments for continuity with an ohmmeter is time consuming, and if care is not taken, the tube may be damaged during the test since the current from the ohmmeter (set on its lowest scale) may be high enough to burn out the filament.

A better test is to measure the voltage across the filament terminals of the tube socket (provided that the bottom of the socket is accessible and all the tubes are left in their sockets). All good filaments in the string will show zero voltage, but the one that is defective (burned out) will have the full voltage that is applied to the filament string, as shown in Fig. 1-16.

Testing solid-state and IC receivers. Unlike vacuum tubes, most transistors, ICs, and solid-state diodes are not easily replaced. Thus, the old electronic troubleshooting procedure of replacing tubes at the first sign of trouble has not carried over into solid-state equipment. Instead, solid-state circuits are analyzed by testing to locate faulty parts.

Testing the active device. For service purposes, the vacuum tube, transistor, IC, and solid-state diode may be considered the *active device* (or common denominator) in any electronic circuit. Because of their key position in the circuit, these devices are a convenient point for evaluating operation of the entire circuit (through waveform, voltage, and resistance tests). Making these tests at the terminals of the active device will usually result in locating the trouble quickly.

Waveform testing. Usually, the first step in circuit testing is to analyze the output waveform of the circuit or the output waveform of the active device (generally the plate of a vacuum tube or the collector of a transistor). Of course, in some circuits (such as power supplies) there is no output waveform. And in some circuits there is no waveform of any significance.

In addition to checking for the presence of waveforms on an oscilloscope, the waveform must be analyzed in detail to check the amplitude, duration, phase, and/or shape. As discussed throughout this book, a careful analysis of waveforms can pinpoint the most likely branch of a circuit that is defective.

Transistor and diode testers. It is possible to test transistors and diodes in circuit, using in-circuit testers. Such testers are discussed further in Chapter 2. These testers are usually quite good for transistors used at lower frequencies, particularly in the audio range. However, most in-circuit testers will not show the high-frequency characteristics of transistors. (The same is true for out-of-circuit transistor and diode testers.) For example, it is quite possible for a transistor to perform well in the audio section, or even in the sweep sections, of a receiver. However, the same transistor will be hopelessly inadequately in the RF, IF, or video sections of the receiver.

Voltage testing. After waveform analysis and/or in-circuit tests, the next logical step is voltage measurement at the active device terminals or leads. Always pay particular attention to those terminals which show an abnormal waveform. These are the terminals most likely to show an abnormal voltage.

When properly prepared service literature is available (with waveform, voltage, and resistance information) the actual voltage measurements can be compared with the normal voltages listed in the service literature. This test will often help isolate the trouble to a single branch of the circuit.

Relative voltages in solid-state receivers. It is often necessary to trouble-shoot solid-state circuits without benefit of adequate voltage and resistance information. This can be done using the schematic diagram to make a logical analysis of the relative voltages at the transistor terminals. For example, with an NPN transistor, the base must be positive in relation to the emitter if there is to be emitter-collector current flow. That is, the emitter-base junction will be forward-biased when the base is more positive (or less negative) than the emitter. The problem of troubleshooting solid-state television circuits on the basis of relative voltages is discussed further in Chapter 3.

Resistance measurements. After waveform and voltage measurements are made, it is often helpful to make resistance measurements at the same point on the active device (or at other points in the circuit), particularly where an abnormal waveform and/or voltage is found. Suspected parts often can be checked by a resistance measurement, or a continuity check can be made to find point-to-point resistance of the suspected branch. Considerable care must be used when making resistance measurements in solid-state circuits. The junctions of transistors act like diodes. When biased with the right polarity (by the ohmmeter battery) the diodes will conduct and produce false resistance readings. This condition is discussed in Chapter 3.

Current measurements. In rare cases, the current in a particular television circuit branch can be measured directly with an ammeter. However, it is usually simpler and more practical to measure the voltage and resistance of a circuit and then calculate the current.

1-11.4   Waveform Measurement

When testing to locate trouble, the waveform measurements are made with the circuit in operation and usually with an input signal (or signals) applied. The signals can be from an external generator, or you can use the regular television broadcast signals, whichever is convenient for the particular measurement. If you use waveform reproductions found in the service literature, follow all of the notes and precautions described in the literature. Usually the literature will specify the position of operating controls, typical input signal amplitudes, and so on.

Note that most television circuit waveforms are measured with the oscilloscope sweep setting at 30 Hz or 7875 Hz. The 30-Hz frequency is one-half the vertical frequency of 60 Hz. Thus, there will be two wave-forms displayed on the oscilloscope when checking any circuit containing vertical pulses or signals (output of video detector, sync separator, vertical sweep). Likewise, the 7875-Hz frequency is one-half the horizontal frequency of 15,750 Hz and is used to display two waveforms when checking circuits containing horizontal signals (output of video detector, sync separator, horizontal oscillator, horizontal output).

Figure 1-17 shows waveform reproductions found in typical television service literature. That is, the approximate shape of the waveform is given on the schematic diagram (rather than in a separate chart), together with the voltage amplitude. There may or may not be a note on the diagram indicating the approximate frequency of the waveform. Note that only the horizontal sync pulses (15,750 Hz) are shown at the input to the sync amplifier Q_1 (at the base). As a practical matter, it is reasonable to assume that there will be vertical sync pulses at the same point. Thus, you could monitor the Q_1 base with an oscilloscope and expect to see two waveforms representing the vertical sync pulses (if the oscilloscope is set to a 30-Hz sweep).

There is a relationship between waveforms and trouble symptoms. Complete failure of a circuit will usually result in the absence of a waveform. A poorly performing circuit will usually produce an abnormal or distorted waveform. Also, exact waveforms are not always critical in all circuits. The representations of waveforms given in television service literature are usually only approximations of the actual waveform. Likewise, the same waveform will not always appear exactly the same when measured with different oscilloscopes.

1-11.5 Voltage Measurements

When testing to locate trouble the voltage measurements are made with the circuit in operation but usually with no signals applied. In some cases, the voltage measurements can be made with the receiver tuned to a television broadcast. If you are using the voltage information found in service literature, follow all of the notes and precautions. Usually the information will specify the position of operating controls, typical input voltages, and so on.

In most television service literature, the voltages are given on the schematic diagram, along with the waveforms, as shown in Fig. 1-18. This system is quite accurate but does require that you find the actual physical location of the terminals where the voltages are to be measured.

Because of the safety practice of setting a voltmeter to its highest scale before making measurements, the terminals having the highest voltage (vacuum-tube plate or transistor collector) should be checked first. (In some solid-state receivers, such as shown in Fig. 1-17, the collector is grounded, and the emitter has the highest voltage.) Then the elements having lesser voltage should be checked in descending order.

If you have had any practical experience in troubleshooting, you know that voltage (as well as resistance and waveform) measurements are seldom

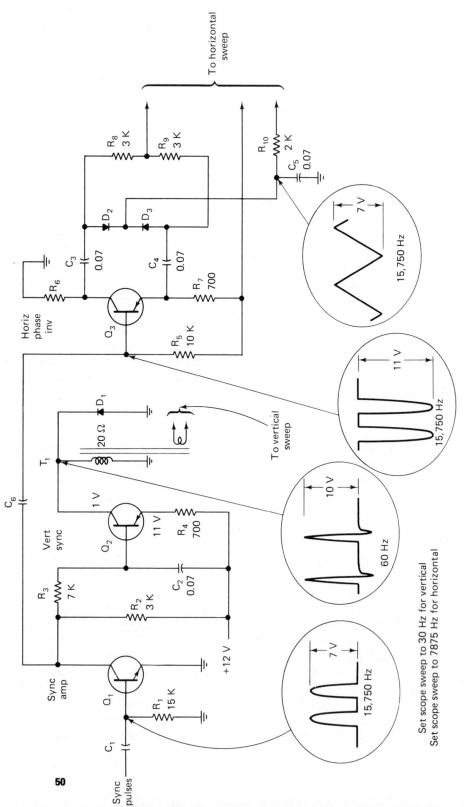

50

Fig. 1-17. Waveform representations found on typical television service schematics

Set scope sweep to 30 Hz for vertical
Set scope sweep to 7875 Hz for horizontal

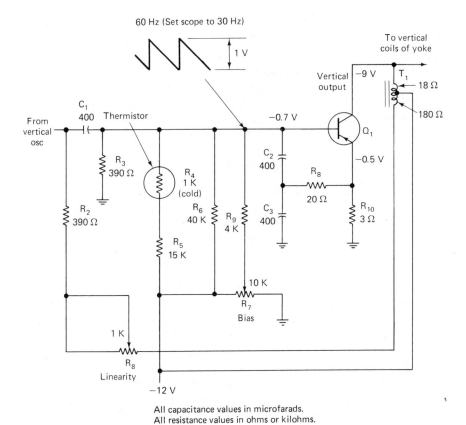

Fig. 1-18. Voltage (and waveform) representations found on typical television service schematics

identical to those listed in the service literature. This brings up an important question concerning voltage measurements: "How close is good enough?" In answering this question, there are several factors to consider.

The tolerances of the resistors, which greatly affect the voltage readings in a circuit, may be 20, 10, or 5%. Resistors with 1% (or better) tolerance are used in some critical circuits. The tolerances marked or color-coded on the parts are therefore one important factor. Transistors and diodes have a fairly wide range of characteristics and thus will cause variations in voltage readings.

The accuracy of test instruments must also be considered. Most voltmeters have accuracies of a few percent (typically 5 to 10%). Precision laboratory meters (generally not used in television troubleshooting) have a much greater accuracy.

For proper operation, critical circuits may require voltage to be within a very close tolerance (at least within 10% and probably closer to 3%). However, many circuits will operate satisfactorily if the voltages are to within 20 or 30%.

Generally, the most important factors to consider in voltage measurement accuracy are the symptoms and the output signal. If no output signal is produced, you should expect a fairly large variation of voltages in the trouble area. Trouble which results in circuit performance that is just out of tolerance may cause only a slight change in circuit voltages.

1-11.6 Resistance Measurements

Resistance measurements must be made with no power applied. However, in some cases, various operating controls must be in certain positions to produce resistance readings similar to those found in the service literature. This is particularly true of controls that have variable resistances.

Always observe any notes or precautions found on the service literature. In any circuit, always check that the filter capacitors are discharged before making resistance measurements. After all safety precautions and notes have been observed, measure the resistance from the terminals of the active device to the chassis (or ground) or between any two points that are connected by wiring or parts.

In most television service literature, resistance information is given on the schematic diagram, along with the waveforms and voltage information, as shown in Fig. 1-18. Do not be surprised if you find television service literature with little or no resistance information. Generally, the only resistances given are the values of resistors and the dc resistance of coils and transformers. Rarely do you find the resistance from all terminals of the active device given in television service literature.

The reasoning for the omission of active device resistances has some logic. If there is a condition in any active device terminal circuit that will produce an abnormal resistance (say an open or shorted resistor or a resistor that has changed drastically in value), the voltage at that terminal will be abnormal. If such an abnormal voltage reading is found, it is then necessary to check out each resistance in the terminal circuit on an individual basis.

Because of the shunting effect of other parts connected in parallel, the resistance of an individual part or circuit may be difficult to check. In such cases, it is necessary to disconnect one terminal of the part being tested from the rest of the circuit. This will leave the part open at one end, and the value of resistance measured is that of the part only.

Keep in mind that when making resistance checks a zero reading indicates a short circuit, and an infinite reading indicates an open circuit. Also remember the effect of the transistor junctions (acting as a forward-biased diode when biased on). The problems of resistance measurements are discussed thoroughly in Chapter 3.

1-11.7   Duplicating Waveform, Voltage, and Resistance Measurements

If you are responsible for service of one type or model of television receiver, it is strongly recommended that you duplicate all of the waveform, voltage, and resistance measurements found in the service literature with test equipment of your own. This should be done with a known good receiver that is operating properly. Then when you make measurements during troubleshooting you can spot even slight variations in voltage. Always make the initial measurements with test equipment that will normally be used during troubleshooting. If more than one set of test equipment is used, make the initial measurements with all available test equipment, and record the variations.

1-11.8   Using Schematic Diagrams

Regardless of the type of trouble symptom, the actual fault can be traced eventually to one or more of the circuit parts—vacuum tubes, transistors, ICs, diodes, resistors, capacitors, coils, transformers, etc. The waveform-voltage-resistance checks will then indicate which branch within a circuit is at fault. Then you must locate the particular part that is causing the trouble in the branch.

This requires that you be able to read a schematic diagram. These diagrams show what is inside the blocks on a block diagram and provide the final picture of the television receiver. Usually, you must service television receivers with nothing but a schematic diagram. If you are fortunate, the diagram will show some voltages and waveforms.

Examples of using schematic diagrams. The following example shows how the schematic diagram is used to locate a fault within a circuit. Although this example involves only selected circuits, the same *basic troubleshooting principles* apply to all circuits of a television receiver.

Assume that the circuit of Fig. 1-17 is being serviced. The reasoning that led to this particular circuit is as follows. The symptom is a complete lack of vertical sync. That is, the television picture rolls vertically and cannot be controlled by the vertical hold. All other functions, including the horizontal sync, are normal. This localizes the trouble to either the vertical sweep circuits or to the sync separator circuits. You isolate the trouble to the sync separator circuits by means of waveform measurement. Your first waveform measurement is at the collector of Q_2, which is (simultaneously) the vertical output of the sync separator and the input to the vertical sweep circuits. The waveform is absent or abnormal, and you place a bad-output bracket at the collector of Q_2. All of the remaining waveforms shown in Fig. 1-17 are normal. Thus, you can assume that Q_1 is functioning normally (it is passing the horizontal sync pulses) and should pass the vertical sync pulses. However, it is better not to assume anything. Instead, confirm that there are vertical sync pulses at both the input (base) and output (collector) of Q_1. Note that neither of these pulses is shown on Fig. 1-17 (which is quite typical for television service literature).

To overcome this lack of information, set the oscilloscope sweep to 30 Hz, and measure whatever waveforms appear at the base and collector of Q_1. If you find two pulses, they are the vertical sync pulses. The amplitude of these pulses should be *approximately* equal to the horizontal pulse amplitudes. That is, the base pulse should be about 7 V, with a 10-V pulse at the collector. Assume that this is the case in making your measurement. Now, it is reasonable to assume that Q_1 is *probably* functioning normally and that you have a good input to Q_2.

With the trouble isolated to Q_2, you must now locate the specific fault in the Q_2 circuit, using voltage and resistance measurements. The parts involved are R_3, R_4, C_2, Q_2, D_1, and T_1. Note that only two voltages are given for the terminals of the active device, Q_2. These are 1 V for the collector and 11 V for the emitter. Thus, there is only a 1-V drop across the winding of T_1, yet the normal waveform shows a pulse of about 10 V in amplitude. This indicates that Q_2 is normally biased to (or beyond) cutoff and is switched on by pulses from the junction of R_3 and C_2 (which form the vertical integrator discussed in Sec. 1-2.5). To bias a PNP transistor to cutoff, the base must be less negative (or more positive) than the emitter. Since a positive 12-V supply is used, the base of Q_2 must be more positive than the emitter (11 V) but less than 12 V. Thus, if you measure some voltage in the 11- to 12-V range from the base of Q_2 to ground, it is probably correct, and R_3 is probably good.

To clear the vertical integrator (R_3 and C_2) from any suspicion, measure the waveforms at the collector of Q_1 and the base of Q_2, using oscilloscope sweeps of 30 Hz and 7875 Hz. If R_3 and C_2 are doing their job, waveforms should appear at the base of Q_2 only when the 30-Hz sweep is used. If no waveforms appear, or if both the horizontal and vertical waveforms are found at the base of Q_2, R_3 and C_2 are suspect.

Now assume that there is a good pulse at the base of Q_2, and you have thus narrowed the problem down to Q_2, R_4, D_1, or T_1. At this point, you could check Q_2 by substitution, or with an in-circuit transistor tester, if convenient. However, you will probably do better to make voltage measurements at all terminals of Q_2. This will pinpoint any obvious part failures. For example, if R_4 or T_1 is open, all of the voltages will be abnormal. If R_4 is shorted, the emitter will be at the supply voltage (12 V) instead of at 11 V.

Keep in mind that voltage measurements alone may not solve the problem. For example, if D_1 is shorted, or leaking badly, it is still possible to get a near-normal voltage reading at the Q_2 collector but a poor pulse waveform. Also, voltage measurements in pulse circuits are sometimes ambiguous. The dc voltage is given as 1 V, yet the pulse (or ac) voltage is 10 V. If the meter is set to measure dc, the pulses may increase the average dc voltage (in some meters) so that an abnormal reading will appear, even though the circuit is functioning normally. If D_1 is open, the dc voltage could be normal, but the pulse waveform will be abnormal. Diode D_1 functions to remove any negative-going pulses at the collector of Q_2. As another example, assume that the primary winding of T_1 has partially shorted turns. This will produce a near-normal dc voltage but would drastically reduce the pulse output to the vertical sweep circuits.

Unless you have pinpointed the problem with waveform and voltage measurements, your last step is to make resistance measurements. As shown, no resistance values are given for any of the Q_2 terminals. Thus, there is no point in measuring from the terminals to ground (as is standard practice in troubleshooting military-type equipment). Instead, you must check the resistance of each part on an individual basis.

To get accurate resistance readings of individual parts, you must disconnect one lead from the remainder of the circuit. If not, the effect of solid-state devices in the circuit can further confuse the troubleshooting process. For example, assume that you are measuring the T_1 primary winding resistance by connecting an ohmmeter across the winding. If the ohmmeter leads are connected so that the positive terminal of the ohmmeter battery is connected to the D_1 anode (ground end), D_1 will be forward-biased, and the ohmmeter will read the combined D_1 and T_1 resistances. This problem can be eliminated by reversing the ohmmeter leads and measuring the resistance both ways. If there is a difference in the resistance values with the leads reversed, check the schematic for possible forward-bias conditions in diodes *and transistor* junctions in the associated circuit. Whenever practical, simply disconnect one lead of the part being measured. The problem of making resistance measurements during troubleshooting is discussed further in Chapter 3.

1-11.9   Internal Adjustments During Troubleshooting

Adjustment of controls (both internal adjustment controls and receiver operating controls) can affect circuit conditions. This may lead to false conclusions during troubleshooting.

For example, the bias on the base of vertical output transistor Q_1 (in Fig. 1-18) is set by vertical bias potentiometer R_7. In turn, the value of the bias determines the portion of the sweep used by Q_1. Thus, the sweep voltages applied to the vertical deflection yoke are set, in part, by the vertical bias control. That is, adjustment of the vertical bias control affects both height and linearity of the vertical sweep. However, the main purpose of the vertical bias control is to compensate when a major part in the circuit (such as the output transistor, output transformer, or vertical deflection yoke) is replaced. The vertical bias control usually does not require adjustment during troubleshooting. On the other hand, the vertical linearity control R_8 has a considerable effect on the linearity of the vertical sweep and often requires adjustment during the troubleshooting process.

These two extremes often lead some technicians to one of two unwise courses of action:

First, the technician may launch into a complete alignment procedure (or whatever internal adjustments are available) once the trouble is isolated to a circuit. No internal control, no matter how inaccessible, is left untouched. The technician reasons that it is easier to make adjustments than to replace parts. While such a procedure will eliminate improper adjustment as a possible fault, the procedure can also create more trouble than is repaired. Indiscriminate internal adjustment is the technician's version of operator trouble.

At the other extreme, a technician may replace part after part, when a simple screwdriver adjustment will repair the problem. This usually means that the technician simply does not know how to perform the adjustment procedure or does not know what the control does in the circuit. Either way, a study of the service literature should resolve the situation.

To take the middle ground, do not make any internal adjustments during the troubleshooting procedure until trouble has been isolated to a circuit, and then only when the trouble symptom or test results indicate possible maladjustment.

For example, assume that the vertical oscillator is provided with a back-panel adjustment control (vertical hold) that sets the frequency of oscillation. If waveform measurements at the circuit show that the vertical oscillator is off frequency (not at 60 Hz), it is logical to adjust the vertical hold control. However, if waveform measurements show only a very low output (but on frequency), adjustment of the vertical hold control during troubleshooting could be confusing (and could cause further problems).

An exception to this rule is when the service literature recommends alignment or adjustment as part of the troubleshooting procedure. Generally, alignment or adjustment is checked after test and repair have been performed. This assures that the repair (parts replacement) procedure has not upset circuit adjustment.

1-11.10   Trouble Resulting From More Than One Fault

A review of all the symptoms and test information obtained thus far will help you verify the part located as the sole trouble or to isolate other faulty parts. This is true whether the malfunction of these parts is due to the isolated part or some entirely unrelated cause.

If the isolated bad part can produce all the normal and abnormal symptoms and indications that you have accumulated, you can logically assume that it is the sole cause of the trouble. If not, then you must use your knowledge of electronics and of the receiver to find what other part (or parts) could have become defective and produce all of the symptoms.

When one part fails it often results in abnormal voltages or currents which could damage other parts. Trouble is often isolated to a faulty part which is a result of an original trouble, rather than the source of trouble.

For example, assume that the troubleshooting procedure thus far has isolated a transistor as the cause of trouble—the transistor is burned out. What would cause this? Excessive current can destroy the transistor by causing internal shorts or by altering the characteristics of the semiconductor material, which is very temperature sensitive. Thus, the problem becomes a matter of finding how the excessive current was produced.

Excessive current in a transistor can be caused by an extremely large input signal, which will overdrive the transistor. This indicates a fault somewhere in the circuitry ahead of the input connection.

Power surges (intermittent, excessive outputs) from the power supply can also cause the transistor to burn out. In fact, power supply surges are a common cause of transistor burnout.

All of these conditions should be checked *before* placing a new transistor in the circuit. Some other typical malfunctions, along with their common causes, include:

1. Burned-out transistors caused by *thermal runaway.* An increase in transistor current heats the transistor, causing a further increase in current, resulting in more heat. This continues until the heat dissipation capabilities of the transistor are exceeded. Bias-stabilization circuits are generally included in most well-designed transistor equipment to prevent thermal runaway.

2. Power supply overload caused by a short circuit in some portion of the voltage distribution network.

3. Burned-out transistor caused by a shorted blocking capacitor.

4. Blown fuses caused by power supply surges or shorts in filtering (power network).

It is obviously impractical to list all of the common faults and their related causes that you may find in troubleshooting a television receiver. Generally, when a part fails the cause is a circuit condition which exceeded the maximum ratings of the part. However, it is quite possible for a part to simply "go bad."

The circuit condition that causes a failure can be temporary and accidental, or it can even be a basic design problem (as indicated by a history of repeated failures). No matter what the cause, your job is to find the trouble, verify the source or cause, and then repair the trouble.

1-11.11   Repairing Troubles

In a strict sense, repairing the trouble is not part of the troubleshooting procedure. However, repair is an important part of the total effort involved in getting the receiver back into operation. Repairs must be made before the receiver can be checked out and made ready for operation.

Never replace a part if it fails a second time without making sure that the cause of trouble is eliminated. Preferably, the cause of trouble should be pinpointed before replacing a part the first time. However, this is not always practical. For example, if a resistor burns out because of an intermittent short, and you have cleared the short, the next step is to replace the resistor. However, the short could happen again, burning out the replacement resistor. If so, you must recheck every element and lead in the circuit.

When replacing a defective part, an *exact* replacement should be used if it is available. If not available, and if the original part is beyond repair, an equivalent *or better* part should be used. Never install a replacement part having characteristics or ratings inferior to those of the original.

Another factor to consider when repairing the trouble is that, if at all possible, the replacement part should be installed in the *same physical location* as the original, with the same lead lengths, and so on. This precaution is generally optional in most low-frequency or dc circuits but must be followed for high-frequency applications.

In high-frequency (RF, IF, video, etc.), circuits of the receiver, changing the location of parts may cause the circuit to become detuned or otherwise out of alignment.

1-11.12   Operational Checkout

Even after the trouble is found and the faulty part is located and replaced, the troubleshooting effort is not necessarily completed. You should make an operational check to verify that the receiver is free of *all* faults and is performing properly again. Never assume that simply because a defective part is located and replaced the receiver will automatically operate normally again. As a matter of fact, in practical troubleshooting of any television receiver, never assume anything; prove it. Check the picture and sound on all channels. This will ensure that one fault has not caused another. If practical, after you have checked the receiver, have the customer check both picture and sound on all channels.

When the operational check is completed, and the receiver is "certified" (by you and the customer) to be operating normally, make a brief record of the symptoms, faulty parts, and remedy. This is particularly helpful when you must troubleshoot similar receivers. Even a simple record of trouble-shooting will give you a history of the receiver for future reference.

If the receiver does not perform properly during the operational check-out, you must continue troubleshooting. If the symptoms are the same as (or similar to) the original trouble symptoms, retrace your steps, one at a time. If the symptoms are entirely different, you may have to repeat the entire troubleshooting procedure from the start. However, this is usually not necessary.

For example, assume that the receiver does not check out because a replacement transistor has detuned the circuit (say, an IF amplifier transistor has been replaced). When this is the case you can repair the trouble by IF alignment rather than returning to the first step of troubleshooting and repeating the entire procedure. Keep in mind that you have arrived at the defective circuit or component by a systematic procedure. Thus, retracing your steps, one at a time, is the logical method.

2

Television Service Equipment

The test equipment used in television service is basically the same as that used in other fields of electronics. That is, most service procedures are performed using meters, signal generators, oscilloscopes, power supplies, and assorted clips, patchcords, etc. Theoretically, all television service procedures could be performed using conventional test equipment, provided that the oscilloscopes had the necessary gain and band-pass characteristics, that the signal generators covered the appropriate frequencies, and so on. However, there are specialized versions of the basic test equipment that have been developed specifically to simplify television service. We shall concentrate on this specialized equipment throughout this chapter.

It is not the purpose of this chapter to promote one type of test equipment over another (or one manufacturer over another). Instead, the chapter is devoted to the basic operating principles of test equipment types in common use. Each type of test equipment is discussed in turn. You can then select the type of equipment best suited to your own needs and pocketbook. Although complicated theory has been avoided, the following discussions cover what each type of test equipment does, what signals or characteristics are to be expected from each type of equipment, and how, in brief, the equipment circuits operate to produce these signals and characteristics. The discussions describe how the features and outputs found on present-day test equipment relate to specific problems in television service.

A thorough study of this chapter will make you familiar with the basic principles and operating procedures for typical equipment used in general television service. It is assumed that you will take the time to become equally familiar with the principles and operating controls for any particular test equipment you use. Such information is contained in the service literature for the particular equipment.

It is absolutely essential that you become thoroughly familiar with your own particular test instruments. No amount of textbook instruction will make you an expert in operating test equipment; it takes actual practice.

It is strongly recommended that you establish a routine operating procedure, or sequence of operation, for each item of service equipment. That will save time and familiarize you with the capabilities and limitations of your own equipment, thus minimizing false conclusions based on unknown operating conditions.

2-1. SAFETY PRECAUTIONS IN TELEVISION SERVICE

In addition to a routine operating procedure, certain precautions must be observed during operation of any electronic test equipment during service. Many of these precautions are the same for all types of test equipment; others are unique to special test instruments such as meters, oscilloscopes, and signals generators. Some of the precautions are designed to prevent damage to the test equipment or to the circuit where the service operation is being performed. Other precautions are to prevent injury to you. Where applicable, special safety precautions are included throughout the various chapters of this book.

The following general safety precautions should be studied thoroughly and then compared to any specific precautions called for in the test equipment service literature and in the related chapters of this book.

1. Many service instruments are housed in metal cases. These cases are connected to the ground of the internal circuit. For proper operation, the ground terminal of the instrument should always be connected to the ground of the receiver being serviced. Make certain that the chassis of the receiver being serviced is not connected to either side of the ac line or to any potential above ground. If there is any doubt, connect the receiver being serviced to the power line through an *isolation transformer,* as discussed in Sec. 2-5.

2. Remember that there is always danger in servicing receivers that operate at hazardous voltages (such as the high voltages used by the picture tube). Remember this especially as you pull off the receiver back panel and apply power through a "cheater" cord. Always make some effort to familiarize yourself with the receiver *before* servicing it, bearing in mind that high voltages may appear at unexpected points in a defective receiver.

3. It is good practice to remove power before connecting test leads to high-voltage points. It is preferable to make all service connections with the power removed. If this is impractical, be especially careful to avoid accidental contact with receiver circuits and objects that are grounded. Working with one hand away from the equipment and standing on a properly insulated floor lessens the danger of electrical shock.

4. Capacitors may store a charge large enough to be hazardous. Discharge filter capacitors before attaching test leads.

5. Remember that leads with broken insulation offer the additional hazard of high voltages appearing at exposed points along the leads. Check test leads for frayed or broken insulation before working with them.

6. To lessen the danger of accidental shock, disconnect test leads immediately after the test is completed.

7. Remember that the risk of severe shock is only one of the possible hazards. Even a minor shock can place you in danger of more serious risks, such as a bad fall or contact with a source of higher voltage.

8. The experienced service technician guards continuously against injury and does not work on hazardous circuits unless another person is available to assist in case of accident.

9. Even if you have considerable experience with test equipment used in service, always study the service literature of any instrument with which you are not thoroughly familiar.

10. Use only shielded leads and probes. Never allow your fingers to slip down to the meter probe tip when the probe is in contact with a "hot" circuit.

11. Avoid vibration and mechanical shock. Most electronic test equipment is delicate.

12. Study the circuit being serviced before making any test connections. Try to match the capabilities of the instrument to the circuit being serviced.

2-2. SIGNAL GENERATORS

The signal generator is an indispensable tool for practical television service. Without a signal generator, you are entirely dependent on signals broadcast from the transmitting station, and you are limited to signal tracing only. This means that you have no control over frequency, amplitude, or modulation of such signals and have no means for signal injection. With a signal generator of the appropriate type, you can duplicate transmitted signals or produce special signals required for alignment and test of all circuits found in a television receiver. Also, the frequency, amplitude, and modulation characteristics of the signals can be controlled so that you can check operation of a receiver under various signal conditions (weak, strong, normal, abnormal signals).

In addition to conventional RF and audio generators, the signal generators designed specifically for use in television service include the *sweep generator, marker generator, analyst generator, pattern generator,* and *color generator.* Often, the functions of these generator types are combined. For example, several manufacturers produce a sweep and marker generator. Likewise, the analyst and color generator functions are often combined in a single instrument. Color generators are described in Sec. 2-3. The purpose, operating principles, and typical characteristics of the remaining generators are as follows.

2-2.1 Sweep/Marker Generator

At one time, sweep generator and marker generators were manufactured as separate instruments. Today, the sweep and marker functions are usually combined into a single generator. The main purpose of the sweep/ marker generator in television service is the sweep-frequency alignment of the receiver. A sweep/marker generator capable of producing signals of the appropriate frequency can be used in conjunction with an oscilloscope to display the band-pass characteristics of all receiver tuned circuits (RF, IF, video,chroma).

The sweep portion of the sweep/marker generator is essentially an FM (frequency modulation) generator. When the sweep generator is set to a given frequency, this is the *center* frequency. The output varies back and forth through this center frequency. Thus, in simplest form, a sweep generator is a frequency-modulated RF oscillator. The rate at which the frequency modulation takes place is typically 60 Hz. Other sweep rates could be used, but since power lines usually have a 60-Hz frequency, this frequency is both convenient and economical for the sweep rate. The sweep width, or the amount of variation from the center frequency, is determined by a control, as is the center frequency.

The marker portion of the sweep/marker generator is essentially an RF signal generator with highly accurate dial markings that can be calibrated precisely against internal or external signals. Usually, the internal signals are crystal controlled. Marker signals are necessary to pinpoint frequencies when making sweep-frequency alignment. Although sweep generators are accurate in both center frequency and sweep width, it is almost impossible to pick out a particular frequency along the spectrum of frequencies being swept. Thus, fixed-frequency "marker" signals are injected into the circuit along with the sweep-frequency generator output. Usually, this is accomplished by means of a built-in *marker-adder.*

In sweep-frequency alignment, the sweep generator is tuned to sweep the band of frequencies passed by the wide-band circuits (RF, IF, video, chroma) in the television receiver, and a trace representing the response characteristics of the circuits is displayed on the oscilloscope. The marker generator is used to provide calibrated markers along the response curve for checking the frequency settings of traps, for adjustment of capacitors and coils, and for measuring overall bandwidth of the receiver.

When the marker signal is coupled into the circuit under test, a vertical marker (often called a "pip") appears on the response curve. When the marker generator is tuned to a frequency within the band-pass limits of the receiver the marker indicates the position of that frequency on the sweep trace. The receiver circuits can then be adjusted to obtain the desired wave-shape, using the different frequency markers as checkpoints.

In addition to the basic sweep and marker outputs, a typical sweep/ marker generator will have a number of other special features. For example, a *variable bias* source is provided on some sweep/marker generators. During alignment of most receivers, it is necessary to disable the AGC circuits. This can be accomplished by applying a bias voltage (of appropriate amplitude and polarity) to the AGC line of the receiver.

A *blanking circuit* is another feature found on some sweep/marker generators. When the sweep generator output is swept across its spectrum, the frequencies usually go from low to high and then return from high to low. With the blanking circuit in operation, the return or retrace is blanked off. This makes it possible to view a zero reference line on the oscilloscope during the retrace period.

2-2.2   Basic Sweep-frequency Alignment Procedure

The relationship between the sweep/marker generator and the oscilloscope during sweep-frequency alignment is shown in Figs. 2-1 and 2-2.

If the equipment is connected as shown in Fig. 2-1, the oscilloscope horizontal sweep is triggered by a sawtooth output from the sweep generator. The oscilloscope's internal recurrent sweep is switched off, and the oscilloscope sweep selector and sync selector are set to external.

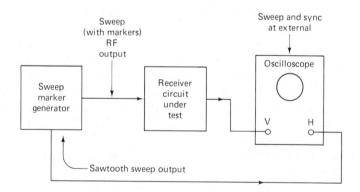

Fig. 2-1. Basic sweep/marker generator and oscilloscope test circuit (using horizontal sweep from generator)

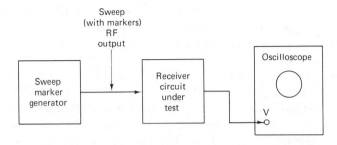

Fig. 2-2. Basic sweep/marker generator and oscilloscope test circuit (using oscilloscope internal horizontal sweep)

Under these conditions, the oscilloscope horizontal sweep should represent the total sweep spectrum. For example, as shown in Fig. 2-3, if the sweep is from 10 to 20 kHz, the left-hand end of the horizontal trace represents 10 kHz and the right-hand end represents 20 kHz. Any point along the horizontal trace represents a corresponding frequency. For example, the midpoint on the trace represents 15 kHz. If you want a rough approximation of frequency, adjust the horizontal gain control until the trace occupies an exact number of scale divisions on the oscilloscope screen (such as 10 cm for the 10- to 20-kHz sweep signal). Each centimeter division then represents 1 kHz.

If the equipment is connected as shown in Fig. 2-2, the oscilloscope horizontal sweep is triggered by the oscilloscope internal circuits (both the sweep selector and sync selector are set to internal). Certain conditions must be met to use the test connections of Fig. 2-2. If the oscilloscope is of the triggered sweep type (where the horizontal sweep is triggered by the

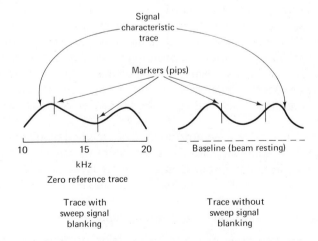

Fig. 2-3. Basic sweep/marker generator and oscilloscope displays

signal applied to the vertical input), there must be sufficient delay in the vertical input, or part of the response curve may be lost. If the oscilloscope is not of the triggered sweep type, the sweep generator must be swept at the same frequency as the oscilloscope horizontal sweep (usually the line power frequency of 60 Hz). Also, the oscilloscope, or the sweep/marker generator, must have a *phasing* control so that the two sweeps can be synchronized.

The method shown in Fig. 2-2 is used where the generator does not have a sweep output separate from the RF signal output or when it is not desired to use the sweep output. The Fig. 2-2 method was popular with older generators, although the method is still in use today. Blanking of the trace (if any blanking is used) is controlled by the oscilloscope circuits. The phase of the sweep is adjusted with the phasing control of the oscilloscope (although some sweep generators have phasing controls). If the phase adjustment is not properly set, the sweep curve on the oscilloscope may be prematurely cut off, or the curve may appear as a double or mirror image. These effects are shown in Fig. 2.4.

Double or Mirror image

Sharp cutoff

Fig. 2-4. Sweep response curves showing effects of improper phasing control adjustment

As shown in Fig. 2-3, the markers are used to provide accurate frequency measurement. On some generators, the marker output frequency can be adjusted until the marker is aligned at the desired point on the trace. The frequency is then read from the marker generator dial. This system has generally been replaced by generators that produce a number of markers at precise, crystal-controlled frequencies. The markers can be selected (one at a time or several at a time) as needed.

The response curve (oscilloscope trace) depends on the receiver circuit under test. If the circuit has a wide band-pass characteristic (as do most tuned circuits in a television receiver, typically 4 to 6 MHz), the sweep generator is set so that its sweep is wider than that of the circuit (typically set to 10 MHz or wider). Under these conditions, the trace will start low at the left, rise toward the middle, and then drop off at the right, as shown in Fig. 2-3.

The sweep/marker generator-oscilloscope method of alignment will tell at a glance the overall band-pass characteristics of the circuit (sharp response, flat response, irregular response at certain frequencies, and so on). The exact frequency limits of the band-pass can be measured with the markers.

2-2.3   Direct Injection Versus Postinjection

There are two basic methods for injection of marker signals into a sweep/marker generator-oscilloscope display.

With *direct injection,* as shown in Fig. 2-5, the sweep generator and marker generator signals are mixed before they are applied to the circuit under test. This method is sometimes known as *preinjection* and has generally been replaced by *postinjection.*

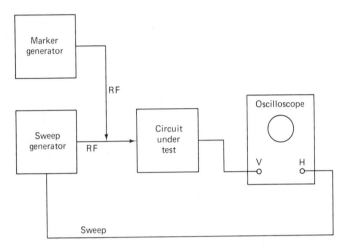

Fig. 2-5. Direct injection (preinjection) of marker signals

With *postinjection,* as shown in Fig. 2-6, the sweep generator output is applied to the circuit under test. A portion of the sweep generator output is also mixed with the marker generator output in a mixer-detector circuit known as a *postinjection marker-adder.* The mixed and detected output from both generators is then mixed with the detected output from the circuit under test. Thus, the oscilloscope vertical input represents the detected values of all three signals (sweep, marker, and circuit output).

Most present-day sweep/marker generators have some form of built-in postinjection marker-adder circuits. The postinjection (sometimes known as *bypass injection*) method for adding markers is usually preferred for television service, because postinjection minimizes the chance of over-loading the circuits under test and permits use of a narrow-band oscilloscope. At one time, postinjection marker-adder units were available as separate units and are still in use today.

2-2.4 Typical Sweep/marker Generators

Figure 2-7 shows a typical sweep/marker generator, with postinjection marker-adder provisions. The instrument has precision output at RF, IF, video, and chroma frequencies. With the generator you can align VHF tuners and IF, video, and chroma amplifiers. The sweep output is at fundamental frequencies on all VHF channels. Also included in continuously

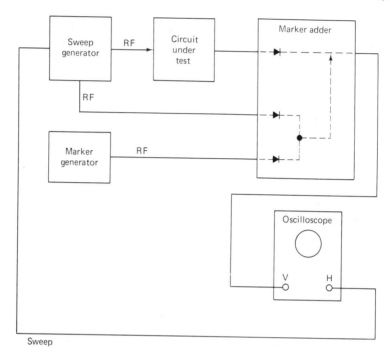

Fig. 2-6. Postinjection (bypass injection) of marker signals

adjustable output from 50 kHz to 50 MHz and from 88 to 108 MHz (the FM broadcast band).

Seven built-in crystal-controlled markers are provided at critical IF and color check points in the IF, video, and chroma band-pass ranges. These include 4.5-MHz sound, 41.25-MHz sound carrier, 41.67-MHz chroma

Fig. 2-7. RCA TV Sweep Chanalyst WR-514A

band pass, 45.75-MHz picture carrier, 47.25-MHz adjacent sound, and 39-to 50-MHz spare oscillator. It is also possible to inject external markers at frequencies from 1 to 250 MHz, and there is a provision for 400-Hz audio modulation. The output voltage is on the order of 0.1 V. The sweep width is continuously adjustable up to about 10 MHz. The output cable terminations are 300-Ω balanced for RF into the antenna terminals and 75-Ω single-ended for IF/video into corresponding circuits.

When the generator in Fig. 2-7 is used with accessories available from the manufacturer, it is possible to inject markers at the VHF channel frequencies. This makes it easier to align the RF tuner. A bias source is also available as an accessory but is not part of the generator.

Figure 2-8 shows another sweep/marker generator, complete with true postinjection marker-adder provisions and three bias supplies. The instrument has a number of features that simplify alignment and troubleshooting of television receivers.

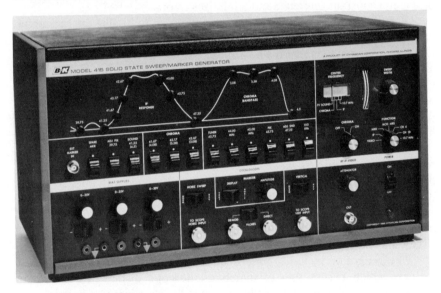

Fig. 2-8. B&K-Precision, Dynascan Corporation Model 415 Sweep/Marker Generators

There are RF outputs, with equivalents of all IF and chroma markers, available for channels 4 and 10. This makes it possible to connect the RF output to the antenna terminals of a receiver and, without further signal input reconnections, to evaluate alignment conditions of all tuned signal-processing circuits of the receiver.

There are 10 crystal-controlled markers with true postinjection marker adding. All 10 markers can be used simultaneously or individually. The fixed markers include the following: 39.75-MHz adjacent channel picture trap, 41.25-MHz sound trap, 41.67-MHz IF chroma side-band carrier, 42.75-MHz IF tuner line, 44.00-MHz IF center frequency, 45.00-MHz IF reference frequency, 45.75-MHz picture carrier, and 47.25-MHz adjacent channel sound trap. (The need for markers at these frequencies is discussed fully in Chapters 3 and 4.)

A continuous chain of 100-kHz markers can be superimposed on any of the internal crystal-controlled markers. This is included primarily for alignment of FM receivers but can be used for alignment of AFT (automatic fine tuning) circuits in television receivers by providing continuous bandwidth reference markers at 100-kHz intervals.

Visual reproduction of idealized alignment curves are provided on the generator front panel to indicate desired marker positions. When the marker switch is pressed the corresponding marker light comes on at the position where the marker should appear (on the oscilloscope waveform). This provides a constant reference and minimizes errors.

A built-in 15,750-kHz filter eliminates the need to disable the horizontal output circuits of the receiver during test and alignment. Some service literature recommends that the horizontal output circuits be disabled during alignment. This is required since the strong horizontal pulses at the output (flyback transformer, yoke, etc.), may be picked up and appear on the oscilloscope display during alignment. The filter, which is placed between the generator pickup probe and the oscilloscope input, removes any horizontal pulses before they can enter the oscilloscope display.

The generator contains *amplifiers and other circuits that compensate for possible poor low-frequency* response of the oscilloscope. One common problem in sweep frequency alignment is an oscilloscope with poor response that makes the receiver circuit appear to be defective or maladjusted. For example, if the oscilloscope has poor low-frequency response (a common problem), the pattern will not by symmetrical, even though the receiver circuit may be good.

The generator includes *pattern polarity reversal and sweep reversal* features which permit you to match oscilloscope displays as shown in manufacturers' alignment procedures. Usually, the alignment patterns are shown as positive polarity, with the low frequency at the left-hand side of the display (as shown in Fig. 2-3). However, you may monitor the waveform at a point where the polarity is negative, or the alignment instruction may show a pattern with the low frequencies at the right-hand side. Confusion is kept to a minimum with polarity and sweep reversal.

The markers can be tilted to horizontal or vertical positions, thus permitting easy identification. For example, if the sides of a band-pass display are steep (vertical) a horizontal marker is easier to identify. A vertical marker will show up better on the flat top (horizontal) of the same pattern.

The video sweep output permits direct sweep alignment of the chroma circuits where specified by the manufacturer. With some generators, it is necessary to use the IF sweep for signal injection and then monitor the video circuits for response.

All of the *crystal-!controlled IF marker signals are available at the generator output* for spot alignment of the traps and prealignment of band-pass circuits. Alignment instructions often specify that a fixed frequency signal be injected into a circuit and a specific control be adjusted for maximum or minimum indication at the output (rather than adjusting the band-pass characteristics for a given pattern or shape, within given frequency limits). Since the fixed frequencies are not available from all sweep generators, it is necessary to use an auxiliary generator for trap alignment.

The generator includes cables for connection to and from the circuit under test, as well as to both the vertical and horizontal inputs of the oscilloscope. After initial hookup to the receiver, IF and chroma response curves can be observed without additional intercabling changes. All changes are done internally with a master switch.

2-2.5   Analyst and Pattern Generators

An analyst generator is used for *signal substitution* (a form of signal injection). A bench-type analyst generator provides outputs that duplicate all essential signals in a television receiver. Such generators have RF, IF, composite video (including sync), sound, audio, separate sync, flyback test, and yoke test signals. As described in Chapters 3 and 4, these signals are used in troubleshooting (rather than alignment) and adjustment. In brief, the signals are injected at appropriate points in the receiver circuits, and the response is noted on the receiver picture tube (or loudspeaker).

The RF, IF, and video signals are usually in the *form of a pattern* (or patterns). On portable-type analyst generators, the patterns are typically lines or bars (horizontal and vertical), dots, crosshatch lines, square pulses, blank rasters, or color bars (color bar generators are discussed in Sec. 2-3). As an example, with a typical portable analyst generator, you can inject a crosshatch pattern (combined horizontal and vertical lines) at some particular VHF channel into the antenna terminals and note the response on the receiver picture tube. If the display is a clear, sharp crosshatch pattern, you know that the receiver is capable of producing a good picture. If desired, you can then inject the same crosshatch pattern into the IF cirucits, or video circuits, separately.

In addition to the basic analyst generator, there are *test pattern generators* available for extensive service work. The test pattern generator differs from the basic pattern generator in that the test pattern generator will reproduce positive transparencies of various pictures that may be inserted into the generator. In effect, the test pattern generator is a miniature television station, capable of reproducing nonmoving pictures.

In operation, the transparency is placed in the generator, the generator output (RF, IF, or video) is connected to the receiver, and the transparency picture is duplicated on the receiver picture tube. A test pattern generator is not necessarily a color generator, but most present-day test pattern generators do incorporate some type of color output. Generally, the color output is the keyed rainbow display (Sec. 2-3), which is superimposed over the black and white test pattern display.

One of the advantages of the test pattern generator is that it will provide a conventional "test pattern" presentation that duplicates the test patterns of television stations. In practically all cases, the test pattern generators also include analyzer functions, because the test pattern generator is primarily a shop or bench instrument for service of "tough dogs."

2-2.6   Typical Analyst/pattern Generators

Figure 2-9 is a typical portable analyst generator. The instrument produces several patterns, including color, at RF, IF, and video frequencies. The color output is the keyed rainbow color-bar display, with adjustable chroma level, discussed in Sec. 2-3. Other outputs include dots (a choice of 3 or 10 horizontal and vertical), crosshatch (a choice of 3 or 10), horizontal or vertical lines (3 or 10), a blank raster, a "superpulse," and a brightness marker bar on three of the color bars (at bars 3, 6, and 9, which represent $90°$, $180°$, and $270°$ phase shift of the color signal).

The superpulse signal is used for checking static convergence, gray-

Fig. 2-9. RCA Master Chro-Bar Generator WR-515A

scale tracking, smearing, and ringing directly on the receiver picture tube. The superpulse can also be used for troubleshooting with an oscilloscope.

The bar marker signals are used to identify the color bars when the keyed rainbow display is used. When a color television receiver is over-scanned, it is difficult to identify the individual color bars. As discussed in Chapter 4, many of the adjustments in a color receiver are based on identi-fying color bars and adjusting for a color match of one bar to another.

The blank raster is used in making color adjustments. As discussed in Chapter 4, the service instructions for most color receivers require that color purity be made with a raster that is free from all signals, noise, snow, etc.

Figure 2-10 is a bench-type analyst generator that produces several patterns, including color, at RF, IF, and video frequencies. The instrument

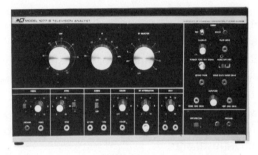

Fig. 2-10. B&K-Precision, Dynascan Corporation Model 1077B Television Analyst

generates all signals normally transmitted by a television station, and those produced within a television receiver for point-by-point signal substition troubleshooting techniques throughout all receiver circuits. For example, the instrument generates

> VHF signals on channels 2, 3, 4, 6, 7, 8, 12, and 13 for testing the RF tuner (and overall performance);
>
> UHF signals on channels 14 through 83 for testing UHF tuners (a portable generator usually excludes UHF);
>
> IF signals from 20 to 48 MHz for testing IF amplifier stages;
>
> Positive and negative composite video signal (including the sync pulses) for injection into video stages;
>
> A keyed color bar pattern which modulates the RF output for troubleshooting and adjusting color circuits;
>
> A color rainbow signal for injection into color amplifiers and demodulators;

A 4.5-MHz sound channel test signal that is frequency-modulated by a 1-kHz audio tone (for adjustment and test of the sound IF circuits);

A 1-kHz audio tone for test of the audio circuits;

A composite sync (horizontal and vertical sync pulses) of positive or negative polarity, adjustable in amplitude, and with variable impedance for troubleshooting sync circuits, picture tubes, blanking circuits, and solid-state keyed AGC circuits;

Vertical grid drive signals for troubleshooting vertical sweep circuits;

Vertical plate drive signal for checking vertical output transformers;

Vertical yoke test signals to determine if vertical yoke windings are defective;

Vertical sweep signals suitable for solid-state receivers;

Horizontal grid drive signals for troubleshooting horizontal sweep circuits;

Horizontal plate drive signal for checking horizontal output (flyback) transformers;

Horizontal sweep signals suitable for solid-state receivers;

A high-level keying pulse for testing keyed circuits (AGC, burst amplifier in color, and blanking); and

A test pattern (dots and lines) for color convergence adjustments, as well as for horizontal and vertical linearity, size, and aspect ratio checks and adjustments.

The instrument also provides for boost voltage indication (checks if picture tube boost voltage is present), high-voltage indication (checks for presence of adequate high voltage to picture tube), test of flyback transformers, horizontal deflection yokes, and width coils for shorts, positive or negative dc bias voltages (variable in amplitude), and a stable signal source for locating intermittent troubles.

The use of these signals in troubleshooting and adjustment is described in Chapters 3 and 4.

2-3. COLOR GENERATORS

Throughout the years, there have been three types of color generators in common use: the unkeyed rainbow generator, the keyed rainbow generator, and the NTSC (National Television Standards Committee) generator. Color generators manufactured today are of the *keyed rainbow type,* and we shall concentrate on these instruments. However, the other two color generator types are worth mentioning since they may still be in use. Also, the unkeyed rainbow principle is the basis for the keyed rainbow generator.

Likewise, the principles involved in the NTSC generator are worth study since they relate directly to how a color receiver converts transmitted signals into color pictures.

2-3.1   Unkeyed Rainbow Generator

The basic rainbow generator is a crystal-controlled oscillator producing a 3,563,795-Hz signal. The signal can be fed directly to the color band-pass amplifier of the receiver, or it can be used to modulate an RF oscillator that operates on the carrier frequency of a selected television channel.

The rainbow generator principle is quite simple. An oscillator that is operated at a frequency of 3,563,795 Hz (the color carrier frequency of 3,579,545 Hz, minus the horizontal sweep frequency of 15,750 Hz) will appear as a 3.58-MHz signal that is constantly changing in phase, when compared to the 3.58-MHz reference oscillator in the television receiver (Chapter 1), so that there is a complete change of phase of 360° for each horizontal sweep. This relationship is shown in Fig. 2-11. Thus, a complete range of colors (rainbow) is produced during each horizontal line. Each

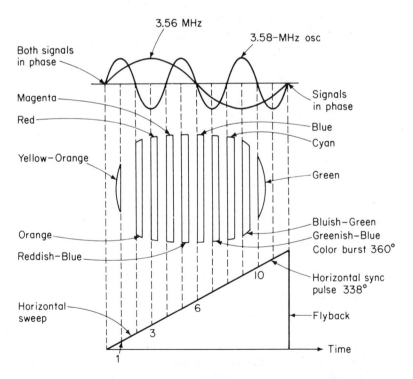

Fig. 2-11. Output signals and display of keyed rainbow generator (10 color bar generator)

line displays all colors simultaneously, since the phase between both oscillators at the beginning of the sweep is always zero. If the phase changes 360° during one sweep, then it will also be zero at the beginning of the next sweep.

The unkeyed rainbow generator has one particular drawback in that there are no "markers" or phase reference points between each of the colors in the rainbow pattern. For that reason, the unkeyed rainbow generator has been discontinued from manufacture and has been replaced by the keyed rainbow generator.

2-3.2 Keyed Rainbow Generator

The keyed rainbow generator is similar to the unkeyed unit, except that each color of the rainbow display is of uniform spacing and width. Also, there are blanks or bars between each color. The color-bar pattern is produced by gating the 3.56-MHz oscillator at a frequency 12 times higher than the horizontal sweep frequency (15,750 Hz × 12 = 189 kHz). This 189-kHz gating produces color bars that are 30° apart all around the color spectrum. When viewed from the picture tube of a normally operating television receiver, these bars appear as shown in Fig. 2-12. The phase relationship of the gated rainbow pattern is shown in Fig. 2-13.

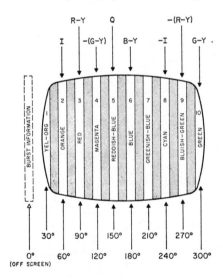

Fig. 2-12. Gated or keyed rainbow display *(Courtesy B&K-Precision, Dynascan Corporation)*

Note that of the 12 gated bursts, only 10 show on the picture tube as color bars, because one of the bursts occurs at the same time as the horizontal sync pulse and is thus eliminated. The other burst occurs immediately after the horizontal sync pulse and becomes the color sync burst, which is used to control the 3.58-MHz reference oscillator in the television receiver.

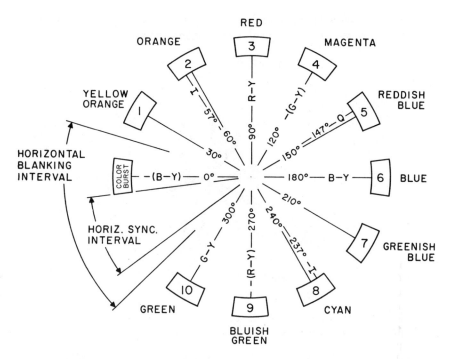

Fig. 2-13. Phase relationship of gated or keyed rainbow pattern *(Courtesy B&K-Precision, Dynascan Corporation)*

Differences in keyed rainbow generators. In its simplest form, the keyed rainbow generator provides a single-channel RF output that can be modulated by color bars, as well as crosshatch, dots, bars, and blank rasters. *Color-gun interrupters* (usually called "gun killers," or some similar term) are found on some keyed rainbow generators. These killers or interrupters consist of switches that ground the corresponding (red, blue, green) color gun of the receiver picture tube through fixed resistors, thus disabling that particular color.

Advanced keyed rainbow generators usually include analyzer functions along with the color signals. Such analyzer functions can include RF, IF, video, and sync signals to check all circuits of a color receiver (as well as black and white).

2-3.3 NTSC Color Generator

The NTSC-type generator differs from the rainbow types in that it produces *single color bars, one at a time.* A typical NTSC-type generator will provide independent selection of six fully saturated colors and white, I, Q, R-Y, and D-Y signals whose phase angles are permanently established in accordance with NTSC standards. The phase relationships and signal amplitudes (produced by a typical NTSC generator) are shown in Fig. 2-14.

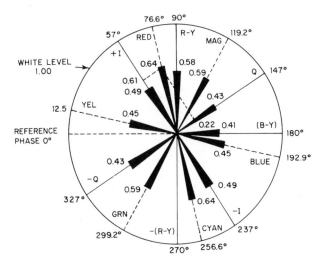

Fig. 2-14. Phase and amplitude relationships of NTSC signals (*Courtesy B&K-Precision, Dynascan Corporation*)

In a typical NTSC-type generator, the signals are selected by taps on a *linear delay line* so that no color adjustments are required. The amplitude of the color signals are also accurately set to NTSC standards, and the color reference burst is placed in its precise NTSC position, closely following the horizontal sync pulse. Thus, the color signal produced by an NTSC-type generator is exactly the same as if the signals were being produced by a television station transmitting color. Figure 2-15 shows the difference between an NTSC waveform and the waveform produced by a typical keyed rainbow generator.

2-3.4 Typical Keyed Rainbow Generator

Figure 2-16 is typical of the portable color-bar generators using the keyed rainbow display. The instrument produces two color signals. One signal is the keyed or gated rainbow shown in Fig. 2-12. The other color signal is a three-bar display as shown in Fig. 2-17. The three bars are red, blue, and bluish green and correspond to bars 3, 6, and 9 of the rainbow display. The three colors also correspond to phase angles of 90°, 180°, and 270°, or the R-Y, B-Y, and —(R-Y) signals of a color transmission. As discussed in Chapter 4, these particular bars are of special use in test and alignment of color circuits.

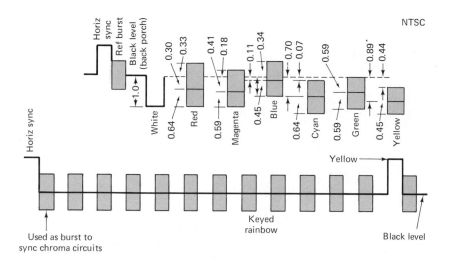

Fig. 2-15. Comparison of NTSC and keyed rainbow generator waveforms *(Courtesy B&K-Precision, Dynascan Corporation)*

The outputs are available on VHF channels 3 and 4. The amplitude of the color subcarrier is set by means of a chroma level control. There are three gun killer switches that allow individual cutoff of the red, blue, or green picture tube guns. Closing the switches shorts the respective color gun to ground through a resistance. A 4.5-MHz switch energizes an oscillator which inserts a 4.5-MHz subcarrier onto the composite video output. This subcarrier blanks the bars in the color outputs (Figs. 2-12 and 2-17) to provide an *ungated or unkeyed rainbow display.* When this sub-

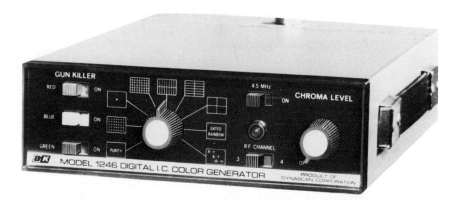

Fig. 2-16. B&K-Precision, Dynascan Corporation Model 1246 Digital IC Color Generator

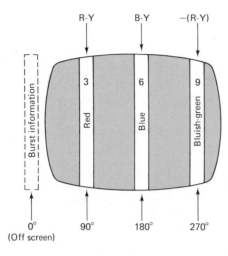

Fig. 2-17. Three-bar display of color generator *(Courtesy B&K-Precision, Dynascan Corporation)*

carrier is applied, a "herringbone" pattern appears on the picture tube display. The herringbone pattern is produced by a 936-kHz signal that is a result of mixing the 4.5-MHz subcarrier with the rainbow subcarrier of 3.56 MHz and is used to aid in accurate fine tuning of the receiver (Chapter 4).

In addition to the two color outputs, seven other outputs can be selected by means of a master switch:

A purity output, which is a sync and an ultra-clean "reference black" level to produce a clear, blemish-free raster. This is used in color purity adjustments or whenever a blank raster is required.

A dot output, which is used in static convergence adjustments. In operation, the dot display is selected and the static convergence magnets of the color picture tube are adjusted so that sharp, clear white dots appear on the screen.

A single dot output, which is used to identify the exact center of the screen.

A crosshatch output used in dynamic convergence, linearity, size, and overscan adjustments.

A single horizontal line and full vertical line output, which is used when performing dynamic convergence at the screen center on the left and right sides.

A single vertical line and full horizontal line output, which is used when performing dynamic convergence at the screen center on the top and bottom.

A crosshatch output, which is used to identify the exact center of the picture tube screen and to "rough in" proper static convergence.

2-4. OSCILLOSCOPES

A conventional oscilloscope can be used for television service. Ideally, the oscilloscope should have a bandwidth of 10 MHz, although you can probably get by with a 5- to 6-MHz bandwidth. The oscilloscope should be capable of a *triggered sweep,* in addition to conventional internal and external sweep synchronization.

With a triggered sweep, the oscilloscope horizontal sweep is synchronized by the signals being applied to the vertical input. The sweep remains at rest until triggered by the signal being observed. This assures that the signals are always synchronized, even when the waveform is of varying frequency. The triggered sweep threshold should be fully adjustable so that the desired portion of the waveform can be used for triggering.

With internal sweep synchronization, the horizontal sweep is set at some fixed rate by selection of controls. With external synchronization, the horizontal sweep is set-controlled by an external trigger. With some oscilloscopes, it is possible to apply an external sweep signal (not just a trigger) directly to the horizontal amplifier. Such provisions are helpful when using the sweep-frequency alignment techniques described in Sec. 2-2.2.

A dual-trace oscilloscope will facilitate some measurements but is not essential for television service. Such refinements as calibrated voltage scales, calibrated sweep rates, z-axis input (for intensity modulation), sweep magnification, and illuminated scales are, of course, always helpful (as they are in any type of electronic service).

Figure 2-18 is an oscilloscope manufactured for general shop and

Fig. 2-18. B&K-Precision, Dynascan Corporation Model 1460 Triggered Sweep Oscilloscope

laboratory use but with special provisions for television service not found in most conventional oscilloscopes. These special features include the *television sync or sweep provision* (usually identified as TV HORIZONTAL and TV VERTICAL, or TVH and TVV, or some similar terms) and a *vectorscope* provision. The following paragraphs are brief descriptions of these features.

2-4.1 TVV and TVH Sync and Sweeps

The TVV and TVH functions are usually selected by means of front-panel switches and permit pairs of vertical or horizontal sync pulses to be displayed on the oscilloscope screen. As shown in Fig. 2-19, when TVV is selected, the oscilloscope horizontal sweep is set to a rate of 30 Hz, and the horizontal trigger is taken from the TVV output of a sync separator in the oscilloscope.

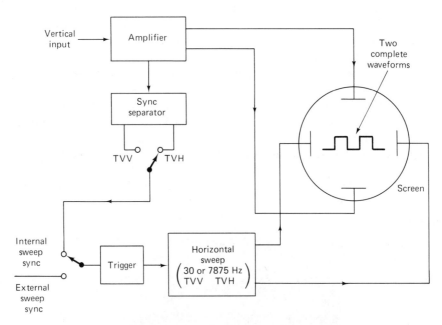

Fig. 2-19. TVV and TVH oscilloscope functions

This sync separator receives both 60-Hz and 15,750-Hz (vertical and horizontal) sync pulses from the receiver signals being monitored. The oscilloscope separator operates similarly to the receiver sync separator and delivers two separate outputs (if both sync inputs are present). If only sync input is present, only one output is available. Either way, the selected output (60 Hz or 15,750 Hz) is used to synchronize the horizontal sweep trigger.

With the horizontal sweep at 30 Hz and the trigger at 60 Hz, there are two pulses for each sweep, and two vertical sync pulses (or two complete vertical displays) appear on the screen. When TVH is selected, the horizontal sweep is set to a rate of 7875 Hz, and the trigger (15,750 Hz) is taken from the horizontal output of the oscilloscope's sync separator. Again, two pulses or displays are presented on the screen since the sweep is one-half the trigger rate.

The TVV and TVH features not only simplify observation of waveforms (you always get exactly two complete displays), but they can also be helpful in trouble localization. As an example, assume that you are monitoring waveforms in the receiver at some point where both vertical and horizontal sync signals are supposed to be present. (This could be anywhere ahead of the sync separator in the receiver.) If you find two steady patterns on the TVV position but not in the TVH position, you know that there is a problem in the horizontal circuits or that horizontal sync pulses are not getting to the point being monitored.

2-4.2 The Vectorscope

A vectorscope is used in service of color television receivers. The vectorscope is normally used in conjunction with the keyed rainbow generator (Sec. 2-3.2) to check a color television receiver's response to color signals. As discussed in Chapter 1, the colors produced by the receiver are dependent on the phase relationship of the 3.58-MHz color signal and the 3.58-MHz color reference burst. The keyed rainbow generator produces 10 color signals, each spaced 30° apart, resulting in a display of 10 corresponding colors (arranged as vertical stripes or a rainbow across the color television screen).

A vectorscope permits the phase relationship (and amplitude) of the 10 color signals to be displayed as a *single pattern* on the oscilloscope screen. The vectorscope monitors the color signals directly at the color-gun inputs (grids or cathodes) of the color picture tube. This makes the vectorscope adaptable to any type of color circuit (any type of color demodulation, vacuum tube, solid-state, etc.). By comparing the vectorscope display against that of an ideal display (for phase relationship, amplitude, general appearance, and so on), the condition of the receiver color circuits can be analyzed.

A conventional oscilloscope can be used as a vectorscope. This requires special connections to the horizontal and vertical deflection system, as described in Chapter 4. Oscilloscopes with vectorscope provisions need no special connections, although some provide a special cable and require special settings of the controls. Likewise, there are vectorscopes used in television broadcast work that have special deflection circuits to provide the vector display. However, these instruments are not needed in television receiver service (although they can be so used).

How a vector display is obtained. A vector pattern is composed of two signals; one is applied to the oscilloscope horizontal plates and the other to the vertical plates. If two sine waves are applied and one is 90° out of phase but at the same frequency, the result will be a circle, as shown in Fig. 2-20.

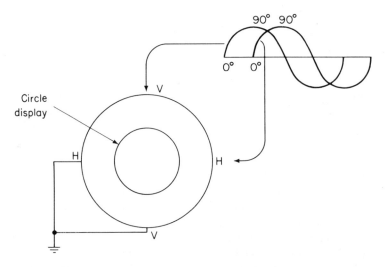

Fig. 2-20. How circular Lissajou pattern is formed (two sinowaves 90° out-of-phase)

If the patterns formed by a standard 10-color keyed rainbow generator are taken from the red and blue grids of a color picture tube and these same signals are applied to the vertical and horizontal deflection plates of an oscilloscope, the result will be a pattern similar to that shown in Fig. 2-21. This pattern is referred to as a vector pattern (or vectorscope pattern).

The phase between the R-Y (red) and B-Y (blue) signals can readily be seen on the oscilloscope. (Oscilloscopes with vectorscope provisions usually have a vector overlay, marked off in degrees, that can be placed over the screen.) A complete circle represents 360°. The screen markings can be superimposed over the display to make phase measurements, in degrees. The length of the arm of the "petals" (patterns produced by each of the 10 color signals) may vary, depending on the signal. Likewise, the entire pattern will rotate to a different position when the receiver "tint" or "hue" or other color control is adjusted. (Such controls change colors by slightly shifting the phase relationship of the color signals to the color burst.)

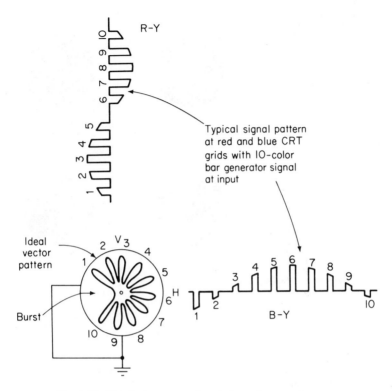

Fig. 2-21. How basic vector pattern is formed on a vectorscope

Practical vectorscope connections. The horizontal and vertical inputs of a vectorscope must be connected to the red and blue grids of the color picture tube, as shown in Fig. 2-22. On some solid-state receivers, the color input is made at the color-gun cathodes instead of the grids. Either way, the green gun need not be connected to present a vector display.

On some vectorscopes, the leads are permanently attached and are color-coded (red lead to red color gun, blue lead to blue gun). Other vectorscopes have connectors for the color-gun leads and a switch that converts the instrument from a conventional oscilloscope to a vectorscope. For some factory work, there are vectorscopes that represent a complete color analyzer system. They contain a color generator as well as the vectorscope.

Figure 2-21 is an "ideal" vectorscope pattern and will rarely be found in practical applications. Figure 2-23 is a more realistic vector pattern. No matter what pattern is produced, interpretation is what counts. The problems of interpreting vectorscope patterns, as well as adjustment and troubleshooting with vectorscopes, are covered in Chapter 4.

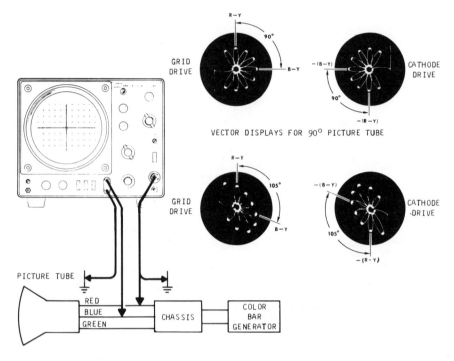

2-5. MISCELLANEOUS TEST EQUIPMENT

In addition to a good oscilloscope (with TVV, TVH, and vectorscope provisions), a sweep/marker generator, and a color generator (preferably with pattern and analyst features), most television service can be accomplished with conventional test equipment found in other electronic service fields. These include meters (digital or moving needle, whichever you prefer), isolation transformers, testers (tube, transistor, and diode),

Fig. 2-23. Typical vectorscope pattern *(Courtesy B&K-Precision, Dynascan Corporation)*

resistance-capacitance substitution boxes, and assorted adapters, clips, and probes.

As a minimum, the oscilloscope and meter should have a high-voltage probe, low-capacity probe, and a demodulator/RF probe. Another item of special test equipment found in many (but not all) television service shops is a picture tube tester/rejuvenator. Probes and picture tube testers are discussed in the following paragraphs.

2-5.1 Oscilloscope and Meter Probes

Practically all meters and oscilloscopes used in television service operate with some type of probe. In addition to providing for electrical contact to the receiver circuit under test, probes serve to modify the voltage being measured to a condition suitable for display on an oscilloscope or readout on a meter. For example, assume that the picture tube high voltage must be measured and that this voltage is beyond the maximum input limits of the meter or oscilloscope. A voltage-divider probe can be used to reduce the voltage to a safe level for measurement. Under these circumstances, the voltage is reduced by a fixed amount (known as the attenuation factor), usually on the order of 10:1, 100:1, or 1000:1.

Basic probe. In its simplest form, the basic probe is a test prod (possibly with a removable alligator clip). Basic probes work well on receiver circuits carrying direct current or audio. If, however, the circuit contains high-frequency alternating current, or if the gain of the oscilloscope or meter is high, it may be necessary to use a special low-capacitance probe. Hand capacitance in a simple probe can cause hum pickup. This condition is offset by shielding in low-capacitance probes. In a more important problem, the input impedance of the meter or oscilloscope is connected directly to the receiver circuit under test by a simple probe. Such input impedance could disturb circuit conditions (as discussed in Chapter 3).

Low-capacitance probes. The low-capacitance probe contains a series capacitor and resistors that increase the meter or oscilloscope impedance. In most low-capacitance probes, the resistors form a divider (typically 10:1) between the circuit under test and the meter or oscilloscope input. Thus, low-capacitance probes serve the dual purpose of capacitance reduction and voltage reduction. You should remember that the voltage indications will be one-tenth (or whatever value of attenuation is used) of the actual value when low-capacitance probes are connected at the inputs of meters or oscilloscopes.

High-voltage probes. Most high-voltage probes are *resistance-type voltage-divider* probes. Such probes are similar to the low-capacitance probe, except that the frequency-compensating series capacitor is omitted. Usually, the conventional resistance-type probe is used when a voltage reduction of 100:1, or greater, is required and when a flat frequency response is of no particular concern.

High-voltage probes for television service must be capable of measuring potentials at or near 30 kV (usually with a 1000:1 voltage reduction). In certain isolated cases, the resistance-type probe is not suitable because of stray conduction paths set up by the resistors. A *capacitance-type* probe can be used in those cases. Such probes contain two (or more) capacitors, with values selected to provide the desired voltage reduction and to match the input capacitance of the meter or oscilloscope.

Special high-voltage probe. Figure 2-24 shows a special high-voltage probe developed specifically for television service. This probe is not connected to an oscilloscope or meter but has a built-in meter for measurement of voltages up to 30,000 V. This permits the picture tube high voltage of both solid-state and vacuum-tube receivers to be measured safely and accurately.

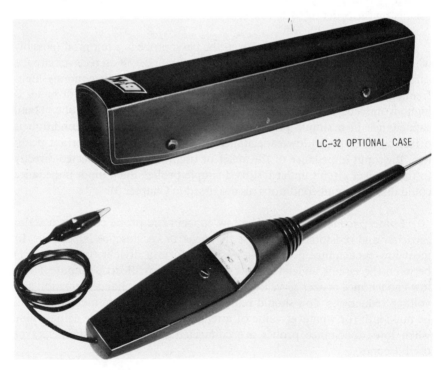

LC-32 OPTIONAL CASE

Fig. 2-24. B&K-Precision, Dynascan Corporation Model HV-40 High Voltage Probe/Meter

Radio-frequency probes. When the signals to be measured are at radio frequencies (such as in the RF tuner and IF amplifier stages) and are beyond the frequency capabilities of the meter or oscilloscope an RF probe is required. RF probes convert (rectify) the RF signals into a dc output voltage that is equal to the peak RF voltage (or possibly equal to the RMS (root mean square) of the RF voltage). The dc output of the probe is applied to the meter or oscilloscope input and is displayed as a voltage readout in the normal manner.

Demodulator probes. The circuit of a demodulator probe is essentially like that of the RF probe. However, the demodulator probe produces both an ac and a dc output. The RF carrier frequency is converted to a dc output voltage equal to the RF carrier. If the carrier is modulated, the modulating voltage appears as ac (or pulsating dc) at the probe output.

In use, the meter or oscilloscope is set to measure direct current, and the RF carrier is measured. The meter or oscilloscope is set to measure alternating current, and the modulating voltage (if any) is measured. In general, demodulator probes are used primarily for signal tracing, and their output is not calibrated to any particular value. However, this is not always true. Likewise, some RF probes are capable of demodulation (producing outputs equal to both carrier and modulation) and may be described in catalogs as RF/demodulator probes.

2-5.2 Picture Tube (CRT) Testers and Rejuvenators

These instruments provide for test and possible rejuvenation of black and white as well as color picture tubes. The units are also known as picture tube renewers or CRT renewers. With these instruments, it is possible to check each element of the picture tube for such factors as shorts, opens, leakage, and proper emission. It is also possible to rejuvenate some picture tubes by application of high voltages, heavy currents, etc., to the proper elements.

Because of the highly specialized nature of tester/rejuvenators, and because you will receive a detailed set of instructions (which must be followed) with the instrument, we shall not describe the units here. However, you should be aware of their use, particularly in service shops specializing in color television.

3

Basic Service Procedures for Black and White Circuits

In this chapter, we shall discuss basic black and white receiver circuit service and troubleshooting. That is, we shall describe how the basic service techniques of Chapter 1 are combined with the practical use of test equipment discussed in Chapter 2 to locate specific faults in various types of black and white receivers and black and white circuits of color receivers.

Since alignment and adjustments, as well as testing, are part of service, we shall also describe basic adjustment procedures for various types of black and white circuits. Throughout this chapter, considerable emphasis will be placed on "universal" test, service, and troubleshooting procedures. These procedures apply to all receivers now in existence and to those that may be found in the future.

3-1. SERVICING NOTES

The following notes summarize practical suggestions for servicing all types of television circuits.

3-1.1 Solid-state Servicing Techniques

Although the following techniques apply primarily to solid-state receivers, they are generally valid for vacuum-tube television also.

Transient voltages. Be sure that power to the receiver is turned off, or that the line cord is removed, when making in-circuit tests or repairs that involve removal and replacement of parts. Transistors (and possibly some diodes) can be damaged from the transient voltages developed when changing components or inserting new transistors (in addition to the possibility of shock or short circuit). In some receivers (the "instant-on" receivers, for example) certain circuits may be *live,* even with the power switch set to OFF. To be on the safe side, pull the power plug.

Disconnected parts. When working on solid-state television, do not operate with any parts, such as loudspeakers or picture tube yokes, disconnected. If the load is removed from some transistor circuits, heavy current will be drawn, resulting in possible damage to the transistor or other part such as an audio transformer.

Sparks and voltage arcs. Avoid sparks and arcs when servicing any type of receiver, especially solid-state television. The transients developed can damage some small-signal transistors. For example, when servicing picture tube circuits, use a meter and high-voltage probe (or the special-purpose high-voltage meters described in Chapter 2) to measure the second anode (high-voltage) potential of the picture tube. Do not arc the second anode lead to the chassis for a spark test, as is often (unwisely) done in vacuum-tube receivers. Such an arc in solid-state television can destroy the high-voltage rectifier and possibly damage the horizontal output transistor.

Intermittent conditions. If you run into an intermittent condition and can find no fault by routine checks, try tapping (not pounding) the parts (vacuum tubes, transistors, diodes, etc.). If this does not work for solid-state parts, try rapid heating and cooling. A small, portable hair dryer and a spray-type circuit cooler make good heating and cooling sources, respectively.

First apply heat, and then cool the part. The quick change in temperature will normally cause an intermittently defective part to go bad permanently. In many cases, the part will open or short, making it easy to locate.

As an alternative procedure, measure the gain of a transistor with an in-circuit transistor checker (Chapter 2). Then subject the transistor to rapid changes in temperature. If the suspected transistor changes its gain *drastically,* or if there is *no change* in gain, the transistor is probably defective.

In any event, do not hold a heated soldering tool directly on, or very near, a transistor or diode case. This will probably destroy the transistor or diode.

If time permits, an intermittently defective transistor can be located by measuring in-circuit gain when the receiver is cold. Then let the receiver operate until the trouble occurs and measure the gain of the transistors while they are hot. Some variation will be noted in all transistors, but a leaky transistor has a much lower gain reading when it is hot.

Operating control settings. If any transistor or vacuum-tube element appears to have a short (particularly the base or grid), check the settings of any operating controls or adjustment controls associated with the circuit. For example, in the audio section of a receiver, a gain or volume control set to zero or minimum can give the same indication as a short from element to ground. This condition is shown in Fig. 3-1.

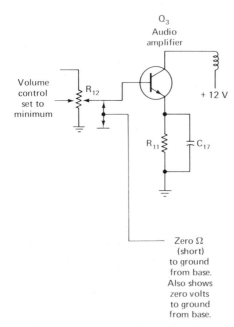

Fig. 3-1. Example of how operating control settings can affect voltage and resistance readings at transistor elements during troubleshooting

Record gain readings. If you must service any particular make or model of receiver regularly, record the transistor gain readings of a good-working unit on the schematic for future reference. Compare these gain readings against the minimum values listed on the service literature.

Shunting capacitors. It is common practice in troubleshooting vacuum-tube circuits to shunt a suspected capacitor with a known good capacitor. This technique is good only if the suspected capacitor is open. The test is of little value if the capacitor is leaking or shorted. In any event, avoid the shunting technique when servicing solid-state television. This is especially true when servicing an electrolytic capacitor (often used as an emitter bypass in solid-state television). The transient voltage surges can damage transistors. In general, avoid any short-circuit tests with solid-state receivers.

Test connections. Most metal-case transistors have their case tied to the collector. Thus, you can use the case as a test point. Avoid using a clip-type probe on transistors. Also avoid clipping onto some of the sub-miniature resistors used in solid-state television. Any subminiature components can break with rough handling.

Injecting signals. When injecting a signal into a circuit (base of a transistor, grid of a vacuum tube, input of an IC, etc.), make sure there is a blocking capacitor in the signal generator output. As discussed in Chapter 2, most signal generators have some form of blocking capacitor to isolate the output circuit from the dc voltages that may appear in the circuit.

In the case of solid-state circuits, the blocking capacitor also prevents the base from being returned to ground (through the generator output circuit) or from being connected to a large dc voltage (in the generator circuit). Either of these conditions can destroy the transistor. If the generator is not provided with a built-in blocking capacitor, connect a capacitor between the generator output lead and the transistor base (or other point of signal injection in the circuit).

3-1.2   Measuring Voltages in Circuit

As discussed in Chapter 1, it is possible to locate many defects in television receivers by measuring and analyzing voltages at the elements of active devices (grid, cathode, and plate of vacuum tubes; base, emitter, and collector of transistors; etc.). This can be done with the circuit operating and without disconnecting any parts. Once located, the defective part can be disconnected and tested or substituted, whichever is most convenient.

Vacuum-tube circuits can be analyzed with a simple VOM or electronic meter. The normal relationships of vacuum-tube elements are generally fixed. For example, the plate is positive, the cathode is at ground or positive, and the grid is (usually) negative.

Transistor circuits are best analyzed with an electronic voltmeter or with a very sensitive VOM. A number of manufacturers produce VOMs designed specifically for transistor troubleshooting. (The Simpson Model 250 is a typical example.) These VOMs have very low-voltage scales to measure the differences that often exist between elements of a transistor (especially the small voltage difference between emitter and base). Such VOMs also have a very low-voltage drop (about 50 mV in the current ranges.

Analyzing transistor voltages. Figure 3-2 shows the basic connections for both PNP and NPN transistors used in television circuits. The coupling or bypass capacitors have been omitted to simplify the explanation. The

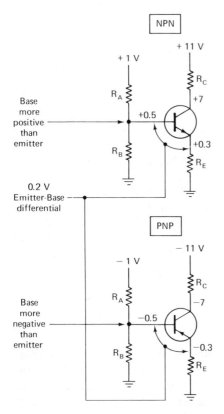

Fig. 3-2. Basic connections for NPN and PNP transistor circuits (with normal voltage relationships)

purpose of Fig. 3-2 is to establish normal transistor voltage relationships. With a normal pattern established, it is relatively simple to find an abnormal condition.

In most circuits, the emitter-base junction must be forward-biased to get current flow through a transistor. In a PNP, this means that the base must be made more negative (or less positive) than the emitter. Under these conditions, the emitter-base junction will draw current and cause heavy electron flow from the collector to the emitter. In an NPN, the base must be made more positive (or less negative) than the emitter for current to flow from emitter to collector.

The following general rules are helpful when analyzing transistor voltage as part of troubleshooting:

1. The middle letter in PNP and NPN always applies to the base.

2. The first two letters in PNP and NPN refer to the relative bias polarities of the emitter with respect to either the base or collector. For example, the letters PN (in PNP) indicate that the emitter is positive with respect to both the base and collector. The letters NP (in NPN) indicate that the emitter is negative with respect to both the base and collector.

3. The collector-base junction is always reverse-biased.

4. The emitter-base junction is usually forward-biased. An exception is a class C amplifier (used in RF circuits).

5. A base input voltage that opposes or decreases the forward bias also decreases the emitter and collector currents.

6. A base input voltage that aids or increases the forward bias also increases the emitter and collector currents.

7. The dc electron flow is always against the direction of the arrow on the emitter.

8. If electron flow is into the emitter, electron flow is out from the collector.

9. If electron flow is out from the emitter, electron flow is into the collector.

Using these rules, normal transistor voltages can be summed up this way:

> For an NPN transistor, the base is positive, the emitter is not quite so positive, and the collector is far more positive.

> For a PNP transistor, the base is negative, the emitter is not quite so negative, and the collector is far more negative.

Measurement of transistor voltages. There are two schools of thought on how to measure transistor voltages in troubleshooting.

Element to element. Some television service technicians prefer to measure transistor voltages from element to element (between electrodes), looking for the *difference in voltage.* For example, in the circuit of Fig. 3-2, an 0.2-V differential exists between base and emitter. The element-to-element method of measuring transistor voltages quickly establishes forward and reverse bias conditions.

Element to ground. The most common method of measuring transistor voltages is to measure from a common to ground to the element. Television service literature usually specifies transistor voltages this way. For example, all of the voltages for the PNP of Fig. 3-3 are positive with respect to ground. This method of labeling transistor voltages is sometimes confusing to those not familiar with transistors, since it appears to break the rules. (Typically, all elements of a PNP are negative.) However, the rules still apply.

In the case of the PNP in Fig. 3-3, the emitter is at + 10V, whereas the

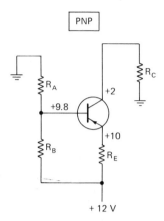

Fig. 3-3. Example of normal voltages for PNP transistor operated from a positive supply

base is at +9.8V. The base is *less positive* than the emitter. Thus, the base is *more negative* than the emitter, and the base-emitter junction is forward-biased (normal).

On the other hand, the base is at +9.8V, whereas the collector is at +2V. The base is more positive (less negative) than the collector, and the base-collector junction is reverse-biased (normal).

3-1.3 Troubleshooting with Transistor Voltages

The following is an example of how voltages measured at the elements of a transistor can be used to analyze failure in solid-state television circuits.

Assume that an NPN transistor circuit is measured and that the voltages found are similar to those of Fig. 3-4. Except in one case, these voltages indicate a defect. It is obvious that the transistor is not forward-biased because the base is less positive than the emitter (reverse bias for an

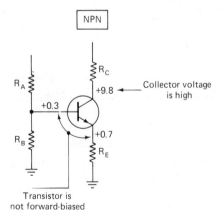

Fig. 3-4. NPN transistor circuit with abnormal voltages (collector voltage is high, base-emitter is not forward biased)

NPN). The only circuit where this might be normal is one that requires a large trigger voltage or pulse to turn it on.

The first troubleshooting clue in Fig. 3-4 is that the collector voltage is almost as large as the collector source (at R_C). This means that very little current is flowing through R_C in the collector-emitter circuit. The transistor could be defective. However, the trouble is more likely caused by a problem in bias. The emitter voltage depends mostly on the current through R_E, so unless the value of R_E has changed substantially (that would be unusual), the problem is one of incorrect bias on the base.

The next step in this case is to measure the bias source voltage at R_A. If the bias source voltage is, as shown in Fig. 3-5, at 0.7V instead of the required 2V, the problem is obvious; the external bias voltage is incorrect. This condition will probably show up as a defect in the power supply and will appear as an incorrect voltage in other circuits.

If the source voltage is correct, as shown in Fig. 3-6, the cause of trouble is probably a defective R_A or R_B or a defect in the transistor.

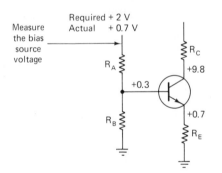

Fig. 3-5. NPN transistor circuit with abnormal voltages (fault traced to incorrect bias source voltage, bias voltage low)

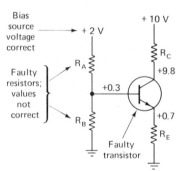

Fig. 3-6. NPN transistor circuit with abnormal voltages (fault traced to bias resistors or transistor)

The next step is to remove all voltage from the receiver and measure the resistance of R_A and R_B. If either value is incorrect, the corresponding resistor must be replaced. If both values are correct, it is reasonable to check the value of R_E. However, it is more likely that the transistor is defective. This can be established by test and/or replacement.

Practical in-circuit resistance measurements. Do not attempt to measure resistance values in transistor circuits with the resistors still connected (even with the power off). This practice may be correct for some vacuum-tube receiver circuits but not with transistor circuits. One reason is that the voltage produced by the ohmmeter battery could damage some transistors.

Even if the voltages are not dangerous, the chance for an error is greater than with a transistor circuit because the transistor junctions will pass current in one direction. This can complete a circuit through other resistors and produce a series or parallel combination, thus making false indications. The condition can be prevented by disconnecting one resistor lead before making the resistance measurement.

For example, assume that an ohmmeter is connected across R_B (Figs. 3-4 through 3-6), with the negative battery terminal of the ohmmeter connected to ground, as shown in Fig. 3-7. Because R_E is also connected to ground, the negative battery terminal is connected to the transistor base, the base-emitter junction is forward-biased, and there is electron flow. In effect, R_E is now in parallel with R_B, and the ohmmeter reading is incorrect. This can be prevented by disconnecting either end of R_B before making the measurement.

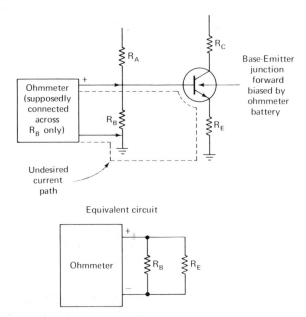

Fig. 3-7. Example of in-circuit resistance measurements showing undesired current path through forward-biased transistor junction

3-1.4   Testing Transistors in Circuit (Forward-bias Method)

Germanium transistors normally have a 0.2- to 0.4-V voltage differential between emitter and base; silicon transistors normally have a voltage differential of 0.4 to 0.8 V. The polarities of voltages at the emitter and base depend on the type of transistor (NPN or PNP).

The voltage differential between emitter and base acts as a forward bias for the transistor. That is, a sufficient differential or forward bias will turn the transistor on, resulting in a corresponding amount of emitter-collector flow. Removal of the voltage differential, or an insufficient differential, produces the opposite results. That is, the transistor is cut off (no emitter-collector flow, or very little flow).

These forward-bias characteristics can be used to troubleshoot transistor circuits, without removing the transistor and without using an in-circuit tester. In the following sections we shall describe two methods of testing transistors in circuit: one by removing the forward bias and the other by introducing a forward bias.

Removal of forward bias. Figure 3-8 shows the test connections for an in-circuit transistor test by removal of forward bias. The procedure is simple. First, measure the emitter-collector differential voltage under normal circuit conditions. Then short the emitter-base junction and note any change in emitter-collector differential. If the transistor is operating, the removal of forward bias causes the emitter-collector current flow to

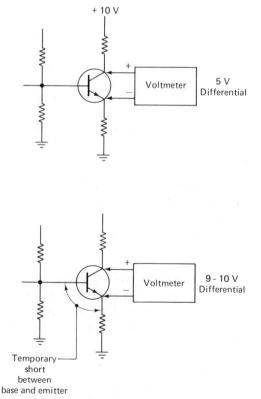

Temporary short between base and emitter

Fig. 3-8. In-circuit transistor test (removal of forward bias method)

stop, and the emitter-collector voltage differential increases. That is, the collector voltage rises to or near the power supply value.

As an example, assume that the power supply voltage is 10V and that the differential between the collector and emitter is 5V when the transistor is operating normally (no short between emitter and base). When the emitter-base junction is shorted, the emitter-collector differential should rise to about 10V (probably somewhere between 9 and 10V).

Application of forward bias. Figure 3-9 shows the test connection for an in-circuit transistor test by the application of forward bias. The procedure is equally simple. First, measure the emitter-collector differential under normal circuit conditions. As an alternative, measure the voltage across R_E as shown in Fig. 3-9.

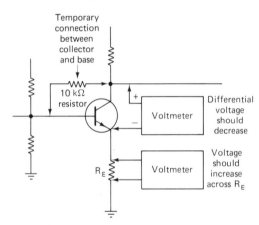

Fig. 3-9. In-circuit transistor test (application of forward bias method)

Next, connect a 10-k Ω resistor between the collector and base, as shown, and note any change in emitter-collector differential (or any change in voltage across R_E). If the transistor is operating, the application of forward bias will cause the emitter-collector current flow to start (or increase), and the emitter-collector voltage differential will decrease, or the voltage across R_E will increase.

Go/no-go test characteristics. The test methods shown in Fig. 3-8 and 3-9 show that the transistor is operating on a go/no-go basis. This is usually sufficient for most dc and low-frequency ac applications. However, the tests do not show transistor gain or leakage. Also, the tests do not establish operation of the transistor at high frequencies.

The fact that these (or similar) in-circuit tests of a transistor do not establish all of the operating characteristics raises a problem in trouble-shooting. For example, the tests do not establish high-frequency characteristics of transistors used in the RF and IF sections of the receiver. Likewise, the tests do not show the rapid pulse-switching characteristics of transistors used in the horizontal sweep circuits.

Some troubleshooters reason that the only satisfactory test of a transistor is in-circuit operation. If a transistor will not perform its function in a given circuit, the transistor must be replaced. Thus, the most logical method of test is replacement.

This reasoning is generally sound, except in one circumstance. It is possible that a replacement transistor will not perform satisfactorily in critical circuits (horizontal sweep and RF/IF circuits), even though the transistor is the correct type (and may even work in another circuit). This can be misleading in troubleshooting. If such a replacement transistor does not restore the circuit to normal, the apparent fault is with another circuit, whereas the true cause of trouble is the new transistor. Fortunately, this does not happen often, even in critical circuits. However, a good troubleshooter should be aware of the possibility.

3-1.5   Testing Transistors Out of Circuit

There are four basic tests required for transistors: gain, leakage, breakdown, and switching time. All of these tests are best made with an oscilloscope using appropriate adapters (curve tracers, switching-characteristic checkers, etc.). However, it is possible to test a transistor with an ohmmeter. These simple tests will show if the transistor has leakage and if the transistor shows some gain. Generally, this is sufficient for most practical troubleshooting applications. However, the only true test of a transistor is in the circuit with which the transistor is to be used.

Testing transistor leakage with an ohmmeter. For test purposes (using an ohmmeter), transistors can be considered as two diodes connected back to back. Each diode should show low forward resistance and high reverse resistance. These resistances can be measured with an ohmmeter, as shown in Fig. 3-10.

The same ohmmeter range should be used for each pair of measurements (base to emitter, base to collector, and collector to emitter). However, avoid using the R × 1 range or an ohmmeter with a high internal battery voltage. Either of these conditions can damage a low-power transistor.

If the reverse reading is low but not shorted, the transistor is leaking.

If both forward and reverse readings are very low or show a short, the transistor is shorted.

If both forward and reverse readings are very high, the transistor is open.

If the forward and reverse readings are the same or nearly equal, the transistor is defective.

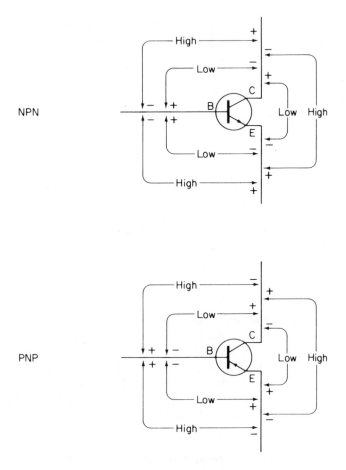

Fig. 3-10. Testing transistor leakage with an ohmmeter

A typical forward resistance is 300 to 700 Ω. However, a low-power transistor might show only a few ohms in the forward direction, especially at the collector-emitter junction. Typical reverse resistances are 10 to 70 kΩ.

The actual resistance values depend on the ohmmeter range and battery voltage. The *ratio* of forward-to-reverse resistance is the best indicator. Almost any transistor will show a ratio of at least 30 to 1. Many transistors show ratios of 100 to 1 or greater.

Testing transistor gain with an ohmmeter. Normally, there will be little or no current flow between emitter and collector until the base-emitter junction is forward-biased. A basic gain test of a transistor can be made using an ohmmeter. The test circuit is shown in Fig. 3-11. In this test, the

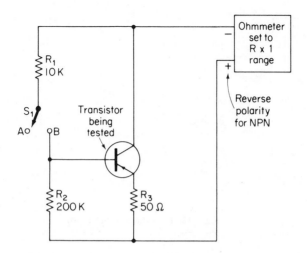

Fig. 3-11. Testing transistor gain with an ohmmeter

R × 1 range should be used. Any internal battery voltage can be used, provided that it does not exceed the maximum collector-emitter breakdown voltage.

In position A of switch S_1, there is no voltage applied to the base, and the base-emitter junction is not forward-biased. The ohmmeter should read a high resistance.

When switch S_1 is set to B, the base-emitter circuit is forward-biased (by the voltage across R_1 and R_2), and current flows in the emitter-collector circuit. This is indicated by a lower resistance reading on the ohmmeter. A 10-to-1 resistance ratio is typical.

3-1.6   Testing Diodes Out of Circuit

There are three basic tests required for diodes. First, the diode must have the ability to pass current in one direction (forward current) and prevent or limit current (reverse current) in the opposite direction. Second, for a given reverse voltage, the reverse current should not exceed a given value. Third, for a given forward current, the voltage drop across the diode should not exceed a given value. If a diode is to be used in pulse or digital work, the switching time must also be tested. These tests are best performed using a scope with appropriate adapters.

Since the elementary purpose of a diode is to prevent current flow in one direction while passing current in the opposite direction, a diode can be tested using an ohmmeter. In this case, the ohmmeter is used to measure forward and reverse resistance of the diode. The basic circuit is shown in Fig. 3-12.

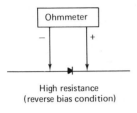

High resistance
(reverse bias condition)

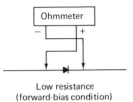

Low resistance
(forward-bias condition)

Fig. 3-12. Basic diode test using an ohmmeter

A good diode will show high resistance in the reverse direction and a low resistance in the forward direction.

If resistance is low in the reverse direction, the diode is probably leaking.

If resistance is high in both directions, the diode is probably open.

A low resistance in both directions usually indicates a shorted diode.

It is possible for a defective diode to show a difference in forward and reverse resistance. The important factor in making a diode-resistance test is the *ratio* of forward-to-reverse resistance (often known as the front-to-back ratio or the back-to-front ratio). The actual ratio depends on the type of diode. As a guideline, a signal diode has a ratio of several hundred to one, whereas a power rectifier diode can operate satisfactorily with a ratio of 10:1.

3-1.7   Testing Integrated Circuits (ICs) in Television Receivers

There is some difference of opinion in testing ICs in circuit or out of circuit during television service. An in-circuit test is the most convenient, since the power source is available and you do not have to unsolder the IC. (Removal and replacement of an IC can be quite a job.)

As a first step with any IC suspected of being defective, measure the dc voltages applied at the IC terminals to make sure that they are available and correct. If the voltages are absent or abnormal, this is a good starting point for troubleshooting.

With the power sources established, the in-circuit IC is tested by applying the appropriate input and monitoring the output. In some cases it is not necessary to inject an input, since the normal input is supplied by the circuits ahead of the IC.

One drawback to testing an IC in the receiver is that the circuits before (input) and after (output) the IC may be defective. This can lead you to think that the IC is bad.

For example, assume that an IC is used as the IF and video detector stages of a television receiver. To test such an IC you inject a signal at the IC input (output of RF tuner) and monitor the IC output (video detector output). Now assume that the IC output terminal is connected to a short circuit. There will be no output indicated, even though the IC and the input signals from the RF tuner are good. Of course, this will show up as an incorrect resistance measurement at a later stage in troubleshooting.

Out-of-circuit tests for ICs have two obvious disadvantages: You must remove the IC, and you must supply the required power. Of course, if you test a suspected IC after removal and find that it is operating properly out of circuit, it is logical to assume that there is trouble in the circuits connected to the IC. This is very convenient to know before you go to the trouble of installing a replacement IC.

IC voltage measurements. While the test procedures for an IC are the same as for the equivalent vacuum-tube or transistor circuits, measurement of the static (dc) voltage applied to the IC is not identical. Many ICs require connection to both a positive and negative power source, although the ICs used as IF amplifier circuits usually can be operated from a single power supply.

Some ICs require equal power supply voltages (such as $+9$ and -9 V). This is not the case with the example circuit of Fig. 3-13, which requires a $+9$ V at pin 8 and a -4.8 V at pin 4.

Manufacturers do not agree on power supply labeling for ICs. For

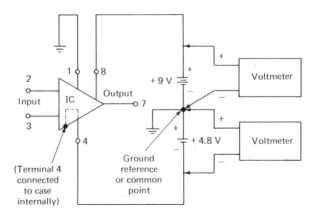

Fig. 3-13. Measuring static (power source) voltages of ICs with IC installed in receiver

example, one manufacturer might use V+ to indicate the positive voltage and V− to indicate the negative voltage. Another manufacturer might use the symbols V_{EE} and V_{CC} to represent negative and positive, respectively. The schematic diagram should be studied carefully before measuring power source voltages during troubleshooting.

No matter what labeling is used, the IC shown in Fig. 3-13 requires two power sources, with the positive lead of one and the negative lead of the other tied to ground. Each voltage is measured separately, as shown.

Note that the IC case (such as a TO-5 type) of the Fig. 3-13 circuit is connected to pin 4. Such a connection is typical for most ICs (but not necessarily pin 4). The case is below ground (or "hot") by 4.8V.

3-1.8   The Effects of Capacitors in Television Service

If you suspect that a capacitor is defective, you can remove it from the receiver and test the capacitor on a checker. This will establish that the capacitor value is correct. If the checker shows the value to be correct, it is reasonable to assume that the capacitor is not open, shorted, or leaking.

From another standpoint, if a capacitor shows no shorts, opens, or leakage, it is also reasonable to assume that the capacitor is good. From a practical standpoint, a simple test that shows the possibility of shorts, opens, or leakage is usually sufficient.

There are two basic methods for a quick check of capacitors. One method involves using the circuit voltages. The other technique requires an ohmmeter.

Checking capacitors with circuit voltages. As shown in Fig. 3-14(a), this method involves disconnecting one lead of the capacitor (the ground or "cold" lead) and connecting a voltmeter between the disconnected lead and ground. In a good capacitor, there should be a momentary voltage indication (or surge) as the capacitor charges up to the voltage at the "hot" end.

If the voltage indication remains high, the capacitor is probably shorted.

If the voltage indication is steady but not necessarily high, the capacitor is probably leaking.

If there is no voltage indication whatsoever, the capacitor is probably open.

Checking capacitors with an ohmmeter. As shown in Fig. 3-14(b), this method involves disconnecting one lead of the capacitor (usually the "hot" end) and connecting an ohmmeter across the capacitor. Make certain all power is removed from the circuit. As a precaution, short across the capacitor to make sure that no charge is being retained after the power is

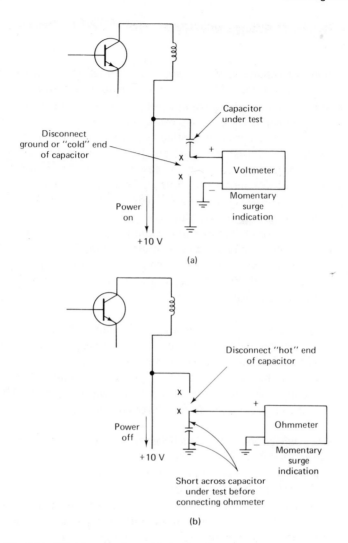

Capacitor
under test

Disconnect
ground or "cold" end
of capacitor

X

+

Voltmeter

X

_

Power
on

Momentary
surge
indication

+10 V

(a)

Disconnect "hot" end
of capacitor

X

+

X

Ohmmeter

Power
off

_

+10 V

Momentary
surge
indication

Short across capacitor
under test before
connecting ohmmeter

(b)

Fig. 3-14. Checking capacitors with circuit voltages (power applied) and with an ohmmeter (power removed.)

removed. In a good capacitor, there should be a momentary resistance indication (or surge) as the capacitor charges up to the voltage of the ohmmeter battery.

If the resistance indication is near zero and remains so, the capacitor is probably shorted.

If the resistance indication is steady at some high value, the capacitor is probably leaking.

If there is no resistance indication whatsoever, the capacitor is probably open.

Functions of capacitors in circuits. The functions of capacitors in solid-state receivers are similar to those of vacuum-tube receivers. However, the results produced by capacitor failure are not necessarily the same. An emitter bypass capacitor is a good example.

The emitter resistor in a solid-state circuit (such as R_4 in Fig. 3-15) is used to stabilize the transistor dc gain and prevent thermal runaway. With an emitter resistor in the circuit, any increase in collector current produces a greater drop in voltage across the resistor. When all other factors remain the same, the change in emitter voltage reduces the base-emitter forward-bias differential, thus tending to reduce collector current flow.

When circuit stability is more important than gain, the emitter resistor is not bypassed. When ac or signal gain must be high, the emitter resistance is bypassed to permit passage of the signal. If the emitter bypass capacitor is open, stage gain is reduced drastically, although the transistor dc voltages remain substantially the same.

Low-gain symptoms. If there is a low-gain symptom in any solid-state amplifier with emitter bypass and the voltages appear normal, check the bypass capacitor. This can be done by shunting the bypass with a known good capacitor of the same value. As a precaution, shut off the power before connecting the shunt capacitor; then reapply power. This will prevent damage to the transistor (due to large current surges).

Coupling capacitors. The functions of coupling (and decoupling) capacitors in solid-state circuits are essentially the same as for vacuum-tube receivers. However, the capacitance values are much larger for solid-state circuits, particularly at *low* frequencies.

Electrolytic capacitors are usually required in solid-state circuits to get the large capacitance values. From a troubleshooting standpoint, electrolytics tend to have more leakage than mica or ceramic capacitors. However, good-quality electrolytics (typically the bantam type found in solid state) have leakage of less than 10μ A at normal operating voltage.

Defects in coupling capacitors. The function of C_1 in Fig. 3-15 is to pass signals from the previous stage to the base of Q_1. If C_1 is shorted or leaking badly, the voltage from the previous stage is applied to Q_1. This forward-biases Q_1, causing heavy current flow and possible burnout of the transistor. In any event, Q_1 is driven into saturation, and stage gain is reduced.

If C_1 is open, there is little or no change in the voltage at Q_1, but the signal from the previous stage will not appear at the base of Q_1. From a

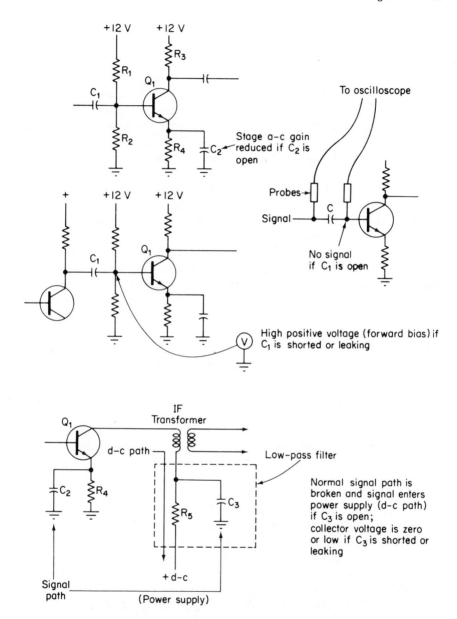

Fig. 3-15. Effects of capacitor failure in solid-state circuits

troubleshooting standpoint, a shorted or leaking C_1 will show up as abnormal voltages (and probably as distortion of the signal waveform). If C_1 is suspected of being shorted or leaky, replace C_1. An open C_1 will show up as a lack of signal at the base of Q_1, with a normal signal at the previous stage. If an open C_1 is suspected, replace C_1 or try shunting C_1 with a known good capacitor, whichever is convenient.

Defects in decoupling or bypass capacitors. The function of C_3 in Fig. 3-15 is to pass operating signal frequencies to ground (to provide a return path) and to prevent signals from entering the power supply line or other circuits connected to the line. In effect, C_3 and R_5 form a low-pass filter that passes dc and very low-frequency signals (well below the operating frequency of the circuit) through the power supply line. Higher-frequency signals are passed to ground and do not enter the power supply line.

If C_3 is shorted or leaking badly, the power supply voltage will be shorted to ground or greatly reduced. This reduction of collector voltage will make the stage totally inoperative or will reduce the output, depending on the amount of leakage in C_3.

If C_3 is open, there will be little or no change in the voltages at Q_1. However, the signals will appear in the power supply line. Also, signal gain will be reduced, and the signal waveform will be distorted. In some cases, at higher signal frequencies, the signal simply cannot pass through the power supply circuits. Since there is no path through an open C_3, the signal will not appear on the collector circuit. From a practical troubleshooting standpoint, the results of an open C_3 depend on the values of R_5 (and the power supply components) as well as on the signal frequency involved.

3-1.9   Effects of Voltage on Circuit Resistance

The effects of shorts on resistors in solid-state circuits are less drastic than in a vacuum-tube circuits, because of the lower voltage used in solid-state. For example, most solid-state receivers operate at voltages well below 25V (except for a few high-power circuits). Typically, solid-state receivers are operated at 12V, or less. A 1-kΩ resistance shorted directly across a 25-V source produces only 25-mA current flow, or about 0.6 W. A 1-W resistor can easily handle this power with no trouble. Even an 0.5-W resistor will probably survive a temporary short of this level.

On the other hand, the same resistance across a 300-V source (typical for vacuum-tube circuits) produces about 0.3-A current flow, or about 90 W. This will destroy all but heavy power resistors.

For these reasons, resistors do not burn out so often in solid-state receivers (compared to similar vacuum-tube receivers). Likewise, solid-state resistance values do not usually change due to prolonged heating. There are exceptions of course, but most solid-state troubles are the result of defects in capacitors (first), transistors (second), and diodes (third).

3-1.10   Effects of Voltage on Poor Solder Joints

The low voltages in solid-state receivers have just the opposite effect on poor solder joints (so-called "cold" solder joints) and partial breaks in printed wiring. Often the high voltages in vacuum-tube equipment can overcome the resistance created by cold solder joints and partial printed circuit breaks.

When there is no obvious cause for a low voltage at some point in the circuit or there is an abnormally high resistance, look for cold solder joints or defects in printed circuit wiring. Use a magnifying glass to locate defects in printed wiring. Sometimes, minor breaks in printed wiring can be repaired by applying solder at the break. However, this is recommended only as a temporary measure. Under emergency conditions, it is possible to run a wire between two points on either side of the break. However, it is recommended that the entire board be replaced as soon as practical.

Finding cold solder joints. Cold solder joints can sometimes be found with an ohmmeter. Remove all power. Connect the ohmmeter across two wires leading out of the suspected cold solder joint, as shown in Fig. 3-16. Flex the wires by applying pressure with the ohmmeter prod tips. Switch the ohmmeter to different ranges, and check if there is any change in resistance.

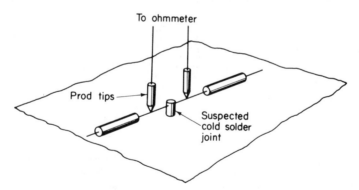

Fig. 3-16. Locating cold solder joints with an ohmmeter

For example, a cold solder joint can appear to be good on the high ohmmeter ranges but as an open on the lower ranges. Look for resistance indications that tend to drift or change when the ohmmeter is returned to a particular scale. If a cold solder joint is suspected, reheat the joint with a soldering tool; then recheck the resistance.

3-1.11   Effects of Transistor Leakage on Circuit Gain

When there is considerable leakage in a solid-state amplifier, the gain is reduced to zero and/or the signal waveform is drastically distorted. Such a condition also produces abnormal waveforms and transistor voltages. These indications make troubleshooting easy, or relatively easy. The troubleshooting problem becomes really difficult when there is just enough leakage to reduce amplifier gain but not enough leakage to seriously distort the waveform or produce transistor voltages that are way off.

Collector-base leakage is the most common form of transistor leakage and produces a classic condition of low gain (in a single stage). When there is any collector-base leakage, the transistor is forward-biased, or the forward bias is increased. This is shown in Fig. 3-17.

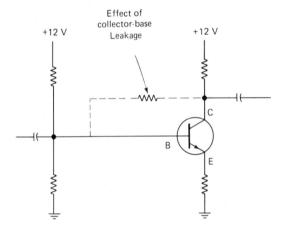

Normal voltages (without leakage)	Voltages with leakage
C = 6 V	C = 4 V
E = 2 V	E = 3 V
B = 2.5 V	B = 3.5 V

Fig. 3-17. Effects of collector-base leakage on transistor element voltages

Collector-base leakage has the same effect as a resistance between the collector and base. The base assumes the same polarity as the collector (although at a lower value), and the transistor is forward-biased. If leakage is sufficient, the forward bias can be enough to drive the transistor into or near saturation. When a transistor is operated at or near the saturation point the gain is reduced (for a single stage). This is shown in Fig. 3-18.

If the normal transistor element voltages are known (from the service literature or previous readings taken when the receiver was operating

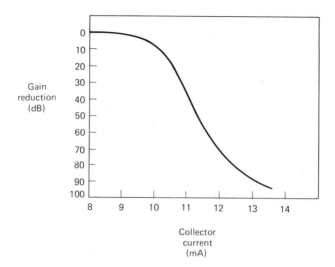

Fig. 3-18. Relative gain of solid-state amplifier at various average collector-current levels (showing effect of collector current increases on gain reduction)

properly), excessive transistor leakage can be spotted easily, since all the transistor voltages will be off. For example, in Fig. 3-17, the base and emitter will be high and the collector will be low (when measured in reference to ground).

If the normal operating voltages are not known, the transistor can appear to be good, since all of the voltage relationships are normal. That is, the collector-base junction is reverse-biased (collector more positive than base for an NPN), and the emitter-base junction is forward-biased (emitter less positive than base for an NPN).

A simple way to check transistor leakage is shown in Fig. 3-19. Measure the collector voltage to ground. Then short the base to the emitter and remeasure the collector voltage. If the transistor is not leaking, the base-emitter short will turn the transistor off, and the collector voltage will rise to the same value as the supply. If there is any leakage, a current path will remain (through the emitter resistor, emitter-base short, collector-base leakage path, and collector resistor). There will be some voltage drop across the collector resistor, and the collector will have a voltage at some value lower than the supply.

Note that most meters draw current, and this current passes through the collector resistor. This can lead to some confusion, particularly if the meter draws heavy current (has a low ohms-per-volt rating). To eliminate any doubt, connect the meter to the supply through a resistor with the same value as the collector resistor. The drop, if any, should be the same as when the transistor is measured to ground. If the drop is much different (lower) when the collector is measured, the transistor is leaking.

As an example, assume that in the circuit in Fig. 3-19 the supply is 12

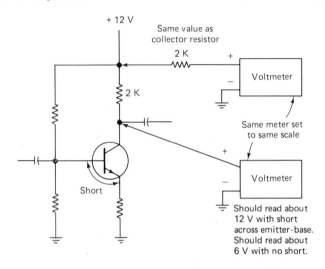

Fig. 3-19. Checking for transistor leakage in amplifier circuit

V, the collector resistance is 2 KΩ, and the collector measures 4 V with respect to ground. This means that there is an 8-V drop across the collector resistor and a collector current of 4 mA (8/2000 = 4mA). Normally, the collector is operated at about one-half the supply voltage (or 6V in this case). However, simply because the collector is at 4 V instead of 6 V does note make the circuit faulty. Some circuits are designed that way.

In any event, the transistor should be checked for leakage, using the emitter-base short test shown in Fig. 3-19. Now assume that the collector voltage rises to 10 V when the base and emitter are shorted. This indicates that the transistor is cutting off, but there is still some current flow through the collector resistor, about 1 mA (2/2000 = 1 mA).

A 1-mA current flow is high for a meter. However, to confirm a leaking transistor, connect the same meter through a 2-kΩ resistor (same as the collector load) to the 12-V supply (preferably at the same point where the collector resistor connects to the supply). Now assume that the indication is 11.7 V through the external resistor. This indicates that there is some transistor leakage.

The amount of transistor leakage can be estimated as follows: 11.7 − 10 = 1.7-V drop; 1.7/2000 = 0.85 mA. However, from a practical troubleshooting standpoint, the presence of any current flow with the transistor supposedly cut off is sufficient cause to replace the transistor.

3-2. BASIC TELEVISION ALIGNMENT PROCEDURES

The general procedure for alignment of split-sound and intercarrier types of television receivers is the same, the major differences being in the number of intermediate frequencies used and the specific frequencies involved. Most modern television receivers (color, black and white) are of the intercarrier type. That is, there is one set of IF amplifiers for both picture and sound signals. Some older television receivers are of the split-sound type, where separate IF circuits are used for picture and sound. Also, most modern television receivers (particularly the larger receivers) use IF circuits that operate in the 40- to 48-MHz range. However, many smaller receivers use IF circuits that operate in the 20- to 30-MHz range. A few older receivers use IF circuits in the 10- to 15-MHz range.

No matter what IF systems and frequencies are used, there are four separate steps for overall alignment of black and white receivers (and the black and white circuits of color receivers): tuner or RF alignment, picture (video) IF alignment, trap alignment, and sound IF and FM detector alignment. The following basic procedures can be used as a guide in performing television alignment procedures. However, always follow the specific procedures recommended in the manufacturer's service literature. In Chapter 5 we shall describe some typical alignment procedures found in manufacturer's data sheets.

3-2.1   RF Tuner Alignment

The primary purpose of alignment is to obtain a response curve of proper shape, frequency coverage, and gain. Most RF tuners merely require "touch-up" alignment in which relatively few of the adjustments are used. For a complete alignment job, follow a specific sequence of adjustments, as recommended in the service literature. However, where only touch-up alignment is required, the sequence of adjustments is usually unimportant.

In principle, complete VHF front-end alignment includes alignment of the antenna input circuits and adjustment of the amplifier and RF oscillator circuits. The antenna input circuits are usually aligned to give a response curve which has a sharp drop-off slightly below channel 2 and which is flat up through channel 13. Alignment of the RF amplifier and oscillator stages includes setting the oscillator frequencies for channels 2 through 13, setting one or more traps to their correct frequencies, and adjustment of tracking with the RF amplifier.

All of these adjustments require that a sweep signal and marker signal be fed into the tuner so that a response curve with markers is reproduced on an oscilloscope screen. The basic sweep/marker generator techniques described in Chapter 2 are used. Alignment is accomplished by setting adjustments so that the waveshape on the oscilloscope screen resembles the waveshape shown in the service literature. Figure 3-20 is a typical RF tuner waveshape obtained with a sweepmarker generator test setup. Some service literature shows separate RF tuner waveshapes for each channel. However, the trend is toward a "typical" tuner waveshape for all channels. In any event, each channel must be aligned separately to obtain the desired curve.

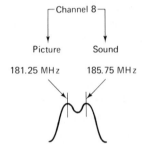

Fig. 3-20. Typical RF tuner wave-shape obtained with a sweep/marker generator and oscilloscope

The marker generator signals are used to provide frequency reference points to aid in shaping the curve. For example, with the sweep generator set to deliver an output on channel 8, markers are injected at 181.25 and 185.75 MHz (picture and sound carrier frequencies for channel 8), as shown in Fig. 3-20. The markers on the curve show the separation between the picture and sound carriers of 4.5 MHz. Since the RF tuner must pass both sound and picture signals, a tuner band pass of approximately 6 MHz is required. (The picture and sound carrier frequencies for all channels are given in Chapter 5.)

If you are working with a suspected defective tuner, leave the tuner set to only one channel position until the trouble is located and cleared. Afterwards, other channel positions can be compared with the initial one for sensitivity, switching noise, and general performance. On the other hand, if you are working with a supposedly good tuner, check the alignment by observing the response curves for each channel. Curves for individual channels should be examined and compared with those shown in the service literature. If a response curve indicates that alignment is required, follow the recommended alignment procedure in the service literature.

RF tuner test setups. The following notes describe typical alignment and test connections for an RF tuner.

Do not start alignment until the receiver, and all test equipment, has warmed up for at least 20 min. If the tuner is to be aligned in the receiver (which is generally the simplest way unless there has been extensive work on the tuner), the alignment may require that the first IF stage be out of operation. In some vacuum-tube receivers, the first IF can be disabled by removing the IF tube. This is not practical in solid-state receivers. Also, in some tuners, resonance in the mixer collector (or plate) circuit may produce undesired reflections. Generally, to remedy these situations, the IF amplifier input must be loaded (with a resistance across the IF input) or the IF transformer primary must be detuned. However, do not make any of these connections or adjustments unless recommended by the service literature.

The tuner oscillator should always be in operation during alignment. It is possible to make some tuner adjustments with the oscillator disabled. However, the lack of oscillator injection voltage at the mixer will alter the mixer bias, resulting in an increase in amplitude of the response curve and distortion of the waveshape. If there has been extensive work on the RF tuner (particularly by untrained "technicians"), it may be necessary to check the frequency of the oscillator on one or more channels with a heterodyne frequency meter. However, this is generally not required.

If you find serious misalignment of the tuner when making the test setup, or if you encounter considerable difficulty or failure in alignment, this usually indicates a defect in the tuner. Likewise, if alignment adjustments fail to produce correct tuner curves, you should submit the tuner to the troubleshooting procedures of Chapter 6.

You can use either the direct signal injection method, or postinjection (bypass alignment) with a marker-adder to align an RF tuner. As discussed in Chapter 2, with the marker-adder system, the marker signal is added to the sweep response curve by the marker-adder unit after the demodulated sweep signal is taken out of the receiver under test. With direct signal injection, in which both the sweep and marker signals are passed through the receiver circuits, overloading or clipping by the receiver circuits can introduce distortion of the marker and distortion of the sweep curve by the marker. The marker-adder system of adding markers to the response curve eliminates this source of distortion. In addition, the marker-adder system permits simple and precise alignment of a variety of trap circuits without marker "suckout." (When marker signals are passed through the circuits being aligned and there are traps tuned to the marker signal frequencies, the markers are removed completely by the traps and do not appear on the oscilloscope pattern.)

1. For marker-adder alignment, connect the equipment as shown in Fig. 3-21. Figure 3-21(a) assumes that the sweep generator, marker generator, and marker-adder are three separate units. As discussed in Chapter 2, the three functions are often combined into one unit. Figure 3-21(b) assumes that a single sweep/marker generator (with adder) is used. With

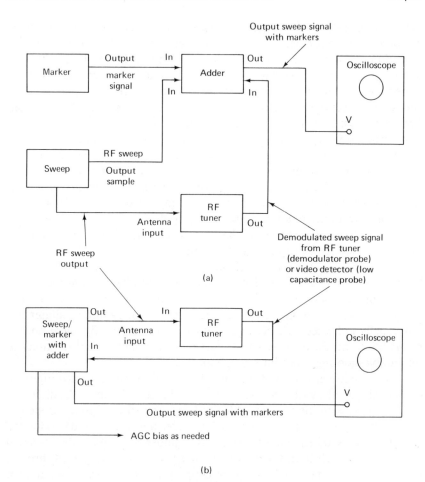

Fig. 3-21. Basic sweep/marker (with adder) postinjection alignment test connections (for RF tuner)

either arrangement, it is necessary to connect the receiver output to the oscilloscope vertical input. The receiver output test point may be across the load resistor of the video detector (IF output) or at the RF tuner output (at the "looker" point). If you connect at the video detector, you will get an "overall" response curve (RF, IF, detector). By connecting directly to the "looker" point, you can check the shape of the tuner curve, independently of other receiver circuits (which is usually the better method if you suspect problems in the tuner). Figure 3-22 shows two typical output test points for an RF tuner. Note that signals at the base test point are at the RF frequency, whereas signals at the collector test point are IF. With either test point, the signals are demodulated (by crystals in the marker-adder) for presentation on the oscilloscope vertical input.

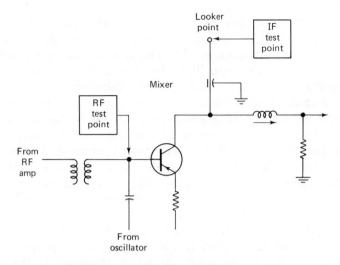

Fig. 3-22. Two typical output test points for an RF tuner

2. For direct injection alignment, connect the equipment as shown in Fig. 3-23. With this arrangement, both the sweep and marker generator outputs are connected directly to the RF tuner input (at the antenna terminals). The receiver output to the oscilloscope can be taken from the video detector (for an overall response curve) or at the tuner looker point. Note that if the video detector output is chosen, a low-capacitance probe

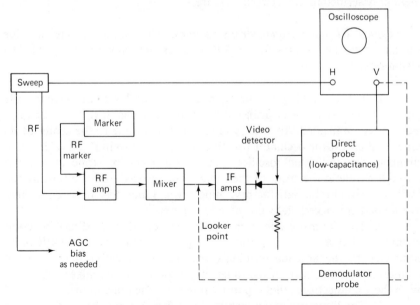

Fig. 3-23. Basic direct injection alignment test connections (for RF tuner)

can be used with the oscilloscope (since the video detector will demodulate both the sweep and marker signals). However, if you monitor the output at the tuner looker point (to get a separate tuner response curve), you must use a demodulator probe with the oscilloscope, because the tuner output is at a frequency (IF) well beyond the band pass of the oscilloscope.

    3. With either marker-adder or direct alignment, it is necessary to disable the AGC line. The diagrams of Figs. 3-21 and 3-23 assume that bias voltages to disable the AGC circuits are available from the sweep/marker generators (which is usually the case with modern generators). Always use the recommended bias voltage (amplitude, polarity, etc.), and apply the bias at the recommended test point. In some receivers, there is a common AGC line for both the tuner and the IF amplifiers. In other receivers, the tuner bias is taken from a separate line.

    4. With the test connections made as shown in Figs. 3-21 or 3-23, adjust the sweep and marker generators to the appropriate channel frequency. Set the sweep width to maximum, or as recommended in the service literature. Typically, the sweep width will be 10 to 15 MHz. Adjust the oscilloscope controls for a trace similar to that shown in Fig. 3-20 (or as shown in the service literature). If the generator sweep output is not connected directly to the oscilloscope horizontal input, it will usually be necessary to adjust the oscilloscope phasing control to obtain the desired trace (as discussed in Chapter 2).

    5. The test setup is now complete and ready for test and/or alignment as described in the service literature.

    Typical RF tuner alignment instructions. The following steps describe the alignment procedures for an RF tuner as they appear in typical television service literature.

    1. Before making any test connections, the literature recommends that the tuner oscillator be adjusted on each active channel. No test equipment is recommended. Instead, you set the fine tuning to the center of its range and adjust the oscillator for "best picture and sound" on each active channel. The individual oscillator slugs are accessible (one at a time) through a hole in the front of the tuner. This procedure for adjusting the oscillator should be satisfactory, unless the tuner has been competely overhauled or worked on by untrained "experts."

    2. The literature does not specify marker-adder or direct injection alignment. However, the literature does specify that the sweep generator be connected to the antenna terminals and that the receiver output to the oscilloscope be taken from the tuner looker point (at the mixer collector). With these connections, the response curve is for the tuner alone.

    3. The literature recommends that a variable bias be applied to the

AGC line at a specific test point. However, the literature does not recommend a specific bias voltage. Instead, you adjust the bias to obtain a response curve "which shows no indication of overloading." That is, the response curve should not change drastically when you vary the generator signal amplitude.

4. The literature recommends that the initial check of tuner alignment be made on channel 13, using a sweep generator frequency of 213 MHz, a sweep width of 10 MHz, and marker signals of 211.25 and 215.75 MHz (picture and sound carrier frequencies).

5. With all test connections made, the bias applied, and the generators properly adjusted, you adjust four specific receiver controls (coil tuning slugs) to obtain "maximum gain and symmetry of the response curve." A sample curve, similar to that in Fig. 3-20, is given in the literature.

6. With channel 13 properly adjusted, you then check all remaining channels (12 through 2), using identical test connections and bias but changing sweep and marker frequencies as necessary. The service literature specifies sweep and marker frequencies for all channels.

7. If all of the channels produce approximately the same response curve, the RF tuner can be considered as properly aligned. If one or more channels produce an abnormal curve, this usually indicates that the oscillator adjustment for that particular channel is incorrect (or that there is a defect in the tuning coils for that channel).

3-2.2   IF Amplifier and Trap Alignment

As in the case of the tuner, the primary purpose of IF amplifier alignment is to obtain a response curve of proper shape, frequency coverage, and gain. The purpose of trap alignment is to remove undesired signals from the IF circuits. Again, most IF amplifiers merely require touch-up alignment. Of course, this may not be true if there has been extensive work in the IF circuits.

If a television receiver is to give wide-band amplification to the television signal, the picture (video) IF system of the receiver must pass a frequency band of approximately 3.5 to 4 MHz. This is necessary to ensure that all the video information is fed through to the picture tube and that the resultant picture has full definition. (The band pass of color receivers must be essentially flat to beyond 4 MHz to ensure that color information contained in the color side bands is not lost.) The desired band pass is obtained by proper alignment of the IF adjustment controls (which are usually coil tuning slugs but may be trimmer capacitors).

All of these adjustments require that a sweep signal and marker signal be fed into the IF amplifiers so that a response curve with markers is reproduced on an oscilloscope screen. The basic sweep/marker generator techniques described in Chapter 2 are used. Alignment is accomplished by setting adjustments so that the waveshape on the oscilloscope screen resembles the waveshape shown in the service literature.

Figure 3-24 is a typical IF amplifier waveshape obtained with a sweep/ marker generator test setup. Note that the frequency relation of the sound carrier to the picture carrier is often reversed in the IF amplifiers because the tuner oscillator usually operates at a frequency higher than that of the transmitted carrier. For example, assume that (for channel 9) a picture carrier is transmitted at 211.25 MHz and that the corresponding sound carrier is transmitted at 215.75 MHz. Further assume that the IF frequency is 26.75 MHz. If the local oscillator is at 238 MHz, this signal will combine with the picture carrier of 211.25 MHz to produce a difference IF signal of 26.75 MHz. However, the sound carrier of 215.75 MHz will also combine with the 238 MHz local oscillator signal to produce a difference IF signal of 22.25 MHz, which is 4.5 MHz *below* picture IF signal. Some sweep/marker generators compensate for this condition by sweeping from high to low frequencies. However, always consult the receiver service literature for proper response curve shape and marker frequencies and the generator service literature for the method of producing the display.

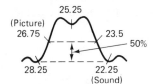

Fig. 3-24. Typical IF amplifier waveshape obtained with a sweep/ marker test setup

No matter how the display is obtained, the following two characteristics of the IF response curve (Fig. 3-24) should be noted: (1) The picture carrier is set at approximately 50% of the maximum response, and (2) the sound carrier frequency must be at 1% (or less) of the maximum response.

The sound carrier is kept at this low level to prevent interference with the video signal. The skirt selectivity (or selectivity at the high and low extremes of the 6-MHz band pass) of the IF response is made sharp enough to reject the sound component of the composite signal. To achieve this selectivity in some receivers, an absorption circuit consisting of a trap tuned to the sound intermediate frequency is used. Some receivers also include additional traps tuned to frequencies of adjacent sound and picture channels. These traps have a marked effect on the shape of the response curve.

The picture carrier is placed at approximately 50% of maximum response because of the nature of the television broadcast transmission. If the IF circuits are adjusted to put the picture carrier too high on the response curve, the effect will be a general decrease in picture quality caused by the resulting low-frequency attenuation. If the picture carrier is placed too low on the curve, there will be a loss of low-frequency video response (that usually shows up as poor definition in the picture). Loss of blanking and poor synchronization can also occur when the picture carrier is too low on the response curve.

Also note that in addition to the picture and sound frequencies, two additional marker frequencies are shown on the curve of Fig. 3-24 (23.50 and 28.25 MHz). These marker frequencies are used to align the IF traps. Most IF amplifiers have one or more traps, depending on the type of receiver. The test equipment setup for alignment of IF traps is essentially the same as for alignment of the IF amplifier stages. However, certain problems may be encountered.

Because the response of the IF amplifier is very low at the trap frequencies, the marker may often be difficult to see on the response curve. The use of a marker-adder is recommended for trap alignment. If extreme difficulty is encountered, the traps can be set with a voltmeter rather than an oscilloscope. With such an arrangement, the voltmeter is connected across the video detector load resistor, the marker generator is set for the trap frequency, and the trap is tuned for a minimum voltage reading on the meter.

The general procedure in aligning IF amplifiers is first to set the traps and then to align the amplifier circuits. Since any adjustment of the amplifier circuits will usually slightly detune the traps, they may have to be touched up during IF amplifier alignment. Always follow the exact procedure given in the service literature.

IF amplifier test setups. The following notes describe typical alignment and test connections for an IF amplifier.

Do not start alignment until the receiver, and all test equipment, has warmed up for at least 20 minutes.

The tuner oscillator should not be in operation during alignment of the IF amplifiers. The oscillator signal may interfere with the sweep and marker signals. In a vacuum-tube receiver, the oscillator tube can be removed. In solid-state receivers, connect a short between the base and emitter of the oscillator transistor. In those receivers where the input to the IF amplifiers (output of RF tuner) is made through a coaxial cable, disconnect the cable and inject the sweep signals directly into the IF amplifier at the connector. This eliminates the need for disabling the tuner oscillator.

If you find serious misalignment of the IF amplifiers when making the test setup, or if you encounter considerable difficulty or failure in alignment, this usually indicates a defect in the IF amplifiers. Likewise, if alignment adjustments fail to produce correct curves, you should submit the IF amplifiers to the troubleshooting procedures of Chapter 6.

You can use either the direct signal injection method or postinjection (bypass alignment) with a marker-adder to align the IF amplifiers. However, as discussed, the marker-adder system is always recommended, particularly where the IF amplifier has traps. The marker-adder system will eliminate the problem of marker "suckout." However, if you do use direct signal injection to align the IF amplifiers, you can adjust the traps by means of a meter connected across the video detector load resistor.

     1. For marker-adder alignment, connect the equipment as shown in Fig. 3-25. Figure 3-25(a) assumes that the sweep generator, marker gener-

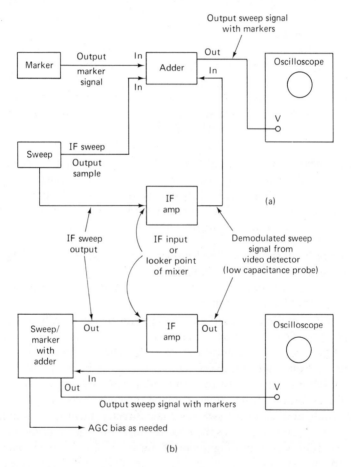

Fig. 3-25. Basic sweep/marker (with adder) postinjection alignment and connections (for IF amplifier)

ator, and marker-adder are three separate units. Figure 3-25(b) assumes that a single sweep/marker generator (with adder) is used. With either arrangement, it is necessary to connect the receiver output to the oscillo-scope vertical input. Usually, the receiver output test point is across the load resistor of the video detector. This provides an overall response curve for the entire IF amplifier circuit as well as the video detector. However, some service literature recommends that you connect the oscilloscope input to the output of individual IF stages. This is discussed in Sec. 3-2.4.

    2. For direct injection alignment, connect the equipment as shown in Fig. 3-26. With this arrangement, both the sweep and marker generator

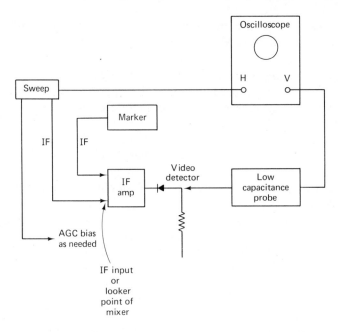

Fig. 3-26. Basic direct injection alignment test connections (for IF amplifier)

outputs are connected directly to the IF amplifier input (at the IF input connector or at the looker point in the mixer of the RF tuner). The receiver output to the oscilloscope is taken from the video detector (across the load resistor).

    3. With either marker-adder or direct alignment, it is necessary to disable the AGC line. The diagrams of Figs. 3-25 and 3-26 assume that bias voltages to disable the AGC circuits are available from the sweep/marker generators. Always use the recommended bias voltage (amplitude, polarity, etc.), and apply the bias at the recommended test point. In some receivers, there is a common AGC line for both the tuner and IF amplifiers. In other receivers, the IF amplifier bias is taken from a separate line.

4. With the test connections made as shown in Fig. 3-25 or 3-26, adjust the sweep and marker generators to the appropriate IF frequency. Set the sweep width to maximum or as recommended in the service literature. Typically, the sweep width will be 10 to 15 MHz. Adjust the oscilloscope controls for a trace similar to that shown in Fig. 3-24 (or as shown in the service literature). If the generator sweep output is not connected directly to the oscilloscope horizontal input, it will usually be necessary to adjust the oscilloscope phasing control to obtain the desired trace (as discussed in Chapter 2).

5. The test setup is now complete and ready for test and/or alignment as described in the service literature.

Typical IF amplifier alignment instructions. The following steps describe the alignment procedures for an IF amplifier (including IF traps) as they appear in typical television service literature.

1. Before making any test connections, the literature recommends that the tuner oscillator be disabled and that the channel selector be set to any noninterfering channel. The receiver involved is one where the RF tuner output is permanently wired to the IF amplifier input (not through a removable coaxial cable).

2. The literature does not specify marker-adder or direct injection alignment. However, the literature does specify that the generator outputs be connected to the IF input at the looker point (mixer of the RF tuner) and that the receiver output to the oscilloscope be taken from the video detector load resistor.

3. The literature recommends that a variable bias be applied to the AGC line at a specific test point. However, the literature does not recommend a specific bias voltage. Instead, you adjust the bias to obtain a response curve "which shows no indication of overloading." That is, the response curve should not change drastically when you vary the generator signal amplitude.

4. The initial step of alignment is made with a marker generator connected to the IF amplifier input, and the receiver output is monitored with a dc voltmeter connected across the video detector output. The marker generator is set to each of three trap frequencies (20.75, 22.25, and 28.25 MHz) in turn. With each marker signal applied, the corresponding trap control (coil tuning slug) is adjusted for a minimum indication on the voltmeter. This action sets the three traps so that they provide maximum attenuation for the three undesired signal frequencies.

5. The next step of alignment is done using the same test connections (marker generator and voltmeter). However, you now adjust four controls (including the mixer collector coil) for maximum indication on the voltmeter when the corresonding one of four marker signals (at frequencies of 23.6, 24.3, 25.5, and 26.3 MHz) is applied. This action stagger-tunes the IF amplifiers so that they pass the desired frequencies.

6. The final step of alignment is made with a sweep generator connected to the IF amplifier input, and the receiver output is monitored with an oscilloscope connected across the video detector output. The marker generator is connected to the IF amplifier input (for direct injection) or through a marker-adder to the oscilloscope (for postinjection). The sweep generator is set to 25 MHz (center frequency) with a sweep width of 10MHz. The marker generator is set to provide signals at frequencies of 22.5, 23.5, 25.25, 26.75, and 28.25 MHz. With all of these connections made and the test equipment operating, you check the response on the oscilloscope.

7. The service literature recommends that you "check for maximum gain and symmetry of response with markers as shown in the diagram (similar to that of Fig. 3-24)." The literature goes on to recommend that "in order to obtain a proper response, it may be necessary to slightly retouch the same four controls adjusted during step 5."

3-2.3   4-5-MHz Sound Trap Alignment

The video amplifier circuits of all television receivers (black and white as well as color) have a 4.5-MHz sound trap. This trap is tuned to reject any sound signals (at 4.5 MHz) that may pass through the video amplifier so as to prevent the signals from appearing on the picture tube. Adjustment of the sound trap is usually not critical in most black and white receivers. However, the problem of sound-in-picture for a color receiver may be quite critical. For this reason, the problem of sound trap adjustment is discussed fully in Chapter 4.

For a typical black and white receiver, the sound trap can be adjusted simply, without test equipment, once the RF and IF stages have been aligned. To adjust the trap, tune in a strong television broadcast signal, and set the receiver contrast control at maximum. Adjust the fine tuning control until a beat pattern (herringbone) is visible on the picture tube screen. Then adjust the sound trap (usually a coil tuning slug) for minimum beat interference. Reset the fine tuning, making sure that the beat interference is completely gone when the fine tuning is set back to normal (midrange, for best picture and sound).

3-2.4 Checking Response of Individual Stages

The response of individual IF amplifier stages, or of two or more stages together, may be checked using a sweep and marker generator. Basically, the sweep signal is fed into the stage immediately preceding the stage being checked. The response curve is checked on an oscilloscope which is connected across the video detector load resistor. As in the case of other IF circuit tests, the AGC line must be disabled for accurate results.

1. Connect equipment as shown in Fig. 3-27.

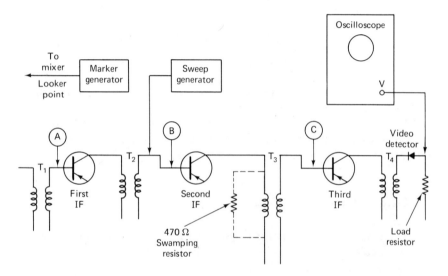

Fig. 3-27. Checking response of individual stages

2. Disable the AGC line using the bias recommended in the service literature.

3. Connect the marker generator (or the marker output of a combined sweep/marker generator) to the looker point of the RF tuner.

4. Connect the sweep generator to the input (base or grid) of the IF amplifier stage ahead of the stage to be checked. This will isolate the test equipment from the stage being checked. The RF output from the sweep generator should not be connected to the input of the stage being checked, because even a slight loading of the input circuit may cause a change in circuit impedance and result in distortion of the normal response characteristics. The test connections of Fig. 3-27 (sweep generator at test point B) are to check response of the third IF amplifier.

5. Connect a resistor of approximately $470\,\Omega$ across the primary of the following IF transformer T_3. This resistor acts to swamp the primary winding and prevents inductive reactance of the winding from affecting the band-pass characteristics of the amplifier being checked (the third IF amplifier). The shape of the response curve on the oscilloscope is determined by the band-pass characteristics of the third amplifier stage and the video detector, with the connections as shown in Fig. 3-27.

6. To check the band-pass characteristics of the video detector only, move the sweep generator from point B to point C, and place the swamping resistor across the primary of T_4.

7. To check the response of the second IF, third IF, and video detector stages together, move the sweep generator to point A, and connect the swamping resistor across the primary of T_2.

3-2.5   Sound IF and Audio Detector Alignment

The procedure used for aligning the sound IF amplifiers in a television receiver are similar to those of an FM receiver. Present-day intercarrier-type receivers use a sound IF of 4.5 MHz. Older split-sound receivers may use either 21.25 or 41.25 MHz.

The procedures for overall alignment of the sound section, including both the IF amplifier and FM detector, may vary depending on the type of FM detector. In general, alignment procedures for nearly all types of amplifiers and detectors involve feeding a sweep signal of the intermediate frequency through the IF amplifier and observing the sweep trace on an oscilloscope connected across the audio detector load resistor.

Procedure (overall response curve)

1. Connect equipment as shown in Fig. 3-28.

2. Connect the sweep and marker generator to test point A. Set the generators to 4.5 Hz (or to another IF if required).

3. Connect the oscilloscope to test point C where a demodulated signal appears.

4. Set the sweep width to about 1 MHz. An S-shaped curve, similar to that shown in Fig. 3-29, should appear on the oscilloscope.

5. Usually, the service literature will specify that the detector should be tuned so that the S-curve peaks are equal in height or distance from the base line and that the extreme right- and left-hand points, as well as the center point between the two peaks, should lie along the base line.

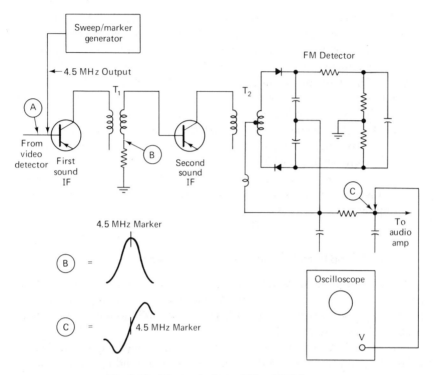

Fig. 3-28. Alignment of sound IF and FM detector

Procedure (alignment of first IF amplifier only)

1. Connect the equipment as shown in Fig. 3-28.

2. Leave the sweep generator and marker generator connected to test point A.

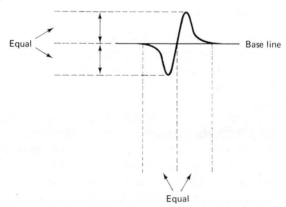

Fig. 3-29. Response curve for FM sound detector (© *1958 by the American Psychological Association. Reprinted by permission.*)

3. Connect the oscilloscope to test point B where a sweep response similar to that shown in Fig. 3-30 should be obtained. The marker indicates

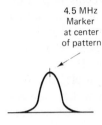

4.5 MHz
Marker
at center
of pattern

Fig. 3-30. Sound IF response curve with marker showing center of passband

the 4.5-MHz center frequency of the curve. If the marker does not appear exactly at the center of the curve, the amplifier should be adjusted as recommended in the service literature until the marker is exactly at the center of the curve. This assures that response is symmetrical and that adequate audio bandwidth is obtained.

4. If a discriminator detector is used, the stage will be preceded by a limiter. The overall response of the IF amplifier is checked by connecting the oscilloscope across the resistor in the input circuit (base or grid) of the limiter. In some receivers, the time constant of the input circuit may be large enough to cause distortion of the pattern when the oscilloscope is connected. If the pattern is distorted, the difficulty may be eliminated by temporarily shunting the resistor with another resistor of a value determined by experimentation.

5. Some service literature recommends that the marker generator be modulated with an audio tone when a discriminator is used as the audio detector. The test connections and procedures are the same as for other types of detectors. However, the modulation will produce a different appearance of the trace. In the band-pass trace (Fig. 3-30), the 4.5-MHz marker will appear wavy or fuzzy. In the overall response curve (Fig. 3-29), the modulation causes a wide trace or waviness of the base line. However, when the center frequency of the detector (discriminator) is tuned exactly to the sound intermediate frequency, the modulation will disappear. This characteristic can be used to set the detector exactly on the center frequency (4.5 MHz).

3-2.6 Checking Overall Video Sweep Modulated Response

It is possible to check the overall response of a television receiver from the RF tuner to the picture tube using a technique known as video sweep modulation. The technique (also known as chroma sweep modulation, or some similar term when color receivers are involved) permits observation of the true overall frequency response, including the effect of the video detector load circuit. When RF, IF, and video circuits are checked separately.

the effect of the video detector load may change, and this change cannot be observed (except by an overall check).

Observation of the overall frequency response is especially important in the alignment of color receivers. For this reason, the video and chroma sweep techniques are discussed further in Chapter 4. In most color receivers, the IF and video sections individually do not have a flat response characteristic over the required bandwidth. The responses of the IF and video amplifiers in these receivers are complementary. For example, the first video stage is usually designed to have a rising characteristic which compensates for the falling characteristic of the IF amplifiers. In such receivers, the color subcarrier is positioned at the 50% response point on the picture IF response curve. However, the rising response characteristic of the video amplifier results in a flat overall IF/video response.

There are many methods used to check response with video sweep modulation. In all of the methods, the video output to the picture tube (at the cathode or grid) is monitored with an oscilloscope. The means of introducing the sweep signal depends on the test equipment available. Two methods are presented here. Other methods are discussed in Chapter 4.

One method for checking overall response with video sweep modulation is shown in Fig. 3-31. This method requires an absorption marker. Absorption markers have been available commercially, although they are not used extensively today. The typical absorption marker schematic is shown in Fig. 3-32. Each tuned circuit introduces a fixed marker into the

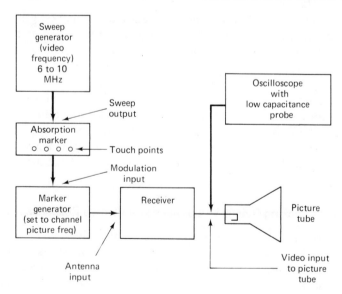

Fig. 3-31. Checking overall video sweep modulation response (absorption marker method)

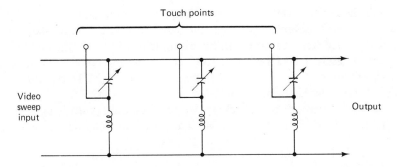

Fig. 3-32. Typical absorption marker circuit

display, usually at such frequencies as 0.5, 1.5, 2.5, 3.08, 3.58, 4.08, and 4.5 MHz. These are the frequencies most often used in checking and aligning chrominance circuitry in color receivers. Each tuned circuit has a "touch terminal." When one or more of the terminals is touched, the corresponding marker or markers on the sweep display will disappear. On some units, all markers (or selected markers) can be removed simultaneously by means of a shorting switch.

With the absorption marker method, the sweep generator output is set for a video sweep frequency (typically 6 to 10 MHz) and is applied to the external modulation input of the marker generator. The marker generator output is applied to the receiver at the RF tuner antenna terminals and is tuned to the picture frequency of a selected channel. The resultant waveform at the video output (picture tube input) is similar to that shown in Fig. 3-33. The markers in the display indicate the approximate bandwidth

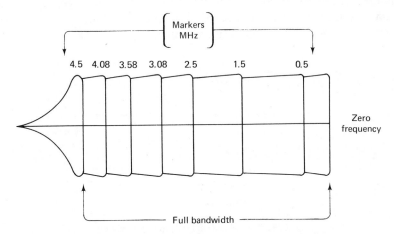

Fig. 3-33. Overall response of RF, IF and video circuits using video sweep modulation (with markers)

through the entire receiver. The zero-frequency point is at the right-hand end of the oscilloscope trace. The full bandwidth is at the left-hand end where the trace starts to slope. In the example of Fig. 3-33, the bandwidth is approximately 4.5 MHz. Most receivers, including color, that have an overall (front end to picture tube) bandwidth of 4 MHz or wider are generally considered good.

Another method for checking overall response with video sweep modulation is shown in Fig. 3-34. This method requires an external

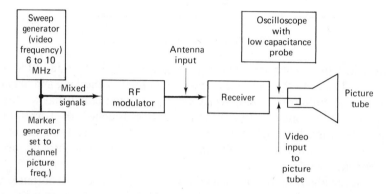

Fig. 3-34. Checking overall video sweep modulation response (external RF modulator method)

modulator, such as shown in Fig. 3-35. The external modulator contains circuits that introduce markers into the display. The external modulator also permits the outputs of the marker and sweep generator to be mixed so that the marker generator output is modulated by the sweep generator. Again, the marker generator is tuned to the picture frequency of a selected channel, and the sweep generator is set to produce a sweep at about 6 MHz.

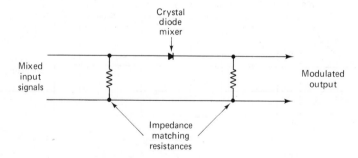

Fig. 3-35. Typical external RF modulator circuit

The resultant pattern is similar to that of Fig. 3-33, and overall response or bandwidth is indicated in the same way.

3-3. TROUBLESHOOTING WITH SIGNAL INJECTION VERSUS SIGNAL TRACING

As discussed in Chapter 1, there are two basic methods for troubleshooting television receivers: signal tracing and signal injection. With true signal tracing, you must rely on the signals broadcast by the television station and on the signals developed by the receiver in response to the broadcast signals. In effect, you trace through all receiver circuits checking that the waveforms, pulses, signals, etc., are as they appear in the service literature. You compare the oscilloscope patterns obtained from the receiver being serviced with the standard patterns (found in the literature) for amplitude, shape, frequency, etc. An absent or abnormal waveform indicates problems in the circuits being measured.

With true signal injection, you must have an analyst generator such as discussed in Chapter 2. These generators duplicate the broadcast signals as well as the signals produced by the receiver. An advanced analyst generator will have such signals as RF, picture IF, sound IF, audio, composite video (including the sync pulse), separate sync pulses, and signals suitable for test of the vertical and horizontal sweep circuits and yoke deflection coils. Other analyst generators provide RF, IF, sound, audio, and a composite video signal. With these generators, if you are troubleshooting circuits beyond the video detector, you must inject the composite video signal and then trace signals with an oscilloscope through the remaining circuits.

We shall not attempt to promote signal tracing or signal injection as the better method. This controversy has gone on for many years and eventually boils down to a matter of choice. Nor shall we try to compare one type of equipment with another.

However, it is generally conceded by most television service technicians that signal injection is the quickest (and thus the most profitable). For that reason, we shall devote most of this chapter to signal injection using analyst generators. In Sec. 3-4 we shall describe the use of an analyst generator without separate sync signals. In Sec. 3-5 we shall cover troubleshooting with an advanced analyst generator. We shall conclude the chapter with a discussion of signal-tracing methods using the composite video signal, the horizontal blanking pulse, and the VITS (vertical interval test signal) broadcast by most television stations. It should also be noted that the comprehensive troubleshooting procedures of Chapter 6 are based primarily on signal tracing.

3-4. BASIC TROUBLESHOOTING WITH AN ANALYST GENERATOR

This section is devoted to basic troubleshooting with an analyst generator such as the RCA WR-515A. With such a generator, you can inject test patterns (lines, dots, crosshatch, pulses, and color signals) at RF, IF, and video frequencies in stage-by-stage troubleshooting techniques. You apply the signals at appropriate test points throughout the circuits and observe the results on the receiver screen.

As a supplemental troubleshooting procedure, you can use an oscilloscope to view the test pattern waveform as it is processed through the receiver. With an analyst generator, you have the advantage of being able to inject the signal directly to the stage being checked. Any of the test patterns can be used in this procedure. However, a pulse waveform (known as the superpulse pattern) is especially helpful to check IF and video stages. This pulse is a simple, sharply defined trace. Thus, distortion or loss of gain in the waveform will be readily apparent.

The method described in this section starts by connecting an RF signal from the generator to the antenna terminals to evaluate the overall performance of the receiver and to note the particular picture problem (no picture, smeary picture, etc.). Then you inject signals back to the video and chroma circuits (in color receivers), working forward through the IF stages until the picture problem appears. In this way, you isolate the defective circuit. Of course, the defective stage can be found just as easily by starting at the mixer input and working back through the receiver until a normal picture is obtained. The technique you use is a matter of individual preference.

3-4.1 Troubleshooting Modular Receivers

As discussed in other chapters, the newer receivers with various stages or parts of stages on plug-in modules are serviced in a manner different from other receivers. Many times the trouble can be corrected quickly and easily simply by replacing the defective module. An analyst generator can be of great help to you in determining what stage or module is defective. The generator is especially helpful in isolating a problem that is not on a plug-in module but is in the master board that the modules plug into.

The troubleshooting procedures with signal-injection and signal-tracing techniques on modular receivers are similar to those used with any other receiver. The basic difference is that to isolate a defective module, the test points specified on the master board, or on the module socket representing the input and output points of the modules, are particularly important. On some modules, the components are accessible and you can use standard troubleshooting techniques to locate defective components. Other modules are sealed, however, and cannot be replaced.

3-4.2 Checking Overall Receiver Performance

To evaluate overall receiver performance, connect the generator to the VHF antenna terminals as shown in Fig. 3-36. Set the receiver tuner to channel 3. Set the generator RF LEVEL and CHROMA controls to mid-range. Release the IF push button (this applies an RF signal).

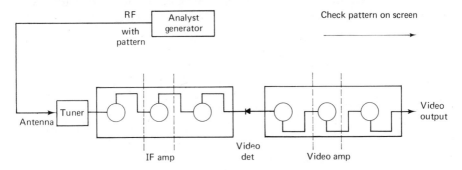

Fig. 3-36. Checking overall receiver performance with analyst generator

1. First press the SUPERPULSE pushbutton, then press the COLOR BAR, and then LINES. Adjust the receiver fine tuning control to obtain the best pattern. Set the receiver brightness, contrast, and other controls as necessary.

2. If the receiver is operating normally and has had proper setup adjustments (AGC, horizontal and vertical oscillators, focus, color killer, etc.), good patterns should be obtained on all functions. If a problem in the picture is noted, proceed to Sec. 3-4.5 (video amplifier check).

3-4.3 Checking AGC

The analyst generator can be used to determine whether the receiver AGC system is operating. Connect the generator to the receiver antenna terminals as described in Sec. 3-4.2. Connect a voltmeter (usually an electronic voltmeter) to the AGC test point at the tuner (or IF stages). If the AGC system is operating, the dc voltage at the test point will vary as the generator RF LEVEL control is tuned through its range.

3-4.4 Checking Sync Lock-in Range

By varying the SUPERPULSE SYNC control (during the overall receiver performance check, Sec. 3-4.2) and observing the superpulse pattern, you can determine how well the receiver holds sync (horizontal and vertical) over a wide range of sync levels. In this way, you can determine if the receiver sync is "touchy" (overly sensitive) to sync level variations.

3-4.5 Checking Video Amplifier

1. Set the receiver tuner to an unused channel or to UHF. Disconnect the generator from the antenna terminals (but do not reconnect the antenna).

2. Attach the generator cables as shown in Fig. 3-37. This applies a video signal from the generator to the video amplifier input.

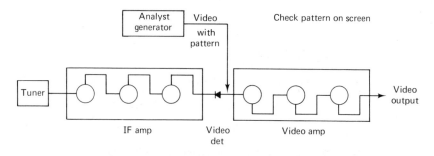

Fig. 3-37. Checking video amplifier with analyst generator

3. Select the desired generator pattern (such as a crosshatch of vertical and horizontal lines), and adjust the VIDEO LEVEL control to " + " (clockwise) or " − " (counterclockwise) as required to produce a good picture. If the VIDEO LEVEL control is turned to the wrong polarity, the picture will be dark, and there will be no sync. Turn the control in the proper direction until the picture begins to "tear," and then back off the control just to the point where the best picture is obtained.

4. If the picture is about normal, the video amplifier (from video detector to picture tube) is good. Proceed to a check of the IF and tuner circuits as described in Sec. 3-4.7 However, if the picture has the original problem discovered during overall response check (Sec. 3-4.2), you have isolated the trouble as being in the video amplifier.

5. You can further isolate the defective circuit within the video amplifier by additional signal injection into test points at each video stage (input and output of each video amplifier). As the signal is applied to various test points in the video amplifier, it will be necessary to readjust the VIDEO LEVEL control. In most receivers, the video sync polarity will be inverted (reversed) at each stage of the amplifier. You can also inject the signal at the video detector and signal-trace through the video amplifier circuits using an oscilloscope, as described in Sec. 3-4.8.

3-4.6   Checking Chroma Amplifier

A rough check of the chroma amplifier of many color receivers can be made by injecting a color-bar pattern at the video frequency into the chroma amplifier input (video amplifier output). The receiver sync may not be too stable when the signal is injected at that point because it is beyond the sync takeoff point. However, the color in the pattern should be fairly good. If trouble in the chroma amplifier is indicated, you can trace the color-bar signal using an oscilloscope. Refer to Chapter 4 for detailed descriptions of color receiver troubleshooting procedures.

3-4.7   Checking IF Amplifier and Tuner

1. Press the IF 45.75-MHz push button and set the LEVEL control to midrange (this applies an IF signal of 45.75 MHz). Connect the generator cable to the mixer input test point on the tuner (looker point) or to a test point at the input to the IF amplifier, as shown in Fig. 3-38. On some receivers, you can unplug the cable connecting the tuner to the IF amplifier and apply the generator signal to the cable.

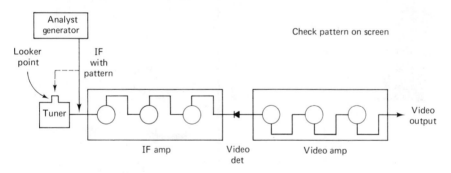

Fig. 3-38. Checking IF amplifier and tuner with analyst generator

2. Adjust the generator LEVEL control for the best picture. Observe various test patterns, including color bars. If you obtain normal patterns and the problem noted in Sec. 3-4.2 (overall response) is not apparent, the IF amplifier stages are operating properly, and you have isolated the problem as being in the tuner. However, if the picture still has the original problem, you have isolated the trouble as being in the IF amplifier (or the video detector diode).

3. You can further isolate the defective area by additional signal injection into test points at each IF stage. Figure 3-39 shows typical test points for both IF and video signal injection. You can also inject the signal at the mixer input (or IF input) and trace the signal using an oscilloscope, as described in Sec. 3-4.8.

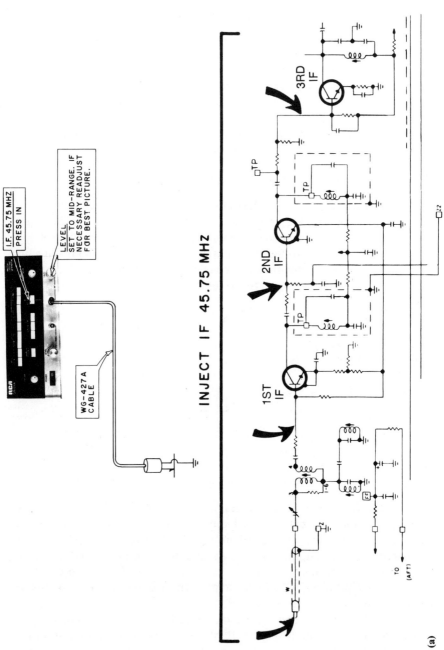

I.F. 45.75 MHZ
PRESS IN

LEVEL
SET TO MID-RANGE. IF
NECESSARY READJUST
FOR BEST PICTURE.

WG-427A
CABLE

INJECT IF 45.75 MHz

1ST IF 2ND IF 3RD IF

TO (AFT)

Fig. 3-39a and b. Test points for IF and video signal injection on a typical television receiver (*Courtesy RCA*)

(a)

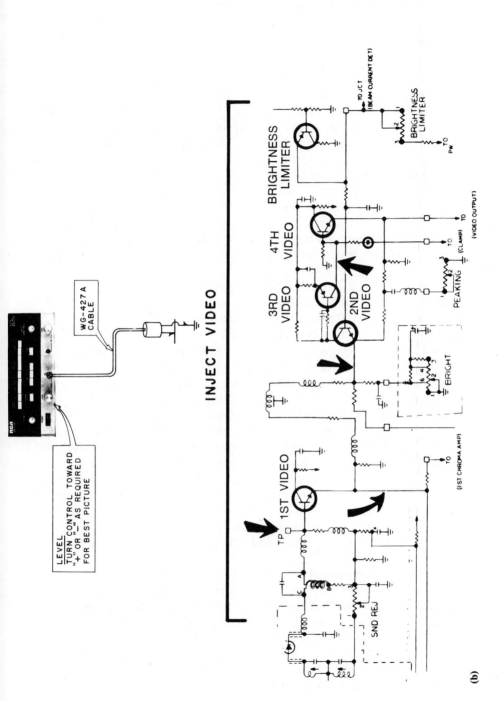

INJECT VIDEO

LEVEL
TURN CONTROL TOWARD
"+" OR "−" AS REQUIRED
FOR BEST PICTURE

WG-427 A
CABLE

1ST VIDEO

SND REJ

2ND VIDEO

3RD VIDEO

4TH VIDEO

BRIGHTNESS LIMITER

BRIGHT

PEAKING

BRIGHTNESS LIMITER

TO JCT
(BEAM CURRENT DET)

TO
PW

TO
(VIDEO OUTPUT)

TO
(CLAMP)

TO
(1ST CHROMA AMP)

(b)

3-4.8   Signal Tracing with an Analyst Generator

When an analyst generator is used in signal tracing, the generator serves as the signal source (in place of television broadcast signals). The basic procedure consists of applying a test signal at the antenna or mixer input, and then using an oscilloscope to view the pattern as it is processed through the IF, video, and chroma stages. When you reach a test point where the waveform has disappeared, is badly distorted, or does not have sufficient gain you have localized the trouble.

The oscilloscope must have a low-capacitance probe (for the signal tracing of circuits after the video detector) and a demodulator probe (for signal tracing ahead of the video detector). A normal pattern for use as a reference can be obtained by connecting the oscilloscope probes directly to the generator output. In this way, you will know what pattern is produced by each of the probes before the signals are processed by the receiver circuits.

It is usually recommended that the generator be set to produce a pulse pattern (such as the SUPERPULSE pattern) when tracing through the IF and video circuits. If you are signal tracing in the chroma amplifier circuit, a color-bar pattern must be used.

For signal tracing a receiver having a faulty picture, it is recommended that the initial point for signal injection be the mixer input or the input to the IF amplifier. You could connect the signal to the antenna terminals, but if the picture is really poor, it will be difficult to obtain correct fine tuning, and this could result in misleading waveforms. Although you can inject the signal at other test points within the IF amplifier or video stages (as described for signal injection), the oscilloscope waveform may have some distortion due to circuit loading or "hash" due to pickup through the test lead.

If the analyst generator has a control that varies the sync pulse amplitude in relation to the video signal, such a provision can be extremely useful in signal tracing a receiver that has poor sync or complete loss of sync. You can evaluate performance of the circuit by observing the sync pulse of the waveform at various test points and noting how the sync pulse appears at different sync levels.

Some service literature recommends that the horizontal circuits of the receiver be disabled during test and troubleshooting. This is done so that the horizontal signals will not interfere with the test signals. When such a recommendation is made, it is necessary to supply a bias voltage to the AGC line (since most receivers use a keyed AGC that requires signals from the horizontal circuits). If the AGC is not operating, the waveforms obtained during signal tracing may be distorted. Always follow the service literature recommendation concerning AGC bias.

Preliminary signal-tracing procedure. See Figs. 3-40 and 3-41 for signal-tracing test points in the IF and video sections, respectively.

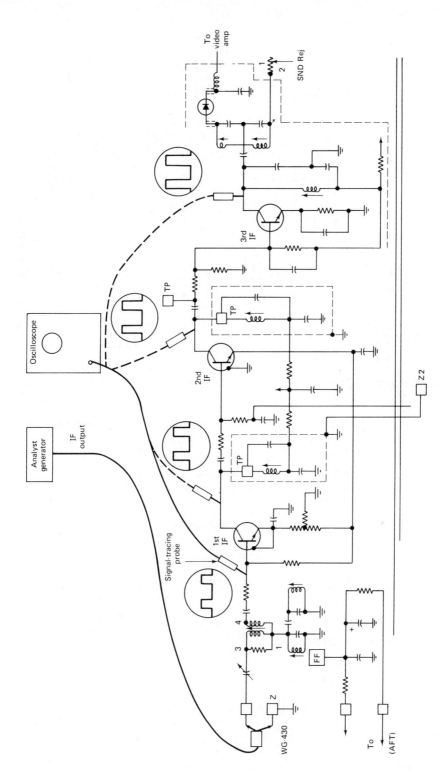

Fig. 3-40. Signal tracing test points in a typical IF amplifier (*Courtesy RCA*)

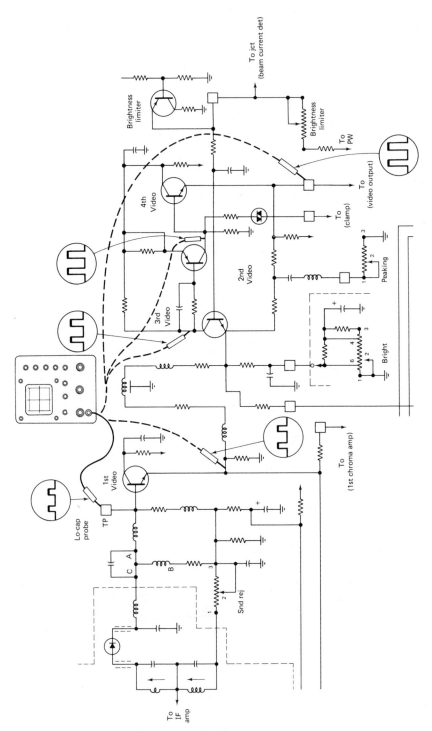

Fig. 3-41. Signal tracing test points in a typical video amplifier *(Courtesy RCA)*

1. Connect cables to the generator. Press the 45.75-MHz push button and set the LEVEL control to midrange (this applies an IF signal of 45.75 MHz). Press the SUPERPULSE push button, and set the SUPER— PULSE SYNC controls to midrange (this applies pulses at a video frequency).

2. As a reference, first connect the oscilloscope signal-tracing probe directly to the generator output. If the oscilloscope has preset TVV and H positions, use the V position. This will produce two pulses on the oscilloscope trace, as shown in Figs. 3-40 and 3-41. If the oscilloscope does not have preset positions, set the oscilloscope sweep range to sweep at about 30 Hz. Adjust the oscilloscope vertical gain, and other controls, for a pattern similar to that of 3-42(c). Set the oscilloscope internal sync switch to " + " or " − " as necessary to lock in a stable pattern of two pulses. This step will familiarize you with the SUPERPULSE waveform and provides a good check of the signal-tracing probe and the oscilloscope.

3. Disconnect the antenna from the receiver, and set the tuner to an unused channel. Connect the generator output to the mixer input or directly to the input of the IF amplifier (Fig. 3-40).

4. If the receiver is operating well enough to provide a viewable picture, adjust the generator LEVEL control to the minimum position that provides a good oscilloscope waveform and at the same time produces a good SUPERPULSE pattern. If the receiver does produce a picture, adjust the brightness, contrast, and vertical and horizontal hold controls for the best picture.

Signal tracing in the IF amplifier. See Figs. 3-40 and 3-42.

1. Connect the oscilloscope signal-tracing probe first to the input and then to the output of each stage in the IF amplifier. Each time you change test points, adjust the oscilloscope vertical gain to maintain trace height.

2. The waveforms should appear as shown in Fig. 3-42(a). An increase in trace height should be noted as you proceed from the input to the output of each stage. Trouble is indicated if the pattern cannot be obtained or has low amplitude at the proper test points [Fig. 3-42(b)] or if the pattern is severely distorted. If all signals appear normal, proceed to the following video amplifier check.

Signal tracing in the video amplifier. See Figs. 3-41 and 3-42.

1. Remove the signal-tracing probe, and use the low-capacitance oscilloscope probe. Connect the probe to the output of the video detector (Fig. 3-41). The patterns should appear similar to that of Fig. 3-42(c).

2. Proceed to check the input and output of each video stage. Each

Notes: Set oscilloscope to TVV or 30 Hz sweep rate. Use a
signal tracing probe for IF signal tracing. Use a low
capacitance probe for video signal tracing.

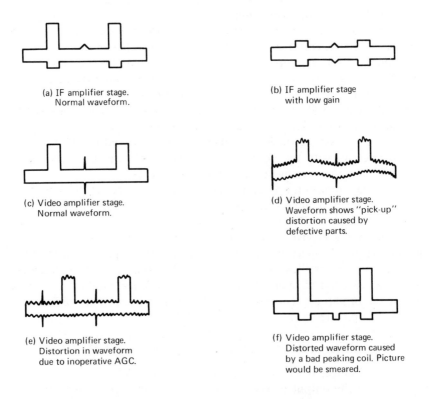

(a) IF amplifier stage.
Normal waveform.

(b) IF amplifier stage
with low gain

(c) Video amplifier stage.
Normal waveform.

(d) Video amplifier stage.
Waveform shows "pick-up"
distortion caused by
defective parts.

(e) Video amplifier stage.
Distortion in waveform
due to inoperative AGC.

(f) Video amplifier stage.
Distorted waveform caused
by a bad peaking coil. Picture
would be smeared.

(g) Superpulse waveform at
WR-515A video output with
"SUPERPULSE SYNC" set to
minimum, simulating a station
signal with a very weak sync.
Receiver pattern would be out
of sync. Note sync pulses extend
downward.

(h) Superpulse waveform as in (g),
except with "SUPERPULSE
SYNC" control set to Maximum.

Fig. 3-42. Superpulse waveforms obtained from RCA WR-515A analyst
generator

time you change test points, adjust the oscilloscope vertical gain to maintain
trace height. In most cases, to maintain stable sync, you will also have to
set the oscilloscope sync polarity switch from "+" to "−," or back, each

time you go from the input to the output of a video stage (since phase is usually inverted by each video stage).

3. The waveforms should appear as in Fig. 3-42(c). If you reach a defective area, the patterns will be missing, have low-gain, or be severely distorted. Figures 3-42(d), 3-42(e), and 3-42(f) are examples of waveforms obtained in a typical video stage that is defective. Figures 3-42(g) and 3-42(h) show patterns when the sync pulses are set to minimum and maximum (simulating weak and strong broadcast signals).

3-5. ADVANCED TROUBLESHOOTING (SIGNAL SUBSTITUTION) WITH AN ANALYST GENERATOR

This section is devoted to advanced troubleshooting with an analyst generator such as the B&K 1077B Television Analyst. With such a generator (or television analyst, as it is called by the manufacturer), you can inject black and white test patterns (similar to those broadcast by television stations—or your own test patterns from photo transparencies) and color-bar patterns at RF, IF, and video frequencies for stage-by-stage troubleshooting. Additional outputs are available for injection into the sound IF and audio circuits of a receiver. You can also inject separate sync signals, as well as signals suitable for test of the vertical and horizontal sweep circuits and yoke deflection coils.

As a supplemental troubleshooting procedure, you can use an oscilloscope to view the test pattern waveforms as they are processed through the receiver. However, the primary advantage of a full analyst generator is signal injection, where the receiver's indicators (picture tube and loudspeaker) are used to troubleshoot the circuits.

The signal-injection technique is also known as signal substitution. In effect, the analyst signals are substituted for the missing (or abnormal) signal in the receiver and are used to restore operation. The signal is usually injected nearest the picture tube, and then moved one stage at a time toward the antenna until signal substitution does not restore operation. The defective stage has then been located. Note that this procedure is opposite to that described for the basic analyst generator (Sec. 3-4), where the signal is initially injected at the receiver front end or antenna.

Every television receiver has many different types of signals present at various stages throughout the circuits. The analyst generates 15 types of signals, most of them adjustable in amplitude, which permits substitution of the correct signal at almost any point in any television receiver.

Figure 3-43 shows a block diagram of a typical black and white television receiver, illustrating 23 typical signal-injection points that may be used for locating a defective stage. To successfully use the analyst, you

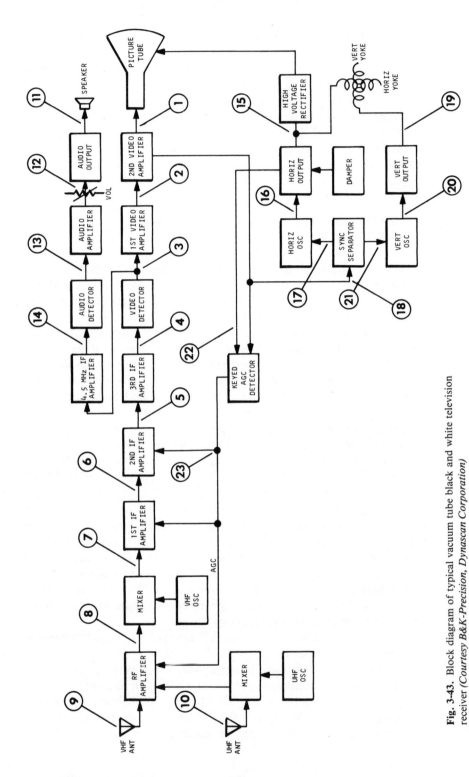

Fig. 3-43. Block diagram of typical vacuum tube black and white television receiver *(Courtesy B&K-Precision, Dynascan Corporation)*

need only to know which signal to inject at each stage and the approximate amplitude of the signal. The steps in the remaining paragraphs of this section give you the specific information for such signal injection or substitution. During most signal substitution procedures, it is permissible to simultaneously use as many outputs from the analyst as necessary to restore operation.

3-5.1 Basic Signal Substitution Techniques

The basic procedure for troubleshooting by the signal substitution method may be summarized as follows:

1. First, you analyze the symptoms to determine the group of stages that should be checked. The relationship between trouble symptoms and areas of the receiver where the trouble is most likely to occur is essentially the same as described in Chapters 1 and 6. However, the sequence used to isolate trouble with signal substitution differs from that when signal tracing is used. Thus, the symptom analysis described in this section is not exactly the same as covered in Chapters 1 and 6.

2. Next, you inject a signal of the proper type into the suspected stage *furthest* from the antenna.

3. If proper operation is restored, you inject a signal into the *next stage nearer* the antenna. You must inject the proper type and level of signal to simulate normal operating conditions in each stage. You continue this stage-by-stage injection until you find a point where signal injection does not restore proper operation. You have now located the defective stage.

4. Next, you inject a signal at each part that is in *series with the signal path* (such as coupling capacitors and transformers) until you have isolated the trouble to as small an area as possible.

5. Finally, you use voltage and resistance checks to locate the specific trouble in the area isolated by the signal-injection process.

6. When you become more familiar with your particular analyst, you may wish to skip stages and check a complete section at a time. An instrument such as the B&K 1077B then becomes even more of a time saver and a more profitable servicing instrument.

Example of signal substitution with an analyst. Assume that you are servicing a receiver with a block diagram similar to Fig. 3-43.

1. You inject a VHF signal at the antenna terminals (test point 9) and analyze the symptoms. The symptoms are no video, audio normal, raster normal. These symptoms tell you that the horizontal and high-voltage sections are operating because the raster is present. Since sound is normal, you can assume that all stages from the antenna to the video detector are operating. The stages that require checking are the picture tube, second video amplifier, and first video amplifier.

2. You inject a *high-amplitude sync signal* directly into the picture tube (point 1) and find that video bars are displayed on the picture tube screen. This proves that the picture tube is good.

3. You inject a *maximum video signal* at the input to the second video amplifier (point 2) and find that a test pattern is displayed on the screen. This proves that the second video amplifier is good.

4. You then reduce the video signal amplitude until the picture is barely visible. (The next step will inject the signal at the first video amplifier. If the first video stage amplifies normally, it should provide a much stronger picture when the signal is applied.)

5. You inject a *low-level video signal* at the input of the first video amplifier (point 3) and find that the test pattern disappears completely. You have found the defective stage. The trouble lies between the input of the first video amplifier and the input of the second video amplifier.

6. Next, you refer to Fig. 3-44, which is a schematic diagram of the

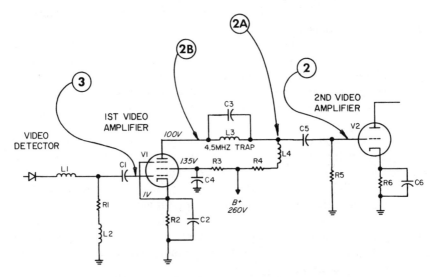

Fig. 3-44. Schematic diagram of typical video amplifier *(Courtesy B&K-Precision, Dynascan Corporation)*

defective stage. You inject a *maximum video signal* at point 2A and find that the test pattern is again displayed. This proves that the coupling capacitor C_5 is good.

7. You inject a *maximum video signal* at the plate of the first video amplifier (point 2B) and find that the test pattern is still displayed. This means that the amplifier tube V_1 is not providing an output signal. (You have already tested the tube and found it to be good.)

8. Finally, you make voltage measurements and find that the plate voltage is normal (100 V) but that the screen voltage is zero. A resistance check shows that C_4 is shorted.

3-5.2   Troubleshooting Transistor Circuits with Signal Substitution

The procedures for troubleshooting transistor receiver circuits with signal substitution are essentially the same as for vacuum-tube receivers. However, there are a few specific differences. The following notes summarize these differences.

Advantages of an analyst. Unlike vacuum-tube circuits, transistors cannot normally be tested by substitution quickly and easily. A complete stage-by-stage troubleshooting is required. An analyst generator such as the B&K 1077B and the signal substitution technique are essential for efficient troubleshooting of transistor stages.

Stage impedances. Transistor stages usually have low input and output impedances compared to vacuum tubes. Present-day analyst generators use low-impedance signals which are well suited for injection into both transistor or vacuum-tube circuits.

Bias. One difference which should be noted that directly concerns signal substitution troubleshooting techniques is the different biasing. Vacuum tubes always use negative bias on the control grid. Transistors use forward bias; that is, the base is positive in respect to the emitter in NPN transistors and negative in PNP transistors. The biasing voltage for transistors, as well as the collector voltage, is normally much lower than in equivalent vacuum-tube stages.

When injecting the bias voltage, always apply it to provide forward bias across the base-emitter junction. When possible, select an injection point that will route the signal through a series resistor into the transistor base. The resistor will absorb much of the voltage difference and prevent possible damage to the transistor. There is normally only a few tenths of a volt of forward bias on most transistors.

Bias reference point. Be very careful of the bias reference point that is used. Usually, it is preferable to use the emitter of the transistor as reference, rather than chassis ground. In both examples shown in Fig. 3-45, chassis ground cannot be used as reference. If used, even a low or zero bias would

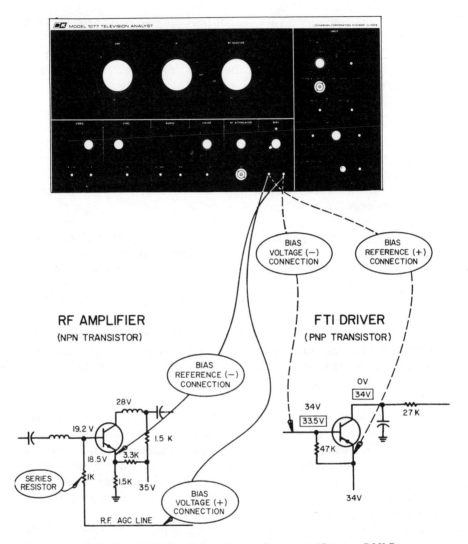

Fig. 3-45. Proper bias connections for transistor stages *(Courtesy B&K-Precision, Dynascan Corporation)*

place a substantial difference of potential (18.5V in one case and 34V in the other case) across the base-emitter junction the instant the test lead is touched to the base of the transistor.

Transient voltages. Transistors are very vulnerable to transient voltage spikes. Careless or random use of test probes during signal substitution can easily cause trouble in this respect. The safest method for preventing transient voltages from damaging transistors is to turn off the receiver while making and removing test connections.

Shorts. Shorts can easily burn out a transistor. Transistors are especially vulnerable to a short between the base and collector. Since the parts in transistor circuits are usually very small compared to equivalent vacuum-tube circuit parts, care must be exercised to prevent shorting transistor leads to other circuits with the tip of the test probe. Use *miniature tips* when servicing transistor circuits with an analyst.

Excessive injection voltages. Remember this one important precaution when using signal injection in transistor stages: *Too much voltage will burn out a transistor.* This is especially true when too much voltage is applied across the base-emitter junction. Such applications, even accidently for only a fraction of a second, will break down the base-emitter junction. As an example of a typical analyst generator, most of the output signals from the B&K 1077B can be used in transistor circuits, even at maximum amplitude. Precautions must be used only with the following output signals:

1. Do not use the horizontal grid drive output in transistor circuits.

2. Use the sync output at low level in transistor circuits. The asterisk (*) on the SYNC control setting represents the highest level (10V peak to peak) that should be used, unless you have checked the service literature and are absolutely sure that higher levels are normally used in the circuit.

3. Use the bias voltage at low levels in transistor circuits. The asterisk (*) on the BIAS control setting represents the highest level (12.5 V) that you should use, unless you have checked the service literature.

Sync signals. When injecting sync signals, keep the SYNC control at zero when not in use. Then if you happen to make a circuit connection before you adjust the level, it will not burn out a transistor. After injecting the sync signal, make sure the amplitude is reduced to a safe level before changing the point of injection. A safe level in one circuit may not be safe in another circuit. When possible, select an injection point that will route the signal through a series resistor into the transistor base. The resistor will help absorb the signal and protect the transistor.

Other defective parts. Transistors are generally dependable parts and should not be replaced until it is proven that they are actually defective. Sometimes another defective part in the circuit will cut off the transistor and it will appear defective. Even when the transistor is proven defective, a check of other parts in the circuit is wise. Many times, a defective part caused the transistor to burn out in the first place. Replacing the transistor without replacing the other defective part will just result in another burned-out transistor.

3-5.3 Troubleshooting Tuned Amplifiers with Signal Substitution

Tuned amplifiers are stages which amplify signals within a specific frequency band and block signals of other frequencies. The stages are usually transformer coupled. Tuned stages within a television receiver are the RF amplifier, IF amplifier, 4.5-MHz sound IF amplifier, and color IF amplifiers. Troubleshooting of all tuned amplifiers by signal substitution is done in essentially the same manner, as outlined in the following steps (See Fig. 3-46).

1. You inject a signal of the correct frequency (RF, IF, etc.). With the B&K 1077B, the VHF signal is of the correct frequency for the channel indicated on the RF SELECTOR. The frequency of the UHF channel is selected by the UHF control. The IF control may be adjusted to provide the best display while injecting the signal. The 4.5-MHz signal and the color signal are stable, fixed-frequency sources that will provide the correct frequency without adjustment.

2. You inject the signal at the input (grid or base) of the suspected stage that is furthest from the antenna. The input of the stage is preferred because the low-impedance signal source of the analyst is less likely to load the tuned circuit, which may affect the circuit's frequency response. If no display is provided, this is probably the defective stage. If the desired results are obtained (test pattern, color, or sound, depending on the stage being tested), reduce the amplitude of the injected signal until the display is barely visible (the sound signal is not adjustable in amplitude).

3. Move the injection point to the input of the next stage nearer the antenna. If that stage is operating properly, the display will be much brighter. The amplification of the stage should produce the brighter display. This method not only locates inoperative stages but weak stages as well. Reduce the signal amplification again to the point where the display is barely visible. Note the difference in the setting of the amplitude adjustment, which is a relative indication of gain. After checking a few television receivers, you will soon learn the normal gain to be expected in various stages. This information will help you spot a weak stage immediately.

4. Continue to move the point of signal injection toward the antenna, one stage at a time, until no display or a poor display is produced. When this occurs you have located the defective stage. Now you must locate the defective part within the stage.

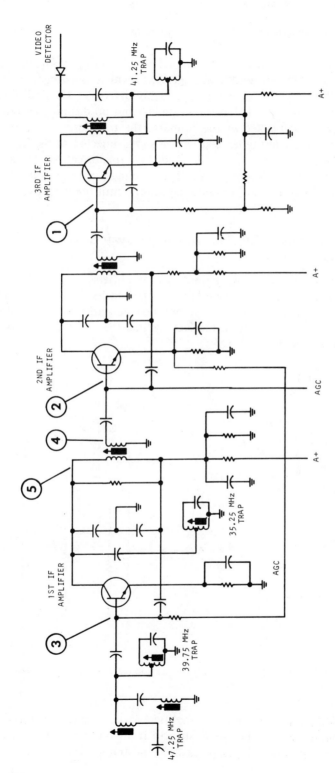

Fig. 3-46. Typical tuned amplifier schematic diagram *(Courtesy B&K-Precision, Dynascan Corporation)*

5. Inject the signal at the secondary and primary of the stage coupling transformers. If no display is produced with a signal injected at the secondary, the coupling capacitor is probably defective. If the display is normal with a signal injected at the secondary but no display is produced when the signal is injected at the primary, the transformer is probably defective. If the display is normal with a signal injected at the primary, check the collector, emitter, and base voltages (plate, cathode, and grid voltages for vacuum-tube receivers). If any voltages are abnormal, make resistance checks of all resistors and capacitors until the defective part is located.

3-5.4   Troubleshooting Untuned Amplifiers with Signal Substitution

Untuned amplifiers differ from tuned amplifiers (Sec. 3-5.3) in that untuned amplifiers amplify a wide band of frequencies. Resistive-capacitive coupling is normally used between untuned amplifier stages. Untuned amplifier stages in a television receiver include the audio amplifiers and video amplifiers (see Fig. 3-47).

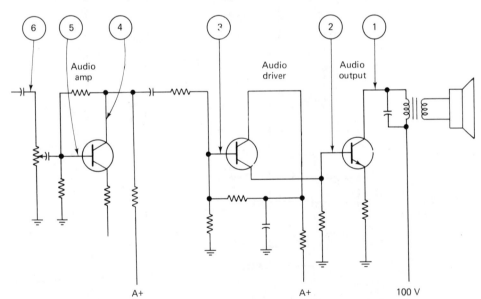

Fig. 3-47. Typical untuned amplifier schematic diagram *(Courtesy B&K-Precision, Dynascan Corporation)*

Troubleshooting of untuned amplifiers is performed by injecting a signal at the output of the stage furthest from the antenna to produce normal operation. For video signals, the signal amplitude is decreased until the display is barely visible; then the point of injection is changed to the input of the stage. If the stage is operative, the amplification should produce a brighter display. For audio, listen for a louder volume when the signal is moved to the input of the stage. (Typically, the sound signal is not adjustable in analyst generators.)

Continue moving the point of injection toward the antenna until no output is obtained. This is the defective stage. Proceed with voltage and resistance checks to locate the defective part.

3-5.5   Troubleshooting Pulse Stages with Signal Substitution

Pulse circuits have a short duration of operation compared to "off" time and are thus often difficult to troubleshoot. Pulse circuits in a television receiver include the sync separator, sync amplifiers, keyed AGC circuit, burst amplifier (in color receivers), and horizontal and vertical sweep circuits (see Fig. 3-48).

In any pulse circuit, waveshape, as well as operation, is important. Not only must the pulses be present at a given point in the circuit, but the pulses must be of the correct amplitude, duration, and frequency. An analyst generator can provide an advantage over signal tracing with an oscilloscope when troubleshooting pulse circuits. With an analyst generator, you can substitute the type of signal actually used in the stage during normal operation. With such pulses, you can easily locate causes of distortion or inoperation because the results are displayed directly on the screen of the television receiver. If normal operation is restored when you apply a pulse of the correct type to some point in the circuit, you know that the stages between that point and the picture tube are good.

As in the case of tuned and untuned amplifiers, you inject signals at every point in the series path of the pulse signal until the point is located at which the display is abnormal. One special precaution to be observed when injecting pulse signals is that *correct polarity* must be observed. As shown in Fig. 3-48, some stages require a positive pulse, while others require a negative pulse. Also, amplifiers usually invert the signal; thus the polarity must be reversed as the point of injection is changed from one stage to the next or from the output to the input of a stage. The video and sync signals from an analyst have reversible polarities.

3-5.6   Analyzing Trouble Symptoms with Signal Substitution

As discussed in Sec. 3-5.1, symptom analysis described in this section is not identical to that covered in Chapters 1 and 6 since the sequence used to isolate trouble with signal substitution differs from that when signal tracing is used. Figure 3-49 groups trouble symptoms and possible causes as they

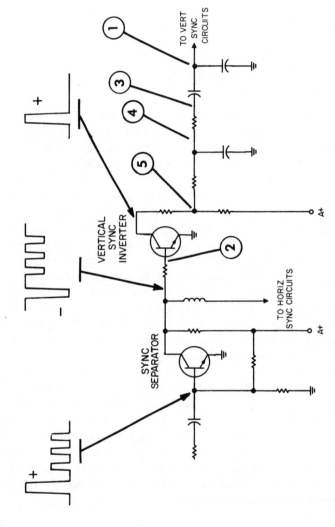

Fig. 3-48. Typical pulse stage schematic diagram *(Courtesy B&K-Precision, Dynascan Corporation)*

relate to signal substitution using an analyst generator. The remaining paragraphs of this section correspond (by title) to the troubles listed in Fig. 3-49. Always observe the following steps before using Fig. 3-49 to isolate trouble.

1. Always start troubleshooting by injecting a VHF test signal at the antenna terminals of the television receiver and checking all symptoms.

2. Use the test pattern from the analyst generator, not a picture from a television station. The picture broadcast by a television transmitter changes continuously, but the test pattern from an analyst generator is reliable. If the display does not match the test pattern slide, it is easily detected. Small irregularities will not be seen easily with a transmitted picture.

3. Analyze the symptoms thoroughly. A careful check of symptoms will direct you to specific stages that could cause the trouble. Do not rely on another person's description of the symptoms. You may gain additional information that will save you time by making your own observations.

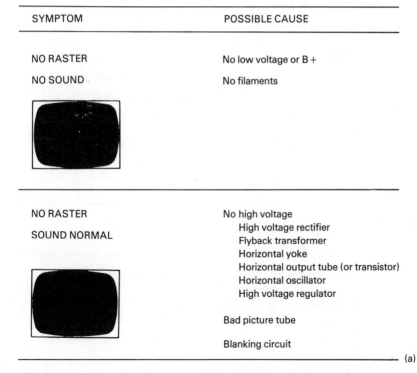

SYMPTOM	POSSIBLE CAUSE
NO RASTER	No low voltage or B +
NO SOUND	No filaments
NO RASTER	No high voltage
SOUND NORMAL	High voltage rectifier
	Flyback transformer
	Horizontal yoke
	Horizontal output tube (or transistor)
	Horizontal oscillator
	High voltage regulator
	Bad picture tube
	Blanking circuit

(a)

Fig. 3-49a, b, and c. Trouble chart using Model 1077B Television Analyst *(Courtesy B&K-Precision, Dynascan Corporation)*

SYMPTOM	POSSIBLE CAUSE

NO VERTICAL DEFLECTION

Vertical yoke

Vertical output tube (or transistor)

Vertical oscillator

Vertical output transformer

NO VIDEO

SOUND NORMAL

Video output

Video driver

CRT

NO VIDEO

NO SOUND

Video detector

1st i-f amplifier

2nd i-f amplifier

3rd i-f amplifier

RF amplifier

Oscillator

Mixer

NO SOUND

VIDEO NORMAL

Speaker

Audio output tube (or transistor)

Audio amplifier

Discriminator

4.5 MHz amplifier

Audio output transformer

(b)

Fig. 3-49 (Cont.)

SYMPTOM	POSSIBLE CAUSE

NO VERTICAL SYNC

Vertical sync inverter

Sync separator

NO HORIZONTAL SYNC

Sync separator

Horizontal AFC

OVERLOADED VIDEO

AGC detector

AGC amplifier

AGC line

NON-LINEAR SWEEP

Requires adjustment

Pincushion amplifier (color)

POOR RESOLUTION

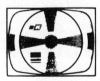

Needs i-f alignment

(c)

Fig. 3-49 *(Cont.)*

Check both video and sound, and if it is a color receiver, check color reception.

4. Learn to analyze the test pattern. Chapter 5 includes a discussion on how to use test patterns to evaluate trouble symptoms. A test pattern shows much more than merely whether the television receiver is operating or not operating. When properly interpreted, the pattern can provide a complete check of overall performance and show whether receiver adjustments are necessary.

No raster, no sound. See Figs. 3-50 and 3-51. The most probable cause of this symptom is the absence of one of the B+ voltages or possibly all low dc voltages. In most television receivers, this is the only portion common to the sound and high-voltage sections. In vacuum-tube receivers, it is also possible that filament voltage is absent or interrupted.

The chart of Fig. 3-50 provides a procedure that will isolate the defective stage for a no-raster, no-sound symptom. The numbers shown in Fig. 3-51 refer to the steps of Fig. 3-50. Step 1 is a visual check of the dial light or tube filaments. For an all-solid-state receiver, without a dial light, start with step 2.

No raster, sound normal. See Figs. 3-52 and 3-53. This symptom appears whenever the picture tube does not have the proper electron beam to illuminate the screen. This could be caused by a bad picture tube or absence of voltage or cathode or grids of the picture tube, but it is usually caused by absence of high voltage on the anode of the picture tube. High voltage, in turn, is dependent on the horizontal sweep signal. There are quite a number of stages in which the trouble could be located; therefore an orderly procedure is required to locate the defective part quickly. The following procedure should lead you to the defect in minimum time. The chart of Fig. 3-52 together with the schematic of Fig. 3-53 and the following steps provide a procedure that will isolate the defective stage for a no-raster, sound-normal symptom.

It is assumed, as with all symptoms, that vacuum tubes have been tested or substituted and the symptom remained unchanged. With this symptom, the picture tube should also be tested (if practical) because it is a prime suspect. *Observe all safety precautions discussed in Chapter 2 concerning high-voltage circuits.*

1. Set the brightness and contrast controls of the television receiver to maximum.

2. Disconnect the plate cap from the high-voltage rectifier tube. Clip the high-voltage probe of the analyst around the plate cap lead. If the lamp in the high-voltage probe lights, high voltage is being developed by the

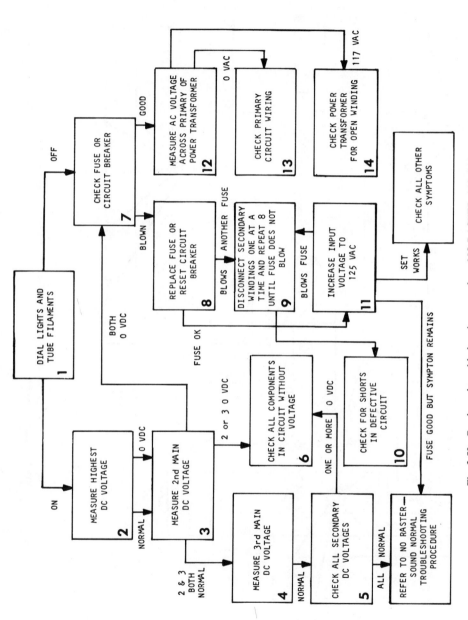

Fig. 3-50. Condensed troubleshooting procedure, NO RASTER, NO SOUND
(Courtesy B&K-Precision, Dynascan Corporation)

163

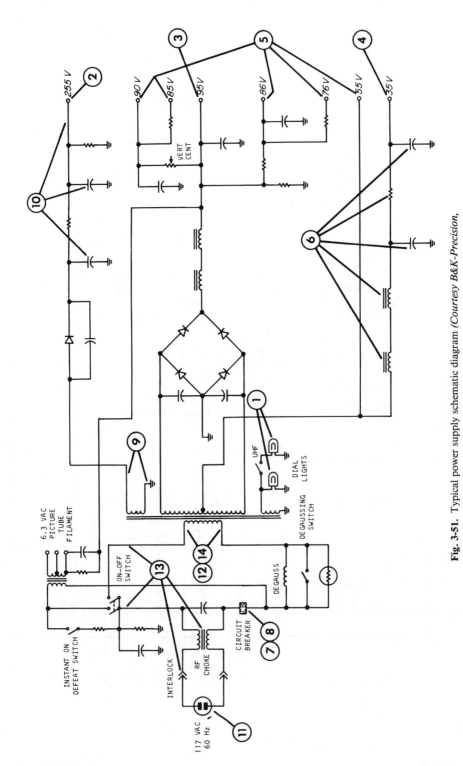

Fig. 3-51. Typical power supply schematic diagram *(Courtesy B&K-Precision, Dynascan Corporation)*

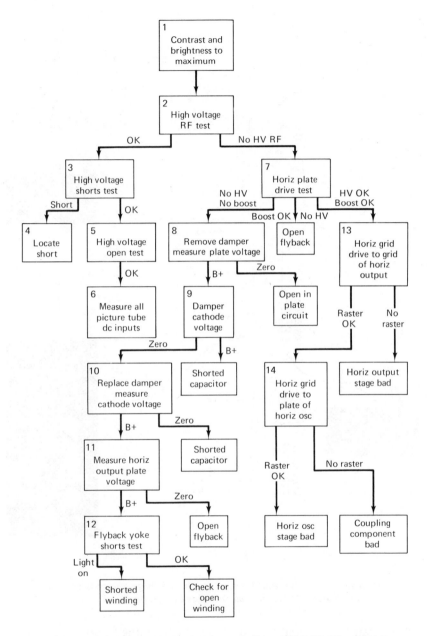

Fig. 3-52. Condensed troubleshooting procedure, NO RASTER, SOUND NORMAL *(Courtesy B&K-Precision, Dynascan Corporation)*

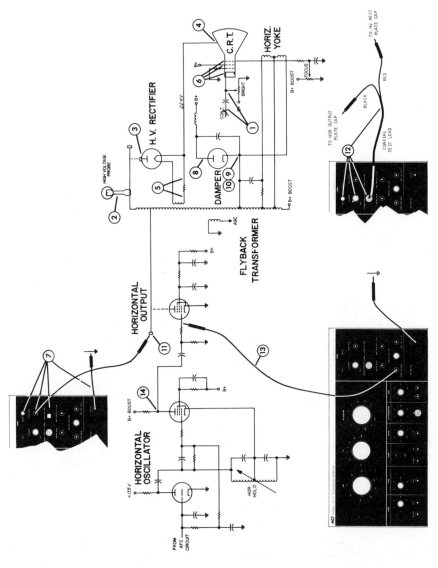

Fig. 3-53. Typical horizontal sweep and high voltage section schematic diagram *(Courtesy B&K-Precision, Dynascan Corporation)*

flyback transformer, and the trouble lies between the transformer and the picture tube. Go to step 3. If the lamp does not light, the trouble lies in the flyback transformer or horizontal sweep circuits. Go to step 7.

3. Replace the cap on the high-voltage rectifier tube. If the lamp in the high-voltage probe goes off, there is a short in the high-voltage circuit. Go to step 4. If the lamp stays on, go to step 5.

4. Disconnect the high-voltage lead from the picture tube anode. If the light comes on, there is an internal high-voltage short in the picture tube. If the lamp stays off, disconnect any filter capacitors, bleeder resistor, or other parts one at a time until the lamp comes on. You have then located the defective part.

5. With the television receiver off, check continuity of all parts in the high-voltage rectifier filament circuit. Make sure a tight connection is made where the high-voltage lead connects to the anode of the picture tube. If good, go to step 6.

6. Check dc voltages at the cathode and all grids of the picture tube. Isolate the reason for any missing voltage.

7. Replace the plate cap on the high-voltage rectifier tube. Inject the horizontal plate drive signal from the analyst at the plate of the horizontal output tube. For solid-state receivers, use the appropriate drive signal as specified by the analyst generator operating instructions.

If the lamp in the high-voltage probe lights, the BOOST INDICATOR lamp of the analyst lights, and the raster is restored on the picture tube, the flyback transformer and associated circuits are proven good. The trouble lies in the horizontal sweep stages. Go to step 13.

If the BOOST INDICATOR lights but the high-voltage lamp does not, there is an open turn in the flyback transformer between the plate of the horizontal output and plate of the high-voltage rectifier (or the corresponding circuits of a solid-state receiver).

If neither lamp lights, the flyback transformer, horizontal yoke, or damper stage is defective. To isolate the defective part, go to step 8.

The flyback transformer and damper stage generate the boosted B+ voltage which is used as plate voltage for the horizontal output stage (sometimes the boost is used for focus voltage on the picture tube and as plate voltage for the vertical output stage and other stages). If B + is not boosted, insufficient plate drive is developed to provide a raster. *These same conditions generally do not hold true for solid-state receivers,* as discussed in Chapter 6. With a vacuum-tube receiver, if the flyback transformer or horizontal yoke is open, no raster will be developed. Likewise, if the flyback transformer or horizontal yoke has shorted windings, it will load the horizontal output and attenuate the flyback pulse to prevent high voltage and boosted B+ from being developed. (The problem of testing flyback transformers without an analyst generator is discussed in Chapter 5.)

8. Remove the damper tube from its socket and measure the dc voltage at the plate pin of the tube socket. B + voltage should be measured. If a zero or unusually low reading is obtained, measure resistances of parts in the plate circuit. If normal, go to step 9. In receivers where a solid-state diode is used for a damper, disconnect the diode before making the measurements.

9. With the damper tube still removed (or the solid-state diode disconnected), measure the dc voltage at the cathode. Zero volts should be measured for vacuum-tube and most solid-state receivers. A possible exception is with some hybrid (combined vacuum-tube and solid-state) receivers, which are discussed in Chapter 6. If B+ is measured in a vacuum-tube receiver, check for a shorted capacitor between the B+ and boosted B+ lines. If normal, go to step 10.

10. Replace the damper tube (or reconnect the solid-state diode) and again measure cathode voltage. B+ or boosted B+ should be measured (for vacuum-tube receivers). If zero or unusually low, check for a shorted capacitor in the cathode circuit. If normal, go to step 11.

11. Measure the dc voltage at the plate lead of the horizontal output tube (or at the collector of the horizontal output transistor). B+ or boosted B+ should be measured. If zero, check for an open circuit in the flyback transformer or other series component between the damper and the horizontal output stages. If normal, go to step 12.

12. Make the flyback transformer and horizontal yoke shorted winding test as described in the analyst generator operating instructions. If you are not using an analyst or your analyst does not have provisions for testing the flyback transformer and yoke windings, refer to Chapter 5 for information on testing yokes and flyback transformers. If no shorts are indicated, make continuity checks of the entire flyback transformer and horizontal yoke to check for an open circuit.

13. Replace the plate cap on the horizontal output tube. Inject the horizontal grid drive signal from the analyst to the grid of the horizontal output tube. For solid-state receivers, use the appropriate drive signal as specified by the analyst operating instructions, and apply the signal to the base of the horizontal output transistor. If the raster is not restored, check all elements of the horizontal output transistor or tube. If the raster is restored, go to step 14.

14. Inject the horizontal grid drive signal (or corresponding solid-state signal) to the plate (or collector) of the horizontal oscillator. If the raster is not produced, check the coupling capacitor (in vacuum-tube receivers). As discussed in Chapter 6, most solid-state receivers will have a coupling transformer and a driver transistor between the horizontal oscillator and horizontal output stages. If the raster is produced, the trouble lies in the horizontal oscillator stage. Make voltage and resistance checks throughout the horizontal oscillator stage until the defective part is located.

No vertical deflection. See Fig. 3-54. This symptom can be caused by a part failure in the vertical oscillator, vertical output stage, vertical output transformer, and associated parts, or vertical yoke. The schematic of Fig. 3-54 provides a procedure that will isolate the defective stage for a no-

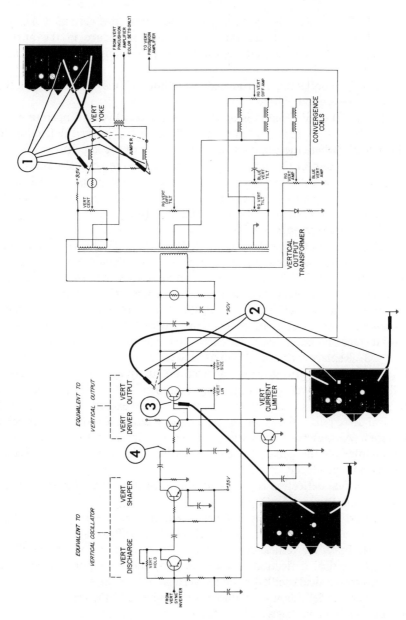

Fig. 3-54. Vertical sweep circuit schematic diagram *(Courtesy B&K-Precision, Dynascan Corporation)*

vertical-deflection symptom. The numbers in Fig. 3-54 refer to the following steps:

1. Disconnect the vertical yoke and inject the vertical yoke test signal from the analyst generator (using the procedure in the analyst instructions). If your analyst does not have provisions for testing yokes, refer to Chapter 5.

If vertical deflection is produced (even though it may be nonlinear), the vertical yoke is good. Go to step 2.

If no vertical deflection is produced, the vertical yoke is open or completely shorted. A few shorted turns will cause the vertical deflection to "keystone" (where there is less height on one side of the screen).

2. Reconnect the vertical yoke and remove the vertical output tube (disconnect the collector lead of solid-state vertical output transistors). Inject a vertical plate drive signal (or appropriate solid-state drive signal) at the disconnected plate (or collector) lead. Adjust the analyst controls until the picture tube screen is filled (vertically).

If vertical deflection is produced, it proves that the vertical output transformer and associated parts that couple the signal from the vertical output stage to the vertical yoke are good. (Note that some solid-state receivers do not have a vertical output transformer.) Go to step 3.

If vertical deflection is not produced, the trouble is in the vertical output transformer or associated circuit. Check B+ voltage to the vertical output stage and make resistance checks to locate the defective part.

3. Replace the vertical output tube (or reconnect the collector lead). Inject the vertical grid drive signal from the analyst at the grid (or base if solid-state) of the vertical output stage and then at the vertical driver stage (if any). In solid-state receivers, be careful not to exceed the peak-to-peak vertical drive pulses shown on the schematic diagrams for the receiver being serviced. Always start by setting the vertical drive control to minimum before injecting the signal. Increase the drive signal slowly to see if vertical deflection is produced.

If vertical deflection is produced, the vertical output stage and vertical driver (if any) are good. Go to step 4.

If vertical deflection is not produced, the trouble is in the vertical output stage or the vertical driver. Check all related voltages and resistances.

4. Inject the vertical drive signal at the output of the vertical oscillator stage. If vertical deflection is not produced, check coupling components between the vertical oscillator and the vertical output stages.

If vertical deflection is produced, the trouble is in the vertical oscillator. Check all related voltages and resistances.

No video, sound normal. See Figs. 3-55 and 3-56. This symptom is normally caused by a defective video amplifier or a bad picture tube. Because the sound and video separation point varies from one receiver to another, it is also possible that the video detector stage is at fault. In other receivers, only the final video amplifier could produce the symptoms. The

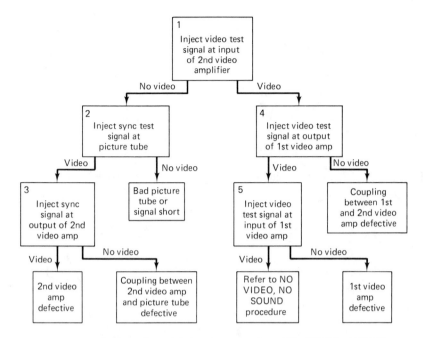

Fig. 3-55. Condensed troubleshooting procedure, NO VIDEO, SOUND NORMAL *(Courtesy B&K-Precision, Dynascan Corporation)*

chart of Fig. 3-55, together with the schematic of Fig. 3-56 and the following steps, will locate the defect no matter where the sound and video separation occurs.

1. Inject a maximum video test signal into the input of the final video amplifier, as described in the analyst instruction manual.

If no video is produced on the picture tube, the trouble lies between the input of the final video amplifier and the picture tube. Go to step 2.

If video is produced (the test pattern appears on the picture tube screen, probably out of sync), the trouble lies before the final video amplifier. Go to step 4.

2. Inject a maximum sync signal direction to the picture tube as described in the analyst instruction manual.

If no diagonal bars are seen, the picture tube is defective or the circuit is shorted to ground. Check for shorts. Note that the sync signal from the analyst should produce diagonal bars on the picture tube but not the test pattern. The video signal from the analyst produces the test pattern, as does the RF and IF.

If diagonal bars are seen, the picture tube is proven capable of displaying video. Go to step 3.

3. Inject maximum sync signal at each series part from the picture

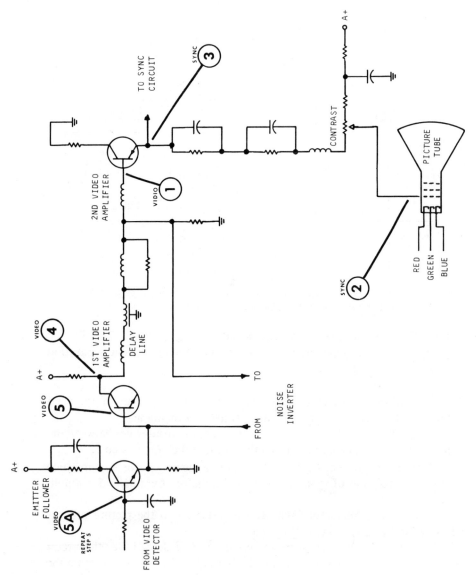

Fig. 3-56. Solid-state video amplifier circuits schematic diagram *(Courtesy B&K-Precision, Dynascan Corporation)*

172

tube to the output of the final video amplifier. If diagonal bars are not displayed for all injection points, check to see at which part the display is lost.

If diagonal bars are still displayed with a sync signal injected at the output of the final video amplifier stage, the stage is inoperative. Check all related voltages and resistances.

4. Inject a maximum video signal at the output of the preceding video amplifier stages. If no video (test pattern) is displayed on the picture tube, check the coupling parts between the first and final video amplifier stages. If video is displayed, go to step 5.

5. Inject a medium-amplitude video signal at the input of the first video amplifier stage. If no video is displayed on the picture tube, the first video amplifier is inoperative. Check all related voltages and resistances.

If video is displayed, sound and video separation occur ahead of the video detector. Follow the procedure outlined in the no-video, no-sound-trouble symptom.

No video, no sound. See Figs. 3-57 and 3-58. This symptom is caused by a malfunction between the antenna terminals and the point where the video and sound are separated. This normally includes the RF tuner (RF amplifier, oscillator, mixer), IF amplifiers, and the video detector. The chart of Fig. 3-57, together with the schematic of Fig. 3-58 and the following steps, will locate the defect.

1. Inject an IF test signal from the analyst into the input of the third IF amplifier (or the last IF amplifier). If no test pattern is displayed, or it is unusually weak, the IF stage or detector is defective. Check all related voltages and resistances.

If the test pattern is displayed properly, the trouble lies toward the antenna. Go to step 2.

2. Move the IF test signal to the input of the second IF amplifier (or the IF stage ahead of the final IF amplifier). Measure the relative gain of the amplifier. If no test pattern is displayed or gain is unusually low, the second IF amplifier stage is defective. Inject the IF signal at each part that is in series with the signal path to locate the point where signal is lost. Check all related voltages and resistances.

If the test pattern is displayed properly, the stage is good. Go to step 3.

3. Repeat the procedures of step 2, but this time inject the signal at the input of the first IF amplifier.

If the test pattern is displayed properly, the stage is good. Go to step 4.

4. If the test pattern is still displayed with the IF signal injected at the first IF amplifier, the trouble is in the RF tuner.

To test the tuner mixer, inject an IF signal into the mixer input. If the

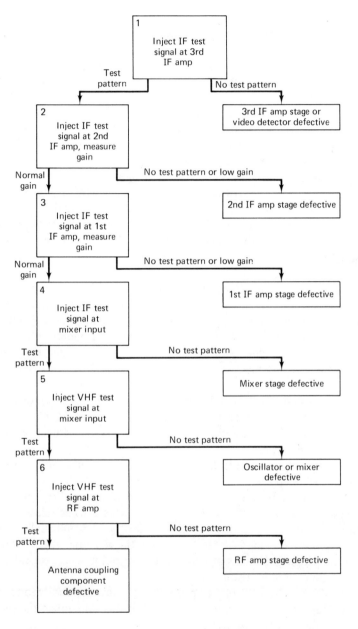

Fig. 3-57. Condensed troubleshooting procedure, NO VIDEO, NO SOUND *(Courtesy B&K-Precision, Dynascan Corporation)*

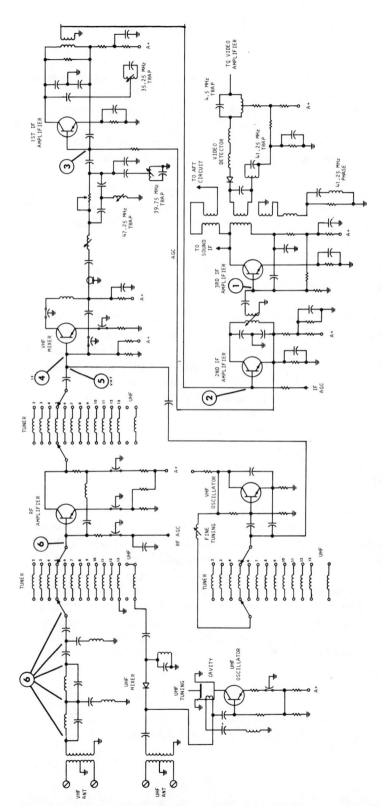

Fig. 3-58. RF and IF circuit schematic diagram *(Courtesy B&K-Precision, Dynascan Corporation)*

mixer is operating, the signal will be passed, and the test pattern will again be displayed. If the mixer is defective, no test pattern will be seen. However, the mixer may pass an IF signal even if the mixer is defective.

5. To test the oscillator, inject an RF test signal into the input of the mixer. If the oscillator is operating, and on the proper frequency, the injected signal and the oscillator signal will mix to produce the IF signal, and the test pattern will be displayed. If the oscillator or mixer is defective, no test pattern will be seen.

6. If the mixer and oscillator are good, the trouble is in the RF amplifier. Inject an RF test signal at various points in the circuit to localize the malfunction. Always inject an RF signal on the same channel to which the television receiver is tuned. Try several channels if the analyst is capable of producing RF test pattern signals on more than one channel.

7. If there is no video and no sound on UHF channels only, the trouble is in the UHF tuner or the UHF position of the channel selector.

No sound, video normal. See Figs. 3-59 and 3-60. This symptom is caused by a malfunction in the 4.5-MHz sound IF stages, discriminator, audio amplifier, or speaker. Using the chart of Fig. 3-59 and the schematic of Fig. 3-60, apply 4.5-MHz sound signals and 1-kHz audio signals to the signal path as described in the analyst instruction manual.

The troubleshooting procedure of Fig. 3-59 applies specifically to the circuit shown in Fig. 3-60. However, the procedures are typical for all television sound systems. Add or omit signal injection points as required so that a signal is injected to test each stage. Check the gain of each audio amplifier by listening for increased volume for each additional stage through which the 1-kHz tone passes.

No vertical sync. See Fig. 3-61. With a loss of vertical sync, the picture will roll no matter where the vertical hold is set. This symptom is caused by the absence of vertical sync pulses at the vertical oscillator. Since the picture has horizontal sync, the trouble should be only in a section of the sync separator stage or in a vertical sync amplifier stage (if used).

Inject a sync signal from the analyst at the points indicated in Fig. 3-61 as described in the analyst instruction manual. The required signal may be positive or negative polarity; try both. Remember to reverse polarity when changing the injection point from the output to the input of a stage. Inject sufficient signal to simulate normal operating values (refer to the receiver schematic for normal peak-to-peak signal values). If too much signal is required to restore proper operation, check for a part that has changed value drastically.

Also check the gain of the sync separator stage or stages. The separator may provide sufficient signal to produce horizontal sync but not vertical

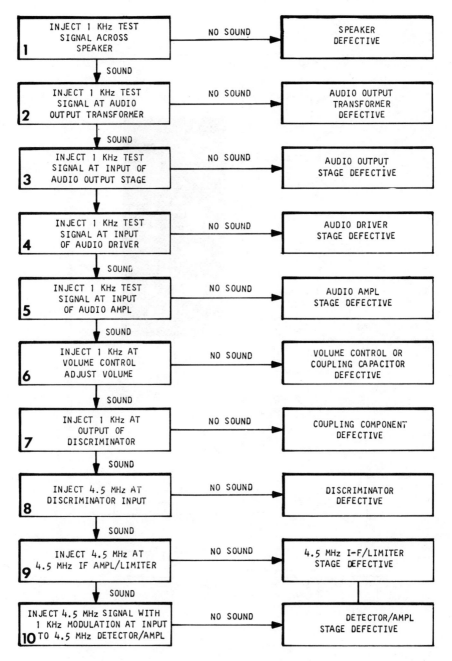

Fig. 3-59. Condensed troubleshooting procedure, NO SOUND, VIDEO NORMAL *(Courtesy B&K-Precision, Dynascan Corporation)*

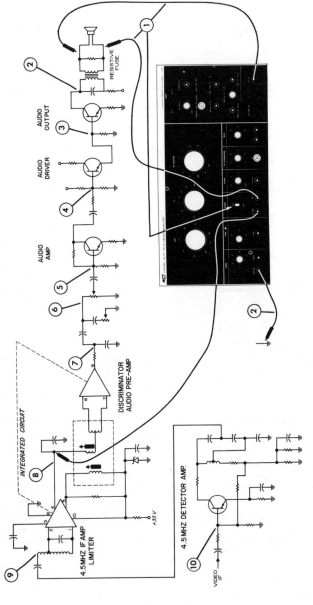

Fig. 3-60. Sound section schematic diagram *(Courtesy B&K-Precision, Dynascan Corporation)*

178

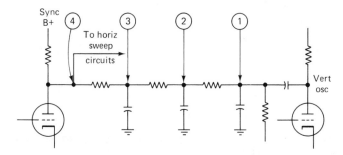

Fig. 3-61. Solid-state and vacuum tube vertical sync circuits schematic diagram *(Courtesy B&K-Precision, Dynascan Corporation)*

sync. Relative gain is checked by injecting the minimum sync signal that will provide vertical sync at the output and input of a stage and noting the difference in the sync control setting.

Normal operation should be restored when the sync signal is injected at point 1. Move the point of injection toward the sync separator stage until vertical sync is lost. The defective part can then be easily checked with voltage and/or resistance measurements.

No horizontal sync. See Fig. 3-62. With this trouble, diagonal lines are seen on the screen, but a picture cannot be locked in with the horizontal hold control. Since no picture is visible, presence or absence of vertical sync cannot be easily detected. However, this is of little consequence, since the following procedure will locate the source of trouble in either case.

The symptom could be caused by a defective sync separator stage or the AFC circuit which precedes the horizontal oscillator.

1. Inject a sync signal from the analyst at the sync input of the horizontal oscillator. The required signal may be positive or negative polarity; try both. Refer to the receiver schematic, and adjust the sync

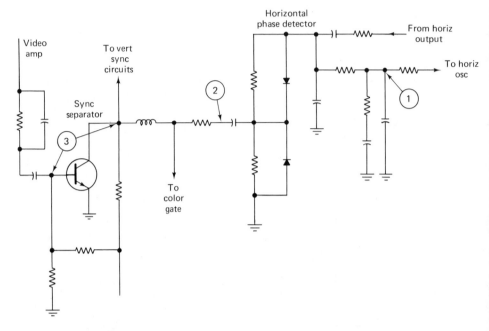

Fig. 3-62. Horizontal sync circuits schematic diagram *(Courtesy B&K-Precision, Dynascan Corporation)*

control for a typical level of sync pulse signal. Horizontal sync should be restored. Readjust the horizontal hold control if necessary.

2. Inject a sync signal into the horizontal phase detector. If sync is not restored, the phase detector is defective. If normal operation is restored, go to step 3.

3. Inject a sync signal at the output and input of the sync separator stage. Check the gain of the stage (if the stage is operational). If no sync is obtained, the sync separator is defective. If sync is restored, check coupling parts of the sync separator stage.

Overloaded video. See Fig. 3-63. Overloaded video causes the picture to appear negative and out of sync at normal and high signal levels injected at the antenna terminals. The normal test pattern (but with considerable snow) will usually appear when you decrease the signal to a very low level. This symptom is usually caused by an RF, IF, or video amplifier that is overdriven, which results in clipping of the signal at normal and high signal levels. A defect in the AGC section can produce such a symptom because a faulty AGC allows the amplifiers to operate at maximum gain.

The first step of the procedure is to determine if the trouble is in the AGC section.

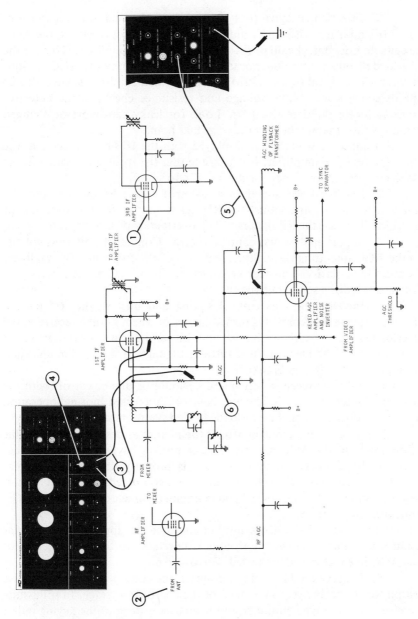

Fig. 3-63. AGC circuit schematic diagram *(Courtesy B&K-Precision, Dynas-can Corporation)*

1. Inject an IF signal (test pattern) from the analyst to the third (or final) IF amplifier. Use a maximum IF signal. In most receivers, the AGC circuits do not control gain of the third (or final) IF amplifier. Thus, if the overloaded video symptom appears, the trouble is not in the AGC section. Inject a video signal (test pattern) into each video amplifier stage to locate the defective stage. Make voltage and resistance checks in the defective stage to locate the defective part. Look for leaky capacitors or changed resistor values that might shift the stage bias point.

If a normal test pattern is displayed with the IF signal injected at the third (or final) IF amplifier, it is possible that the trouble is located in the AGC circuits, although not conclusive. Go to step 2.

2. Inject an RF signal (test pattern) from the analyst into the antenna terminals, and adjust the RF signal control for the symptom of overload. Increase the RF signal until an overloaded picture appears.

3. Set the analyst bias control to zero. Connect the bias test voltage to the AGC line using the cathode (emitter) of the first video IF amplifier. Connect the polarity to provide negative bias for a vacuum-tube receiver or forward bias for a solid-state receiver.

4. Increase the bias control setting and observe the test pattern display. As the bias control is turned higher on vacuum-tube receivers, the overloaded video symptom should diminish, and normal operation should be restored. If the bias is turned further, the picture should dim and then disappear as the IF stage is cut off.

On solid-state receivers, the picture should be dim or cut off with the bias control set to zero. As the bias control is increased, normal operation should be restored; then the overloaded video symptom should return.

If varying the bias control affects the picture as described, the trouble is definitely in the AGC section. The amplifiers have demonstrated that they can be controlled if AGC voltage is properly developed. The bias supply of the analyst has such low impedance compared to the AGC circuit that the analyst will "take over" and maintain complete control of the AGC function. Go to step 5.

If the overloaded video symptom continues to appear while the bias control is adjusted, either the trouble is unrelated to the AGC section or the AGC line is completely shorted. Go to step 6.

5. Remove the bias voltage connections and inject the keying pulse output from the analyst to the plate of the keyed AGC stage. In solid-state receivers, use the sync output from the analyst instead of the keying pulse. If this step produces normal operation, check for trouble in the AGC winding of the flyback transformer or in the coupling parts between the flyback transformer and the keyed AGC amplifier stage.

6. Disconnect the grid (or base) of the second IF amplifier from the AGC circuit and connect the bias voltage source in place of the AGC. If there is a short in the AGC circuit, the short will now be disconnected from the amplifier circuit. Vary the bias control. If the bias control affects the picture as described in step 4, repeat this procedure for the first IF amplifier and the RF amplifier. If the overloaded video symptom does not appear continuously at any of these stages, it is conclusive that the trouble is in the AGC circuit. Disconnect the AGC filter capacitors, one at a time, and repeat the test procedure (steps 3 and 4) until the defective capacitor is located.

If the overloaded video symptom reappears and the bias control has no effect on the picture in any of the stages (second IF, first IF, or RF amplifier), check for a defective part in that stage. Be alert for a leaky capacitor that may shift the bias on the amplifier.

3-6. ANALYZING THE COMPOSITE VIDEO WAVEFORM

Probably the most important waveform in television servicing is the composite waveform consisting of the video signal, the blanking pedestals, and the sync pulses. Figures 3-64 and 3-65 show typical oscilloscope traces when observing composite video signals synchronized with horizontal sync pulses and vertical blanking pulses. Note that the illustrations show both theoretical and practical waveforms.

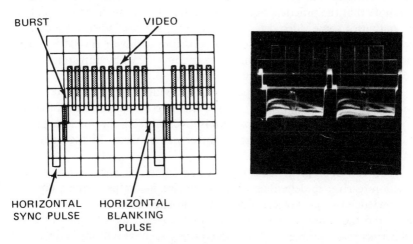

Fig. 3-64. Theoretical and practical composite video waveforms, oscilloscope set to TVH *(Courtesy B&K-Precision, Dynascan Corporation)*

If Fig. 3-64, the oscilloscope is set to provide a sweep of 7875 Hz (the oscilloscope controls are set to HORIZONTAL, or TVH, or some similar position) so that two complete horizontal fields are shown. The theoretical pattern assumes that a color signal from a keyed rainbow generator is

VIDEO

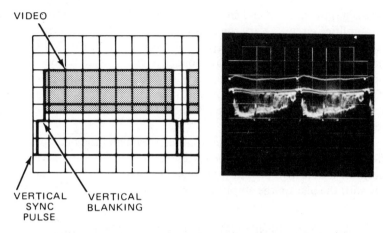

VERTICAL   VERTICAL
SYNC   BLANKING
PULSE

Fig. 3-65. Theoretical and practical composite video waveforms, oscilloscope set to TVV *(Courtesy B&K-Precision, Dynascan Corporation)*

applied to the receiver antenna terminals. Thus, the pattern contains a color burst on the horizontal blanking pulse "back porch," next to the sync pulse. The video portion of the pattern consists of 10 pulses, one for each of the 10 color bars produced by the keyed rainbow generator. The practical pattern assumes that the receiver is tuned to a black and white broadcast. Also note that the practical pattern is inverted from the theoretical pattern. The output of some video detectors is positive, whereas other video detectors produce negative outputs. Also, it is possible to invert the pattern for a positive or negative display (whichever is convenient) by adjustment of the oscilloscope controls (on most oscilloscopes).

In Fig. 3-65, the oscilloscope is set to provide a sweep of 30 Hz (oscilloscope set to VERTICAL, TVV, etc.) so that two complete vertical fields are shown. Again, the theoretical pattern is inverted from the practical pattern. Also note that there is considerable difference between the theoretical and practical patterns.

You can observe the composite video signals at various stages in the television receiver to determine whether the circuits are performing normally. A knowledge of the waveform makeup, the appearance of a normal waveform, and the causes of various abnormal waveforms will help you locate and correct many problems. You should study such waveforms in a television receiver known to be in good operating condition, noting the waveform at various points in the video amplifier. Always observe the oscilloscope operating instructions when making such waveform studies.

Figures 3-64 and 3-65 assume that the composite video waveform is being measured at the video detector output. This is the usual starting point. However, the composite video waveforms may also be checked at other points in the video circuits by moving the probe to those points. On most oscilloscopes, you must reset the vertical gain control (VOLTS/CM control, etc.) each time you move the probe to a new test point. Otherwise, the display may go off-scale. Also, you may have to readjust the triggering level control to maintain stabilization, and the polarity control (since the observed waveform may be reversed when moving from one point to another).

3-7. ANALYZING THE HORIZONTAL SYNC PULSE

The IF amplifier response of a television receiver can be evaluated to some extent by careful observation of the horizontal sync pulse waveform. The appearance of the sync pulse waveform is affected by the IF amplifier band-pass characteristics. Some typical waveform symptoms and their relation to IF amplifier response are indicated in Fig. 3-66. Sync pulse waveform dis-

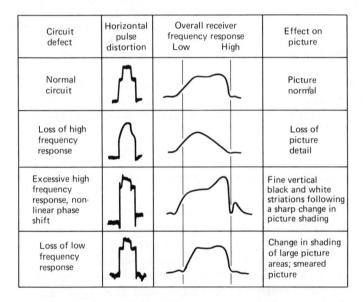

Fig. 3-66. Typical sync pulse waveforms and their relation to IF amplifier response (*Courtesy B&K-Precision, Dynascan Corporation*)

tortions produced by positive or negative limiting in IF overload conditions are shown in Fig. 3-67. An expanded horizontal sync pulse and blanking pedestal of a color signal is shown in Fig. 3-68. Note the color burst information on the "back porch."

Note the change in horizontal pulse waveforms shown in Fig. 3-66. For example, the tips of the sync pulse and blanking pedestal are rounded

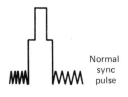

Fig. 3-67. Sync pulse waveform distortions produced by positive or negative limiting in IF amplifiers

Normal sync pulse

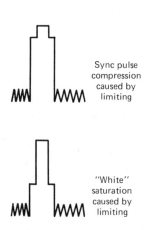

Sync pulse compression caused by limiting

"White" saturation caused by limiting

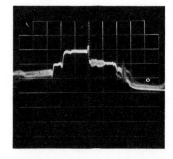

Fig. 3-68. Expanded horizontal sync pulse and blanking pedestal of a color signal *(Courtesy B&K-Precision, Dynascan Corporation)*

off when there is a loss of high-frequency response in the IF stages (or in the overall response of the receiver). When there is excessive high-frequency response, or a general loss of low-frequency response, the pulse tips become sharp.

3-8. VITS (VERTICAL INTERVAL TEST SIGNAL)

Most network television signals contain a built-in test signal (the VITS) that can be a very valuable tool in troubleshooting and servicing television receivers. The VITS can localize trouble to the antenna, tuner, IF, or video sections and shows when realignment may be required. The basic purpose of the VITS is for use by the broadcast engineers. They monitor these test signals to be sure the quality of their transmission meets established standards. The following procedures show how the television service technician can analyze and interpret oscilloscope displays of the VITS to troubleshoot television receivers.

The VITS is transmitted during the vertical blanking interval. On the television receiver, the VITS can be seen as a bright white line (and/or series of dots) above the top of the picture. However, the vertical hold, linearity, or height controls may require adjustment to view the vertical blanking interval. On some receivers with an internal vertical retrace blanking circuit, the vertical blanking pulse must be disabled to see the VITS. The procedures are covered in the following paragraphs. On other receivers, without internal vertical blanking, it is only necessary to advance the brightness control. For troubleshooting purposes, it is not necessary to see the VITS on the picture tube screen. Instead, you monitor the VITS signals on an oscilloscope after they pass through the receiver.

3-8.1   Composition of the VITS Pattern

The transmitted VITS is a precision sequence of specific frequency, amplitude, and waveshape, as shown in Figs. 3-69 and 3-70. The first frame

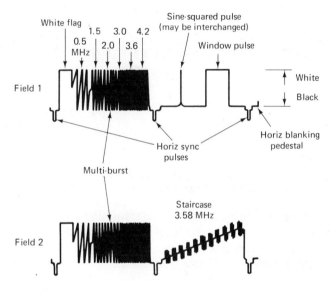

Fig. 3-69. VITS signal, fields 1 and 2 *(Courtesy B&K-Precision, Dynascan Corporation)*

of the VITS (line 17) begins with a "flag" of white video, followed by sine-wave frequencies of 0.5, 1.5, 2.0, 3.0, 3.6 (3.58), and 4.2 MHz. This sequence of frequencies is called the "multiburst." The first frame of field 2 (line 279) also contains an identical multiburst. This multiburst portion of the VITS is the portion that can be most valuable to the service technician.

The second frame of the VITS (lines 18 and 280), which contain the sine-squared pulse, window pulse, and the staircase of 3.58-MHz bursts at progressively lighter shading, are valuable to the network but have less value to the technician.

As seen on the television screen, field 1 is interlaced with field 2 so that line 17 is followed by line 279 and line 18 is followed by line 280. The entire VITS appears at the bottom of the vertical blanking pulse and just before the first line of video.

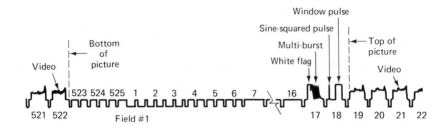

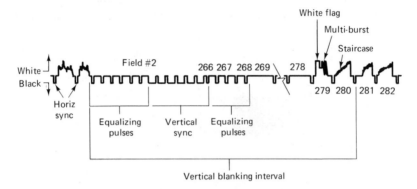

Fig. 3-70. Vertical blanking interval, showing VITS information *(Courtesy B&K-Precision, Dynascan Corporation)*

Each of the multiburst frequencies is transmitted at equal strength. By observing the comparative strengths of these frequencies after the signal is processed through the receiver, you can check the frequency response of the receiver at the normally used video frequencies.

3-8.2 Typical VITS Test Connections and Measurement Procedures

The procedures for monitoring the VITS signals as they are processed by the receiver will vary, depending on the receiver circuits. However, the following steps can be modified as necessary to accommodate all types of receivers.

1. Connect the oscilloscope probe to the output of the video detector or other desired point in the video section of the receiver. Start with the video detector output.

2. Set the oscilloscope sweep controls for a triggered sweep.

3. If the receiver has a vertical retrace blanking circuit, it must be disabled. Figure 3-71 shows a typical picture tube circuit with vertical retrace blanking pulses applied to the control grid. These pulses can be

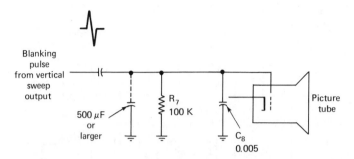

Fig. 3-71. Disabling the vertical retrace blanking pulses to the picture tube

bypassed to ground by placing a large capacitance across capacitor C_8 as shown. On receivers without vertical retrace blanking, simply advance the brightness control.

4. Operate the oscilloscope controls as necessary to obtain a VITS pattern. Figure 3-72 shows a typical pattern on a single-trace oscilloscope. Note that field 1 is superimposed on field 2. Figure 3-73 shows typical patterns on a dual-trace oscilloscope, where field 1 and field 2 are shown separately.

5. In typical operation, you must advance the oscilloscope intensity control full on. Then adjust the sweep sync control slowly to the point where triggering barely starts and a horizontal pattern is formed. Make very slight adjustment of the sweep sync control until the VITS pattern is

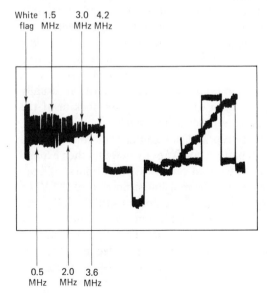

Fig. 3-72. Typical VITS pattern on single-trace oscilloscope (*Courtesy RCA*)

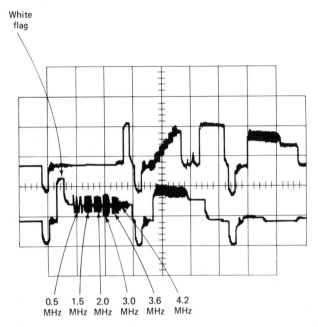

White
flag

0.5   1.5   2.0   3.0   3.6   4.2
MHz   MHz   MHz   MHz   MHz   MHz

Fig. 3-73. Typical VITS pattern on dual-trace oscilloscope *(Courtesy B&K-Precision, Dynascan Corporation)*

obtained. Because of the constantly varying video signal, sync lock-in of the VITS can be quite critical. (On some oscilloscopes, additional fine adjustment of the sync level can be obtained by slight readjustment of the vertical gain control.) It is normal for strong video signals to override the VITS pattern occasionally. This effect will vary with different televised scenes and may be more noticeable on certain channels. However, with careful adjustment of controls, a good, constant VITS pattern can usually be maintained. On some receivers, you can get more solid lock-in of the VITS pattern by using external sync, as described in step 8.

6. Using the oscilloscope sweep and horizontal gain controls, you can select the portion of the VITS pattern you want to view and then expand it as desired. Generally, you are interested only in the multiburst. Figure 3-74 is an expanded trace showing just the multiburst portion of the VITS pattern. It should be noted that the multiburst portions of the VITS are contained on only 2 horizontal lines of the 525-line raster and that the "staircase" and other portions are on only 1 line. The VITS will appear substantially dimmer than the rest of the composite video pattern. The bright vertical line pattern (shown to the right in Fig. 3-74) may be distracting.

0.5   2.0   3.6
MHz   MHz   MHz

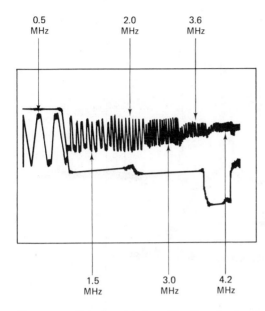

1.5   3.0   4.2
MHz   MHz   MHz

Fig. 3-74. Expanded trace showing just the multiburst portion of the VITS pattern *(Courtesy RCA)*

You can minimize this by adjusting the horizontal position control to place the bright area just beyond the right side of the screen edge.

7. As an alternative, you can display the VITS along with vertical blanking pulses. Figure 3-75 is such a display. The display of Fig. 3-75 is an expanded view of the pattern shown in Fig. 3-76. To produce such a display, set the oscilloscope sweep controls as you would to monitor the vertical composite video waveform (Sec. 3-6). Usually this involves setting the oscilloscope sweep control to VERTICAL or TVV or to a similar position. Then you operate the oscilloscope sweep expansion controls to obtain a display similar to Fig. 3-75 (which includes the blanking pulses along with the VITS).

8. As another alternative, you can display the VITS alone (as with a triggered sweep, steps 1 through 6) but with an external sync to lock in the VITS. On some receivers, improved sync stability can be obtained by using external sync, as follows. Locate a point in the receiver where a negative-going (or positive if the receiver has positive sync pulses) vertical sync pulse is available. One convenient point on most receivers is the appropriate wire in the convergence board that leads to the vertical winding of the deflection yoke. Be sure the receiver vertical and horizontal hold controls are set to provide a stable, in-sync picture. Connect the vertical pulse from the receiver to the external sync input of the oscilloscope. Operate the oscilloscope controls to lock in the VITS pattern and proceed as described in steps 1 through 6.

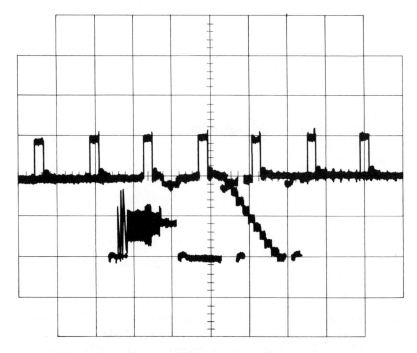

Fig. 3-75. VITS pattern displayed along with vertical blanking pulses *(Courtesy B&K-Precision, Dynascan Corporation)*

3-8.3   Analyzing the VITS Waveform

All frequencies of the multiburst are transmitted at the same level but are not equally passed through the receiver circuits due to the response curve of those circuits (RF, IF, and video). Figure 3-77 shows the desired response for an "ideal" color receiver, identifying each frequency of the multiburst and showing the allowable amount of attenuation for each.

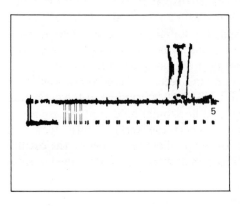

Fig. 3-76. Vertical blanking interval with VITS on right side *(Courtesy RCA)*

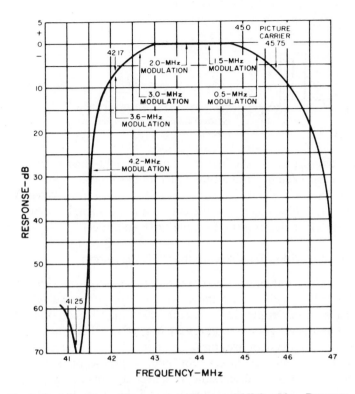

Fig. 3-77. Color IF amplifier response *(Courtesy B&K-Precision, Dynascan Corporation)*

Remember that − 6 dB equals half the reference voltage (the 2.0-MHz modulation should be used for reference).

Figure 3-77 represents the IF amplifier response curve of a color receiver. Thus, the curve frequencies are given in terms of typical IF range (41 to 47 MHz), and the VITS modulation frequencies are superimposed at corresponding points. For example, the 3.6-MHz VITS modulation occurs at the chroma or color frequency of 42.17 MHz.

In an ideal color receiver (and a good black and white receiver) the 2.0- and 1.5-MHz modulation signals should be at about the same level, the 0.5- and 3.0-MHz signals should be about 3 dB down from the 2.0-MHz reference level, the 3.6-MHz (chroma) signal should be about 6 dB down from (or about one half) the 2.0-MHz reference level, and the 4.2-MHz modulation signals should be about 30 dB down from the 2.0-MHz reference. Compare this with Fig. 3-72, which is a normal VITS pattern in a good receiver. Then compare both patterns with that of Fig. 3-78, which is a VITS pattern (measured at the video detector) from a receiver with a defective tuner.

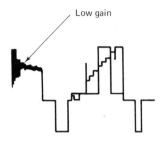

Fig. 3-78. VITS pattern (at video detector) from receiver with defective tuner *(Courtesy RCA)*

Note the loss of gain on the higher-frequency (right-hand side) multiburst signals. In this example, the receiver picture has no color. Since the VITS shows low gain in the chroma frequencies (3.6 MHz) at the video detector, you know the problem is not in the color band-pass amplifier but must be ahead of the video detector (in the tuner or IF amplifier).

3-8.4   Localizing Trouble with the VITS Waveform

To localize trouble, start by observing the VITS at the video detector. This will localize trouble to a point either before or after the detector. If the multiburst is normal at the video detector, check the VITS on other channels. If some channels are good but others are not, you probably have a defective tuner or antenna-system troubles. Do not overlook the chance of the antenna system causing "holes" or poor response on some channels. (Refer to Chapter 5 for discussions on checking antenna systems.) If the VITS is abnormal at the video detector on all channels, the trouble is in the IF amplifier or tuner. If the VITS is normal at the video detector for all channels, the trouble will be in the video amplifier. Look for open peaking coils, off-value resistors, bad solder connections, partial shorts, etc.

As an example of localizing trouble with the VITS, assume that you have a receiver with a poor picture (weak, poorly defined) but that the receiver is fully operative and that the antenna system is known to be good. Further assume that the VITS pattern at the video detector is about normal, except that the 2.0-MHz burst is low compared to the bursts on either side. This suggests that an IF trap is detuned, chopping out frequencies about 2 MHz below the picture carrier frequency.

Switch to another channel carrying VITS. If the same condition is seen, then your reasoning is good, and the IF amplifier requires adjustment. If the poor response at 2.0 MHz is not seen on other channels, possibly an FM trap in the tuner input is misadjusted, causing a drop in signal on only one channel. Other traps at the input (RF tuner) of the receiver could similarly be misadjusted or faulty.

One final note concerning VITS signals. The VITS patterns broadcast by all stations are not identical, although they all resemble those patterns shown here. This may cause a problem when switching from channel to channel during the trouble localization process. Be careful that a different VITS pattern does not lead you to believe a defect exists on one or more channels. Compare the VITS patterns with the picture. For example, if you get an odd VITS pattern on one channel and the picture is poor on that channel, you have definitely localized trouble to the tuner.

4

Basic Service Procedures for Color Receivers

In this chapter we shall discuss basic color receiver service and trouble-shooting. That is, we shall describe how the basic service techniques of Chapter 1 are combined with the practical use of test equipment discussed in Chapter 2 to locate specific faults in various types of color receivers.

Since alignment and adjustments, as well as testing, are part of service, we shall also describe basic adjustment procedures for various types of color receivers. This includes color setup procedures such as purity, convergence, and linearity adjustments as well as testing color sync, chroma demodulator, and matrix circuit characteristics. Throughout this chapter, considerable emphasis is placed on "universal" test, service, and troubleshooting procedures. These procedures apply to all color receivers now in existence, including vacuum tube, and to those that may be found in the future.

4-1. PURITY, CONVERGENCE, AND LINEARITY ADJUSTMENTS

All color receivers require some form of purity, convergence, and linearity adjustments. These adjustments must be made in addition to the adjustments for black and white receivers (height, width, centering, linearity, etc.), as described in Chapter 5. Generally, the adjustments are made when the receiver is first placed in operation and should be checked when the receiver has been subjected to extensive repair, such as replacement of the picture tube and deflection coils. Typically, this setup and checkout process for color receivers consists of degaussing the color picture tube, followed by purity, linearity, and convergence adjustments.

It is recommended that the manufacturer's service instructions be followed exactly whenever possible. Each type of color receiver has its own set of controls which may or may not be quite different from controls on other receivers. In the absence of manufacturer's instruction, and to show you what typical color receiver setup procedures involve, in the following paragraphs we shall describe complete setup for typical color receivers, as recommended by the manufacturers.

An RCA-type and a Zenith-type receiver are covered. Note that these "types" are actually composites of many receivers and do not necessarily apply to any specific receiver. Likewise, you will note that the results of the procedures are similar but that the controls are quite different. With examples of both types of controls shown, you should be able to relate the procedures to a similar set of controls on almost any receiver that may be found. Throughout the procedures, you will find such instructions as "connect the color generator to the receiver antenna terminals, and set the generator to produce vertical lines." It is assumed that you have read Chapter 2 regarding color generators and that you have studied your particular color generator. In Sec. 4-2, we shall describe color setup procedures using specific color generators.

4-1.1   Black and White Adjustments

In the following instructions, it is assumed that the black and white circuits of the receiver are in good working order and properly adjusted. The procedures for basic black and white adjustment are covered in Chapter 5. In some color receivers (and earlier black and white receivers), the picture tube focus is controlled electronically. On such receivers, the focus adjustment is a variable inductance in series with a winding on the horizontal output transformer, and the plate of a focus-rectifier tube, and is located at the rear of the high-voltage cage. No matter what system is used, if focus adjustment is required, set the receiver brightness and contrast controls at the approximate desired viewing level. The following procedures cover only those black and white circuits that are directly related to the color circuits.

4-1.2   Preliminary Convergence Adjustments (RCA)

1. Connect the color generator output to the receiver antenna input.

2. Set the color generator controls to produce a dot pattern. Adjust the receiver controls until the pattern is visible on the screen.

3. Adjust the red, green, and blue magnets and the lateral magnet to attain convergence of the dots in the center of the picture tube. The direction of dot movement using the magnets is shown in Fig. 4-1. Red and green movement is opposite that for the blue. The red and green dots move diagonally, whereas the blue dot moves horizontally or vertically. Location

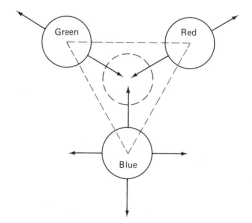

Fig. 4-1. Direction of dot movement for static convergence *(Courtesy B&K-Precision, Dynascan Corporation)*

of convergence and lateral beam magnets on a typical picture tube are shown in Fig. 4-2.

4. If a greater range of adjustments is necessary, the magnets may be reversed (on some picture tubes). To do this, slide the plastic magnet holder out of its metal clip and rotate the holder 180°. Replace the holder in the clip, making sure that the magnet is reinserted in the clip. Keep the receiver in focus when making this adjustment.

4-1.3   Color Purity Adjustments (RCA)

1. The picture tube and associated parts should be subjected to a strong demagnetizing field before any purity adjustments are made. Use a demagnetizing coil (degaussing coil) and slowly move the coil around the

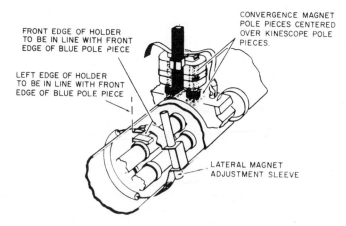

Fig. 4-2. Location of convergence and lateral beam magnets *(Courtesy B&K-Precision, Dynascan Corporation)*

picture tube and around the sides and front of the receiver. (*Do not degauss the magnets around the picture tube neck.*) Then slowly withdraw the coil at least 10 ft. from the receiver before deenergizing the coil. If the receiver has a built-in degaussing system, operate the system as described in the service literature.

2. Remove the third IF tube so that the raster will be blank. On solid-state receivers, disable the final IF stage by placing a short between emitter and base. If the color generator has a "blank raster" or "purity" output, use the output to produce a blank raster.

3. Loosen the screw on the yoke clamp and slide the yoke as far to the rear as possible. Figure 4-3 shows the location of the purity magnets on a typical picture tube.

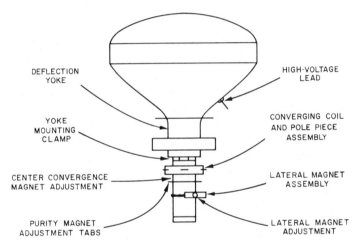

Fig. 4-3. Picture tube adjustments and components *(Courtesy B&K-Precision, Dynascan Corporation)*

4. Shunt the blue and green picture grids to ground through individual 100,000-Ω resistors. If the color generator is provided with color-gun interrupters (gunkillers), connect the leads from the color generator to the corresponding grid leads of the picture tube socket. The location of color-gun leads for an assortment of color picture tubes is shown in Fig. 4-4. Generally, it is not necessary to remove the socket from the color picture tube or make any direct connection to the socket pins. Most color-gun interrupter leads are provided with special alligator clips that will pierce the insulation on the picture tube lead and make contact with the wire.

5. Rotate the purity magnet around the neck of the picture tube and, at the same time, adjust the tabs on the magnet to produce a uniform red screen area (solid red spot) at the center of the picture tube.

6. Slide the yoke forward until the screen becomes completely red.

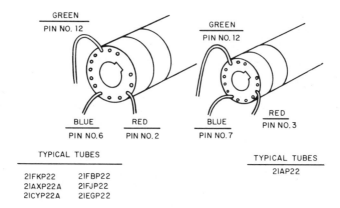

Fig. 4-4. Location of color gun leads for some color picture tubes *(Courtesy RCA)*

7. After checking the red field, it is helpful (but not absolutely essential) to check the green and blue fields (individually) for purity and the white field for uniformity. The green field can be checked by shunting the red and blue picture tube guns to ground through individual 100,000-Ω resistors (or use the color-gun interrupters on generators so equipped). The blue field can be checked by shunting the red and green grids. With all grids shunted, there should be no color, and the white should be uniform across the screen. (If the individual fields of red, blue, and green are present and a white field is present with all guns off, you know that the picture tube is good and is capable of producing all required colors.)

8. Set the generator controls to produce a crosshatch pattern. Check that the height, width, centering, focus, overscan, etc., are properly adjusted by comparing the crosshatch pattern with that specified for the receiver.

9. It is sometimes possible to get good purity only by pushing the yoke too far forward. This reduces the width and height such that the set does not overscan properly. Color generators produce a crosshatch pattern that consists of a fixed number of vertical and horizontal bars. Accordingly, the crosshatch function provides a convenient method for adjusting the receiver overscan to ensure that the proper portion of the raster is extended beyond the edge of the receiver mask.

10. The service literature for color receivers usually specifies a recommended amount of overscan at the left and right and a different amount of overscan at the top and bottom. The recommended overscan varies in different receiver models. Because a color generator provides a fixed number of vertical and horizontal bars, it is relatively easy to judge the amount of overscan. The appearance of the crosshatch pattern with correct overscan (9 vertical and 13 horizontal, in this case) in a representative model color receiver is shown in Fig. 4-5.

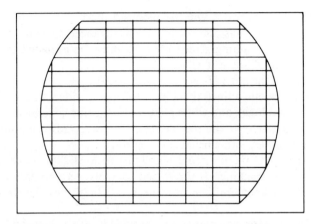

Fig. 4-5. Typical crosshatch pattern *(Courtesy RCA)*

4-1.4 Picture Tube Temperature Adjustments (RCA)

The picture temperature adjustments can be made without an external color generator. Thus, the color generator can be disconnected and/or turned off if its signal interferes with the procedure.

1. Set the picture tube bias (often called the kine bias) and screen controls (red, blue, and green screens) counterclockwise (full off). Figure 4-6 shows the location of the temperature adjustment controls on a typical receiver.

2. Set the normal-service switch (Fig. 4-6) to service.

3. Advance the individual screen controls so that each control just produces a horizontal line on the picture tube. When one or more of the

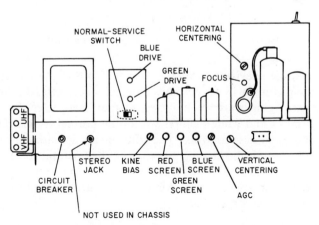

Fig. 4-6. Typical rear chassis color adjustments *(Courtesy B&K-Precision, Dynascan Corporation)*

controls fails to produce a line the picture tube bias (kine bias) must be advanced. After the bias control has been advanced to make the missing line appear, the remaining screen controls must be adjusted to the point where the horizontal line just appears.

4. Return the normal-service switch to normal.

5. Alternately adjust the blue and green video drive controls to produce a normal black and white picture. The color generator can be used to produce a pattern, or the receiver can be connected to an antenna and the controls adjusted with a broadcast picture.

6. Check the picture from highlights to lowlights, at all brightness levels, for proper tracking. If the screen controls were accurately adjusted as previously outlined, proper tracking at all brightness levels should be obtained.

7. Some color generators are provided with a *gray-scale tracking display output.* As shown in Fig. 4-7, a typical gray scale provides six levels of wide vertical gray bars, plus pure black and pure white.

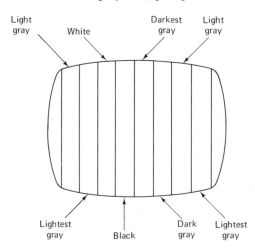

Fig. 4-7. Typical gray scale output of color generator

The compatible television system is designed so that the reception of black and white pictures will be reproduced on color receivers. In the color receiver, the three guns must be adjusted so that these telecasts will be reproduced as black and white pictures within the normal usable range of the contrast and brightness controls. The adjustments are also called *white balance* in addition to screen temperature, since they pertain to the reproduction of various luminance values from black and white. Only a receiver correctly adjusted for gray-scale tracking can, in turn, reproduce proper color when tuned to a color telecast.

Alternative screen temperature adjustments. An alternative method for adjusting picture tube temperature that sometimes provides greater accuracy involves just extinguishing the horizontal line after it has just appeared. This method is illustrated in Fig. 4-8, and the procedure is as follows:

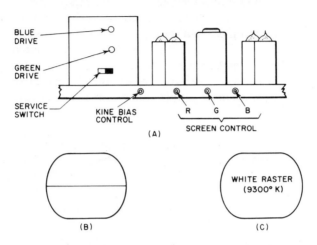

Fig. 4-8. Alternate method for picture tube temperature (white balance) set-up *(Courtesy B&K-Precision, Dynascan Corporation)*

1. Set the service switch to service position (this collapses the vertical sweep).

2. Set the kine bias control and all screens to minimum.

3. Adjust the red, green, and blue screens to the point where beams just cut off (increase kine bias only as required to support the weakest screen adjustment). [See Fig. 4-8(b).]

4. Reset the service switch to normal.

5. Adjust the green and blue video drive for a normal black and white picture (from a generator or broadcast).

4-1.5   Center Convergence Adjustments (RCA)

The convergence of dots at the picture tube center should be checked after purity and temperature adjustments, as described in Sec. 4-1.2. If the temperature and purity adjustments have affected convergence, repeat the convergence procedure using the dot output of the color generator. If the convergence magnets have been touched, the purity and temperature controls must be readjusted. When the center convergence, purity, and temperature controls require no further adjustment, proceed with the vertical and horizontal convergence adjustments.

4-1.6 Vertical Convergence Adjustments (RCA)

In the following procedure, instances may be encountered where adjustments of the vertical tilt controls will not produce optimum convergence. On certain receivers, the range of the tilt controls may be broadened by changing the ground jumper connection of the tilt winding on the vertical output transformer from the tap to the end of the winding. Changing the connections in this manner will provide all the required vertical tilt range for optimum convergence. Consult the schematic and/or service literature for the particular receiver regarding the vertical output transformer connections. Generally, this connection is not practical on solid-state receivers.

1. Connect the color generator output to the receiver antenna input.

2. Set the color generator to produce vertical lines. Use a crosshatch pattern if the generator will not produce separate horizontal and vertical lines.

3. Adjust the receiver controls until the pattern is visible on the screen.

4. Gain access to the convergence controls. On some receivers, the convergence controls are mounted on a board assembly. This board is designed to permit adjustments from the front of the receiver. If not, a mirror will be required.

5. Referring to the vertical line at the center of the picture tube screen, adjust the vertical R-G master amplitude control to converge the center line at the bottom. [See Figs. 4-9(a) and 4-10.] Adjust the vertical R-G master tilt control to converge the center line at the top of the screen. Touch up both adjustments for best convergence along the entire center vertical line. You may find it easier to perform this setup with the blue gun disabled. The blue gun is not necessary to this adjustment since the R-G controls affect only red and green.

6. Set the color generator controls to produce horizontal lines.

7. Referring to the center line of the screen, converge the horizontal line at the bottom of the screen with the vertical R-G differential amplitude control. [See Figs. 4-9(b) and 4-10.] Adjust the vertical R-G differential tilt control to converge the top horizontal line at the center of the screen. Touch up both adjustments for best convergence of all lines at the vertical center line of the screen.

8. Set the color generator controls to produce a dot pattern.

9. Check convergence of the dots at the center of the screen. If necessary, readjust the convergence magnets as described in Sec. 4-1.2.

10. Set the color generator controls to produce horizontal lines.

11. Advance the vertical blue amplitude control to produce displacement of the lines at the top and bottom of the screen at the center line.

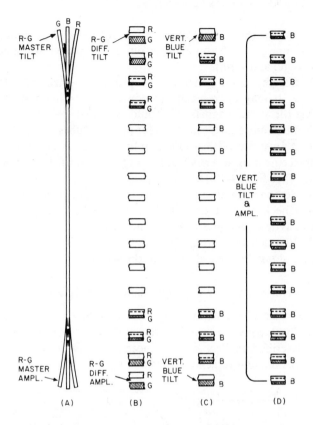

Fig. 4-9. Vertical convergence pattern *(Courtesy B&K-Precision, Dynascan Corporation)*

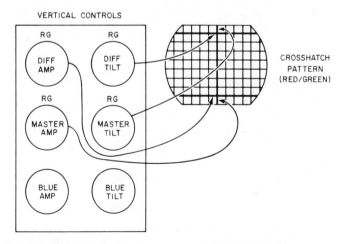

Fig. 4-10. Use of vertical master and differential controls *(Courtesy B&K-Precision, Dynascan Corporation)*

12. Adjust the vertical blue tilt control to produce equal displacement of the lines at both top and bottom of the screen at the center line. [See Figs. 4-9(c) and 4-10.]

13. Set the color generator to produce horizontal lines. Check convergence. It may be necessary to converge the blue with the red and green using the static magnets. It should be noted that the interaction of controls is much more severe in 90° color pictures tubes.

14. Adjust the vertical blue amplitude and tilt controls to produce equal displacement of all lines from top to bottom of the screen along the center line. [See Figs. 4-9(d) and 4-10.]

15. Set the color generator controls to produce a crosshatch pattern.

16. Check convergence of the dots at the center of the screen. Reconverge as described in Sec. 4-1.2 if necessary.

17. Retouch the vertical blue tilt and amplitude controls for best convergence along the vertical center of the screen.

4-1.7   Horizontal Convergence Adjustments (RCA)

1. Set the color generator controls to produce a crosshatch pattern.

2. Check convergence of the dot at the center of the screen. Reconverge as described in Sec. 4-1.2 if necessary. Refer to Fig. 4-11 when making horizontal convergence adjustments.

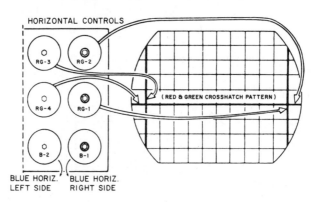

Fig. 4-11. Use of horizontal R-G convergence controls *(Courtesy B&K-Precision, Dynascan Corporation)*

3. Adjust coil B-1 to make the blue line at the right center of the picture (center to right-hand edge) a straight line.

4. Adjust control B-2 to make the blue line at the left center of the picture tube (center to left-hand edge) a straight line.

5. Adjust coil RG-1 to make vertical lines at the right side converge.

6. Adjust coil RG-2 to make horizontal red and green lines at the right side converge. Readjust coil B-1 as necessary to make the blue line at the right center fall on the converged red and green lines. Retouch RG-1 for convergence of vertical lines at the right side.

7. Adjust control RG-3 to make red and green vertical lines at the left side converge.

8. Adjust control RG-4 to make red and green horizontal lines at the left side converge.

9. After adjusting RG-4, repeat adjustment of RG-3 to compensate for any interaction. Readjust control B-2 to make the blue line at the left center fall on the converged red and green lines. The pattern should now show proper convergence on all areas of the screen.

10. After completion of vertical and horizontal convergence adjustments, check and (if necessary) repeat the picture tube purity and temperature adjustments (Secs. 4-1.3 and 4-1.4).

4-1.8   Color Purity Adjustment (Zenith)

The first step toward adjusting a color receiver for good purity is to demagnetize (degauss) the picture tube. If the picture tube shadow mask has become slightly magnetized, good purity may be difficult to get.

Degaussing procedure. The degaussing procedure may be performed with the receiver either in or out of the cabinet. Most technicians prefer to degauss in the cabinet, since this will also degauss any metal in the cabinet (such as the shadow mask).

1. Place the receiver in the same position as for viewing. With the receiver on, set the contrast control to minimum and the brightness control to near maximum.

2. Position the degaussing coil in a parallel plane against the face of the picture tube. Energize the coil.

3. Slowly rotate the coil around the face of the picture tube for approximately 1 min. If the receiver is turned on during the degaussing procedure, the raster will show a "swirling" rainbow-type pattern.

4. Slowly withdraw the coil while continuing to rotate the coil parallel to the picture tube face.

5. Withdraw to a distance of 10 ft or more and deenergize the coil. If it is impossible to back off 10 ft, this can be simulated by withdrawing the coil to a distance of about 5 ft and then gradually reducing the coil supply voltage by use of a variac or similar device.

Purity control adjustment procedure. Locations of the convergence and lateral beam magnets on a typical picture tube are shown in Fig. 4-12. Locations of the associated tracking controls are shown in Fig. 4-13.

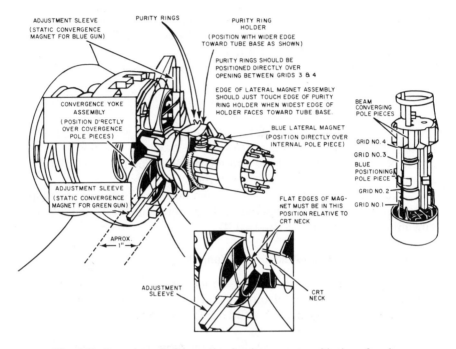

Fig. 4-12. Rear view of picture tube showing correct positioning of neck components *(Courtesy B&K-Precision, Dynascan Corporation)*

1. Connect the color generator output to the receiver input.

2. Set the color generator controls to produce a dot pattern. Adjust the receiver controls until the dot pattern is visible on the screen.

3. Adjust the red, green, and blue magnets and the lateral magnet to get convergence of dots in the center of the picture tube. If a greater range of adjustment is necessary, the magnets may be reversed. To do this slide the adjustment sleeve out of its metal clip and rotate the sleeve 180°. Replace the sleeve in the clip, making sure that the magnet is reinserted in the clip. Keep the receiver in focus when making this adjustment.

4. Place the receiver in the same position as for viewing. With the receiver on, set the contrast control to minimum and the brightness control to near maximum.

5. Disable the blue and green picture tube guns by turning their screen adjustments to minimum (Fig. 4-13) or by using the color-gun interrupter switches on the generator.

6. Rotate each purity ring (spreading the tabs apart) until you get a pure red raster over all of the screen area. If you cannot fill the entire screen area, move the deflection yoke forward or backward until the entire screen is pure red. Purity in the central area of the raster is controlled by

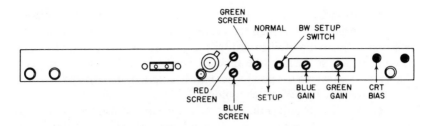

Fig. 4-13. Rear view of chassis showing location of black-and-white tracking adjustments *(Courtesy B&K-Precision, Dynascan Corporation)*

the purity ring adjustment. The purity of the outer area is controlled by the deflection yoke positioning.

7. Check the green raster by turning the red screen adjustment to minimum and turning up the green screen adjustment. The raster should appear green over the entire screen area.

8. Check the blue raster by turning the green screen adjustment to minimum and turning up the blue screen adjustment. The raster should appear uniform blue over the entire screen area.

4-1.9   Black and White Tracking (Zenith)

If a color receiver is to have good black and white tracking, the receiver must produce black and white pictures within the normal usable range of both the contrast and brightness controls. If it is necessary to set the contrast and brightness at extreme positions to get good black and white pictures, blooming may occur and off-colors may exist when watching color broadcasts.

The three screen grid adjustments, plus the brightness and contrast controls, are used for adjusting black and white tracking. (Some manufacturers refer to this as gray-scale tracking; refer to Sec. 4-1.4.) During the procedure, the voltages on the cathodes, control grids, and screen grids of the picture tube guns are adjusted to produce black and white pictures throughout the usable range of the brightness and contrast controls.

1. If the color generator has a gray-scale output, use that output. If not, disconnect the color generator from the receiver antenna input.

2. Tune in a black and white picture that displays an adequate range of light levels, light and gray objects, etc. Set the brightness and contrast controls for a normal picture. Do not bloom the picture (too much contrast and/or brightness).

3. Set the BW switch to setup position (Fig. 4-13). In this position, the vertical sweep is removed to facilitate adjustment.

4. Advance each screen adjustment to just produce a white horizontal line of low brightness through the center of the screen. If one or more of the screen adjustments fail to produce a line, adjust the picture tube bias setting to make the missing line appear. Then adjust the remaining picture tube screen controls to the point where the horizontal line just appears.

5. Return the BW switch to normal.

6. Alternately adjust the blue and green gain adjustments to produce a normal black and white picture. If you have difficulty in getting a good black and white picture, check the service instructions for any special procedures.

4-1.10   Dynamic Convergence Adjustments (Zenith)

This procedure can be performed with individual horizontal and vertical lines. However, it is better to use a crosshatch or some similar pattern. Connect the color generator output to the receiver antenna input, and select the appropriate pattern.

Of the 12 dynamic convergence adjustments, 3 are used for convergence at the top of the raster, 3 for the bottom of the raster, 3 for the left side, and 3 for the right side. (See Fig. 4-14.) The numbers appearing on the dynamic convergence controls are for reference to the convergence pattern of Fig. 4-14 and do not appear on the controls.

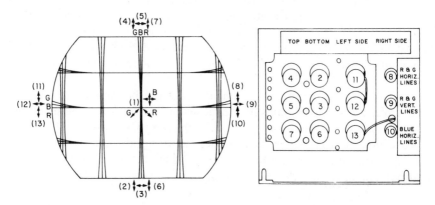

Fig. 4-14. Typical dynamic convergence adjustments *(Courtesy B&K-Precision, Dynascan Corporation)*

1. Check convergence of the pattern at the center of the screen. If necessary, adjust for proper convergence using the permanent magnets (with generator dot pattern), as described in Sec. 4-1.8. Then return the color generator to a crosshatch pattern and check center convergence. Do not proceed with dynamic convergence until the center of the display is properly converged with the static magnets.

2. Merge the red and green horizontal lines at bottom center of the pattern using R & G HORIZ. LINES, bottom adjustment.

3. Merge the red and green vertical lines at bottom center of the pattern using R & G VERT. LINES, bottom adjustment.

4. Merge the red and green horizontal lines at top center of the pattern using R & G HORIZ. LINES, top adjustment.

5. Merge the red and green vertical lines at top center of the pattern using R & G VERT. LINES, top adjustment. Repeat adjustments 1 through 5 to get the best vertical red-green convergence from top center to bottom center.

6. Merge the blue horizontal line with the red and green lines at bottom center of the pattern using BLUE HORIZ. LINES, bottom adjustment.

7. Merge the blue horizontal line with the red and green lines at top center of the pattern using BLUE HORIZ. LINES, top adjustment. Repeat adjustments 6 and 7 to get convergence of the blue line with the red and green lines from top center. If necessary, repeat step 1 to converge the blue beam with the red and green beams.

8. Merge the red and green horizontal lines at the right center of the pattern using R & G HORIZ. LINES, right side adjustment.

9. Merge the red and green vertical lines at right center of the pattern using R & G VERT. LINES, right side adjustment.

10. Merge the blue horizontal line with the red and green lines at right center of the pattern using BLUE HORIZ. LINES, right side adjustment.

11. Merge the red and green horizontal lines at left center of the pattern using R & G HORIZ. LINES, left side adjustment.

12. Merge the red and green vertical lines at left center of the pattern using R & G VERT. LINES, left side adjustment.

13. Merge the blue horizontal line with the red and green lines at left center of the pattern using BLUE HORIZ. LINES, left side adjustment.

14. If necessary, repeat adjustments 8 through 13 to get the best horizontal convergence from left side center to right side center. Although the degree to which color receivers can be converged will vary, you should be able to get good convergence with an area out to about 2 in. from the edges, top, and bottom of the raster. You may have some misconvergence at extreme edges of the raster. This is normal and will pass unnoticed at a normal viewing distance.

4-2. COLOR SETUP PROCEDURES USING A COLOR GENERATOR WITH PATTERN/ANALYST FEATURES

In the following paragraphs we shall describe how typical color generators can be used to set up a color receiver. Note that the procedures cover essentially the same areas described in Sec. 4-1 but use specific color generators discussed in Chapter 2 (the RCA WR-515A and the B&K 1246). By comparing the procedures of these generators to the basic setup procedures of Sec. 4-1, you should be able to match the outputs of your generator to any color receiver.

4-2.1   Color Setup Using RCA WR-515A*

The chart of Fig. 4-15 lists the generator patterns recommended for the setup procedures. Details on how to use the generator, along with many helpful hints and techniques for making setup adjustments, are given in the following paragraphs. It should be noted that not all color receivers have adjustments for video peaking, AGC, and centering, as listed in Fig. 4-15.

ADJUSTMENT OR PROCEDURE	RECOMMENDED WR-515A PATTERN
AGC	Lines, 10 V, 11 H
Video peaking	Superpulse
H and V centering	Lines; 3 V, 3 H
Purity (including degaussing)	Blank raster
Overscan	Lines; 3 V, 3 H
Linearity	Lines: 10 V, 11 H
Pincushion	Lines; 11 H
Gray scale tracking	Superpulse
Convergence	
(static)	Dots, lines (3 V, 3 H) or Superpulse
(dynamic)	Lines; 3 V, 3 H
AFPC (tint)	Color bars, with "mark"

Fig. 4-15. Recommended WR-515A patterns for color setup procedures *(Courtesy RCA)*

Preliminary procedure. Use the following procedure on all receivers.

1. Connect the output of the generator to the receiver antenna input terminals. (Disconnect any antenna that may be connected to the receiver, including internal "rabbit ears." Set the generator RF LEVEL and the CHROMA LEVEL controls to midrange. Release the IF push button to OUT. (This disables the IF signal from the generator.)

2. Turn on the WR-515A, and press the COLOR push button. [This turns on the generator and provides a color bar (keyed rainbow) output.] Turn on the receiver, and tune to channel 3. If the receiver has automatic fine tuning (AFT), turn it off. Adjust the receiver fine tuning, contrast,

See Figure 2-9.

brightness, horizontal and vertical hold controls, and the generator RF LEVEL control to obtain the best picture possible.

AGC adjustment. Use the following procedure on receivers with any form of AGC.

1. Press the LINES, VERT 10, and HORIZ 11 push buttons to obtain a full crosshatch pattern. (See Fig. 4-16.) Adjust the generator

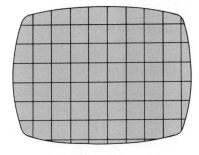

Fig. 4-16. Full crosshatch pattern (10V, 11H) with lines missing because of overscan

V LINE/DOT BRIGHTNESS control to obtain suitable brightness of the vertical lines. Turn the receiver AGC control clockwise until the pattern just begins to bend or distort; then readjust the control counterclockwise to the point slightly below where the distortion is eliminated. Note that on some receivers the AGC control may have very little effect. If so, set the control to obtain the best picture. Also, there are 11 horizontal and 10 vertical lines generated. However, of these lines, only 9 or 10 lines will normally be seen (because of picture tube overscan).

2. Turn the generator RF LEVEL control fully counterclockwise. If the pattern distorts, or loses sync, readjust the receiver AGC control as explained in step 1, setting the control just below the point where the trace distorts.

3. Turn the generator RF LEVEL control throughout its entire range. Readjust the AGC control as necessary to prevent the picture from distorting or losing sync through the widest possible range of the RF LEVEL control.

Video peaking. Use the following procedure on receivers with any form of video peaking adjustments.

1. Press the SUPERPULSE push button. Adjust the SUPERPULSE SYNC for the best pattern. (See Fig. 4-17.) Adjust the receiver video peaking coil (or coils) so that the leading (or left) edge of the white rectangle is crisp and sharp. If you adjust the control too far in one direction, it may

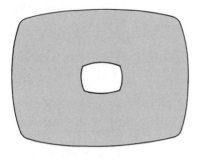

Fig. 4-17. Superpulse pattern (white center on gray background)

cause "ringing" (traces of the white edge repeating across the pattern). If you adjust the peaking too far in the other direction, the white edge may become soft or hazy.

Horizontal and vertical centering. Use the following procedure on receivers with any form of horizontal or vertical centering controls.

1. Press the LINES, VERT 3, and HORIZ 3 push buttons. The picture tube pattern should be as shown in Fig. 4-18. Adjust the horizontal

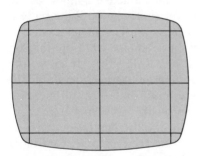

Fig. 4-18. Line pattern with 3 horizontal and 3 vertical lines

and vertical centering controls on the receiver so that the two middle vertical and horizontal lines intersect at the center of the screen and the outer bars are equally spaced from the edge of the picture tube mask.

Purity. Use the following procedure on all receivers.

1. Before adjusting purity, the receiver may need a rough check of static convergence, as described in *convergence adjustment.*

2. Press the BLANK raster push button. Obtain a red screen by turning off the receiver blue and green screen controls or by using the blue and green GUN KILLER switches. (The gun killer cable must be connected to the picture tube leads.)

3. Loosen the picture tube yoke and move it as far to the rear as possible on the neck of the picture tube, without moving the purity magnet assembly.

4. Adjust the purity magnets so that the area in the center of the screen is uniformly red with no dark or discolored areas.

5. Push the yoke forward until the entire screen is uniformly red. Tighten the yoke.

6. Turn on the blue and green guns. The raster should be uniformly white. If not, repeat the adjustment procedure.

7. If good purity adjustment cannot be obtained, part of the receiver has become magnetized and will require *degaussing*. Even receivers that have automatic degaussing can have a magnetized chassis or cabinet. If necessary, use a degaussing coil to demagnetize the affected area.

Overscan. Use the following procedure on all receivers.

1. Press the LINES, VERT 3, and HORIZ 3 push buttons to obtain a three-line crosshatch pattern (Fig. 4-18).

2. Adjust the receiver width and height controls so that the raster extends beyond the edge of the picture tube mask by the amount specified for the particular receiver in its service literature. The recommended amount of overscan varies. However, Fig. 4-18 shows the type of pattern obtained on a typical receiver with proper overscan adjustment.

Linearity. Use the following procedure on all receivers.

1. Press the LINES, VERT 10, and HORIZ 11 push buttons to obtain a full crosshatch pattern (Fig. 4-16).

2. Adjust the height and vertical linearity controls so that the horizontal bars of the crosshatch are equally spaced (particularly at the extreme top and bottom of the screen).

3. Adjust the horizontal linearity controls (if any) so that the vertical bars are equally spaced.

Pincushion adjustment. Use the following procedures on receivers with pincushion controls.

1. Press the LINES and HORIZ 11 push buttons. Release both the VERT 3 and 10 push buttons so that only horizontal lines appear, as shown in Fig. 4-19.

2. Adjust the top and bottom pincushion controls until all horizontal lines are straight. Figure 4-19 shows a pattern with correct pincushion adjustment.

Color temperature adjustment (gray-scale tracking). Correct color temperature applies to all receivers and is necessary to assure that all three color guns are operating at a level to cause the red, blue, and green phosphors to glow at the same level. The beam currents of the three guns must main-

Fig. 4-19. Horizontal line pattern

tain the ratio throughout the excursions of the video signal and throughout the range of the brightness control. Proper color balance must be maintained from black (low lights), through the grays, to white (bright lights).

The SUPERPULSE pattern (Fig. 4-17) permits both high-light and low-light areas to be observed at the same time. This results in an easier and more effective gray-scale tracking adjustment.

1. Press the SUPERPULSE push button and set the SUPERPULSE SYNC control to midrange.

2. Turn the receiver screen controls to minimum, and turn the picture tube bias control (kine bias) to a high bias position (usually this is full counterclockwise). Set the normal-service switch in the receiver to service. Turn the brightness control to maximum.

3. Advance the red screen control until a dim, barely visible horizontal red line can be seen. Advance the blue and green screen controls until each one produces a dim horizontal line. If any of the three screen controls cannot be adjusted to produce a visible line, advance the picture tube bias (kine bias) control as necessary to make the lines visible. Set the kine bias so that the weakest color line is just visible. Then set the remaining screen controls as necessary to produce a dim horizontal line.

4. Set the normal-service switch to normal. Adjust the brightness and contrast controls to provide a superpulse pattern (Fig. 4-17), with the background raster around the white rectangle dimly visible.

5. Adjust the red (if any), blue, and green video drive controls to produce a neutral white superpulse rectangle, with no predominant hue either in the low-light background or high-light superpulse (inner) rectangle.

6. Vary the SUPERPULSE SYNC control setting. The white rectangle and the background area should not take on a color or hue as the raster darkens and becomes black.

Convergence adjustment. Misconvergence of a color receiver can be noted by observing the crosshatch or dot pattern on the screen. If the receiver is properly converged, the lines or dots will be clear and sharp, with no color fringing (one or more colors appearing on the edges of the dots or crosshatch lines).

1. Before adjusting convergence, you should check other adjustments, especially purity, if specified in the service literature. These adjustments may include horizontal tuning, horizontal drive, high-voltage regulation, height, width, linearity, focusing, etc.

2. Press the push button to provide the desired pattern for static convergence. The dot pattern (Fig. 4-20) can be used, but many technicians

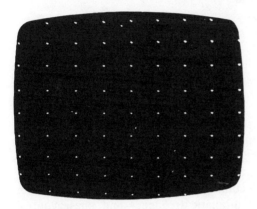

Fig. 4-20. Dot pattern (white dots on black background)

prefer the three-line V and H pattern (Fig. 4-18) since it provides single horizontal and vertical lines at the center of the screen. Superpulse (Fig. 4-17) is another convenient pattern for static convergence, since the four edges of the pattern provide a good reference for observing color fringing.

3. Adjust the receiver fine tuning, brightness, and contrast controls to get the best pattern. Adjust the permanent magnets on the picture tube neck to converge the red, blue, and green beams at the screen center. The pattern is covered when it is white and has the least amount of color fringing.

4. Press the LINE, 3 VERT, and 3 HORIZ push buttons or, if preferred, use a full crosshatch pattern. Adjust the dynamic convergence controls to converge the entire screen, particularly the edge areas. (Refer to Sec. 4-1 for typical procedures.)

5. Repeat static and dynamic convergence procedures as required until the complete crosshatch pattern consists of crisp white lines, having the least amount of color fringing.

Checking the color circuits. Use the following procedures on all color receivers.

1. Press the COLOR BAR and MARK push buttons and set the generator CHROMA controls to midrange.

2. Switch out any special receiver color controls, such as "Accutint." Set the receiver hue or tint control and the color intensity control to midrange. Adjust the receiver fine tuning, brightness, and contrast controls to get the best color-bar patterns. The pattern should appear similar to Fig. 4-21. However, in some receivers all 10 bars may not show due to the receiver's overscan.

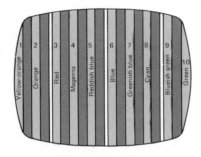

Fig. 4-21. Color bar pattern with bar marks on the 3rd, 6th, and 9th bars *(Courtesy RCA)*

3. Checking the hue (or tint) control. Identify the eighth bar using the MARK function. [When you press in the MARK push button you will notice a bright vertical line on the third (red), sixth (blue), and ninth (green-blue) bars.]

4. Adjust the receiver tint control until the eighth bar is cyan (bluish-green). If the receiver AFPC (automatic frequency and phase control) circuit (if any) is operating properly, the tint control should be somewhere near its midrange position when the correct color-bar pattern is obtained. The tint control should have enough range to shift the color at least one bar in either direction. For further data on checking and aligning the AFPC circuits, refer to Sec. 4-5.

5. *Checking the "Accutint" or similar control to enhance flesh tones.* Switch on the receiver "Accutint" or similar control. The color-bar pattern should change, with several of the bars taking on a more reddish color. The range of the tint control will now be restricted in some systems and will cause only a slight shift of color in each direction. The color saturation or intensity control may also be restricted, so that the color cannot be taken completely out of the pattern.

6. *Checking color sync lock.* The CHROMA LEVEL control may be used to check the color sync lock ability of the receiver. To check color sync lock action, turn the CHROMA LEVEL control slowly counterclockwise (minimum chroma signal). The color should become pale and finally disappear. Since some receivers are equipped with an automatic color control (ACC) circuit, the rate of fading will depend on the receiver under test. Most receivers may lose color sync just before the color disappears. This will be evidenced by diagonal running of the colors.

Both of these conditions indicate *normal operation* of the color sync circuits. If a slight reduction of the chroma amplitude causes the color to fall out of lock, however, it indicates that the color synchronization ability of the receiver may be inadequate. As a further check, you can turn the RF level control to reduce the signal strength, noting the effect of color sync.

When the CHROMA LEVEL control is turned beyond the midrange position, the relative amplitude of the color subcarrier sync is increased. At the full clockwise position, the amplitude is approximately doubled. This additional range is helpful in diagnosing receiver trouble in cases where the response of the RF/IF band-pass amplifiers may be subnormal or in cases when the sync lock action is faulty.

7. *Checking color "fit."* The bars in the color-bar pattern should "fit" into proper position and not lap over into the blank spaces between bars. Narrow brightness pulses are added to the edges of each bar, making it easy to observe color registration. Improper fit may be caused by incorrect delay in the video amplifier or by incorrect alignment of the band-pass amplifier.

4-2.2   Color Setup Using B&K 1246*

In the following paragraphs we shall describe how to use the generator to perform the color setup procedures, previously covered in Secs. 4-1 and 4-2.1. The detailed color setup procedures will not be duplicated here. Instead, we shall concentrate on how to use this type of generator to get the required signals.

Basic operation. Use the following procedure for all types of receivers.

1. Disconnect all antenna lines from the receiver VHF antenna terminals, and attach the generator RF cable to these terminals.

2. Tune both the receiver and generator to either channel 3 or 4.

3. Turn on the generator by rotating the CHROMA LEVEL control clockwise.

4. Rotate the pattern selector to GATED RAINBOW and set the CHROMA LEVEL control to its midpoint position. Advance the receiver color level control approximately one-quarter from minimum. Adjust the contrast and brightness to provide a comfortable viewing intensity.

5. Slide the 4.5-MHz switch to ON. All color bars should disappear.

6. Rotate the receiver fine tuning until an ungated rainbow with a light herringbone pattern appears. This effect is due to a beat between the sound and color subcarriers. Adjust the fine tuning once again for color with *minimum* herringbone. At this point, the receiver IF sound trap has

See Figure 2.16.

attenuated the 4.5 MHz signal, and the receiver is properly tuned. Switch off the 4.5-MHz switch. Refer to Sec. 4-6 for details on setting the 4.5-MHz sound trap in color receivers.

With certain receivers, it may be impossible to either locate or tune out the herringbone. This may be due to a poorly aligned sound trap. An alternative method of fine tuning is presented as follows.

Alternative method of fine tuning

1. Set the pattern selector to crosshatch, switch off the 4.5-MHz switch, and fine-tune the receiver until a reasonably good display is obtained.

2. Reduce the contrast and brightness controls until both the horizontal and vertical lines are barely visible; one may be brighter than the other, but both should be visible.

3. Carefully readjust the fine tuning for brightest *vertical* lines; then reset the contrast and brightness for a comfortable viewing intensity. Never operate the receiver with excessive contrast and brightness during convergence or color adjustments (and thus avoid blooming of the receiver display).

4. Rotate the pattern selector to GATED RAINBOW, and set the CHROMA LEVEL control to its midpoint position.

5. Advance the receiver color level control until color appears. When color does not appear at all, or only with the control near maximum, carefully readjust the fine tuning. It should only require a slight amount of rotation; excessive rotation indicates tuner or IF misalignment. If this last step fails to produce color, it is likely that a malfunction exists somewhere in the receiver, and it must be corrected before proceeding.

When the receiver is properly tuned, it is then ready for convergence and color adjustments.

Convergence. As discussed in previous sections of this chapter, convergence adjustments consist of purity, static convergence, and dynamic convergence steps.

Purity. The purity output of the generator is essentially a blank raster. That is, purity output provides sync and a reference signal free of all video information. A blank raster is an advantage over older methods (selecting an unused channel, disabling the IF stages, etc.) when adjusting for purity. The older methods usually produce "snow" on the screen (which can be annoying and cause inaccuracy in setup). Using the purity (blank raster) position, you can be assured that the adjusted purity condition will be maintained when the convergence procedure is initiated.

Static convergence. Static, or dc, convergence is always performed before and after purity adjustments. The generator provides two dot patterns. The single center dot pattern is most convenient because it automatically pinpoints screen center and is quickly located when working from behind the receiver, viewing the screen at any angle. In many instances, it is easier to "rough in" static convergence with the crosshair pattern. However, for final touch-up, always use the center dot.

Dynamic convergence. Crosshatch is the recommended pattern for performing dynamic convergence. Some technicians prefer to use dots throughout the entire procedure. This is a matter of personal preference. However, misconvergence is most easily seen with horizontal and vertical lines.

The single horizontal with full vertical lines pattern (Fig. 4-22) is

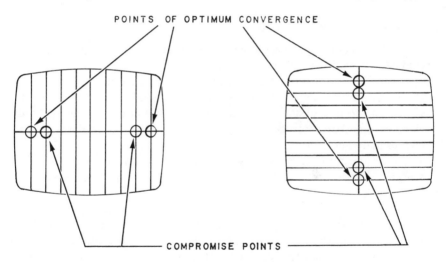

Fig. 4-22. Crosshatch pattern with single horizontal, or single vertical, lines for convergence adjustments *(Courtesy B&K-Precision, Dynascan Corporation)*

especially useful when performing dynamic convergence on the left and right sides (at center vertically). Elimination of all but the center horizontal line removes any confusion as to the correct points to converge. On a typical receiver, the vertical lines closest to the screen's left and right edges would be the best choice. You can reach a compromise on difficult convergence problems by using the second closest line.

The single vertical line with full horizontal lines pattern (Fig. 4-22) is useful when performing convergence on the top and bottom (at horizontal center). Elimination of all but the center vertical lines again removes confusion as to the correct convergence points. On a typical receiver, the horizontal lines closest to the screen's top and bottom edges are the best choices. Use the second lines to compromise on difficult convergence problems.

Use the full crosshatch pattern to verify convergence after the static and dynamic adjustments have been made. Also, if you notice defocusing, blooming, or "kinks" at the crosshatch intersections, this indicates that brightness and contrast are excessive. For greatest accuracy during convergence adjustments, never increase brightness and contrast beyond that point where the patterns are sharp and clear.

Color adjustments. Always check the service literature procedures when adjusting color circuits. The following procedures describe a general technique for adjusting the color circuits of any receiver and can be used when manufacturer's service literature is not available.

Hue (tint) setting and range

1. Select the gated rainbow pattern, and set the CHROMA LEVEL control to its midpoint position.

2. Adjust the receiver's saturation, brightness, and contrast controls to produce a pleasing color pattern. Ten individual color bars should be visible on the face of the screen (see Fig. 2-12). On some receivers, only eight or nine bars may be displayed. This is due to excessive overscan and/or blanking.

3. Select the R-Y, B-Y, – (R-Y) pattern. Three color bars, representing the third, sixth, and ninth bar of the gated rainbow, should now be visible on the screen.

4. Adjust the receiver's hue (or tint) control to display a red, blue, and bluish-green bar (in this order) from left to right, as shown in Fig. 2-17. If this color-bar arrangement cannot be obtained with any setting of the hue control, then internal adjustment of the hue range is necessary. It can be assumed that the color circuits are operating properly if these steps produce the correct results.

Demodulator alignment check. The following steps provide a quick check of the color demodulators, without the use of a vectorscope.

1. Select the R-Y, B-Y, – (R-Y) pattern, and set the CHROMA LEVEL control to its midpoint position.

2. Turn the receiver contrast control to minimum and adjust the brightness control for the brightest display possible without blooming. Set the receiver saturation control to a very low level—just enough to produce color.

3. Disable the red and green guns of the picture tube. Adjust the hue (tint) control so that color and shading to the area left and right of the first bar (left-hand side of the screen) matches the center section of that bar.

4. Disable the blue and green guns, and leave only the red gun active. If the color demodulators are properly aligned, the bar in the center of the screen will match the color and shading of the area to either side of it. A large amount of error (difference in color) usually indicates the need for demodulator alignment.

Color sync locking. The CHROMA LEVEL control varies the amplitude of the color subcarrier from 0 to 200%. Use of CHROMA LEVEL control can help determine if the receiver will adequately lock on a color signal.

1. Select the gated rainbow pattern, and set the CHROMA LEVEL control to its midpoint position. This represents normal or average color subcarrier amplitude.

2. Adjust the receiver color controls to produce a recognizable color pattern.

3. Slowly rotate the CHROMA LEVEL control counterclockwise until the colors become pale and finally disappear. The rate of fading will depend entirely on the receiver under test. (Disable any automatic color control circuits on the receiver.) Most receivers will maintain color sync throughout the entire range of the CHROMA LEVEL control. However, some receivers may lose sync just before the color disappears. This is evidenced by diagonal running of the colors. Both of these conditions indicate normal operation of the sync circuits. However, if a slight reduction of the chroma amplitude from normal causes the color to fall out of sync (diagonal running of colors), receiver color sync is poor. In the full clockwise position of the CHROMA LEVEL control, the amplitude of the subcarrier is 200% of sync amplitude. This additional range is helpful in diagnosing receiver conditions, such as RF/IF misalignment or chroma circuit malfunction.

Color fit. The color bars produced by the generator are raised on a luminance pedestal so that spaces between the colors are black (for reference). When displaying either the 10-bar or 3-bar patterns, the colors should only be seen in the luminance area. If the colors overlap or spill into the black region, there is incorrect delay in the video amplifier or incorrect alignment of the band-pass amplifier.

Color killer. The color killer threshold can be set while displaying the crosshatch pattern. Simply adjust the receiver color killer control until the vertical lines start to tear with color. Back off the color killer control until the tear is removed and then give the control a slight turn more to provide a safety margin.

Deflection system tests. A quick check of receiver scanning can help disclose any abnormal or borderline situations which might exist in the electrical or mechanical parts of the deflection system. When evaluating any results from the following tests, always use the service literature as a criterion.

Overscan

1. Select a crosshatch pattern.
2. Adjust the receiver contrast and brightness to display sharp, thin lines against a black background.
3. Count the number of vertical and horizontal lines. A typical receiver will display nine vertical and nine horizontal lines. A tenth line is outside the normal picture area and should be ignored when performing adjustments.
4. Certain receivers have an inherent tendency toward a greater amount of overscan and/or blanking. This may result in an 8 by 8 crosshatch instead of a 9 by 9 one. The same effect will produce an 8 or 9 color-bar pattern instead of 10.

Linearity, size, and centering. The repetitive spacing of the crosshatch pattern provides a stable source with which to perform these tests and adjustments. Abnormal conditions such as pincushion distortion, deflection nonlinearity, and excessive 60-Hz hum become immediately obvious.

Vertical size and linearity should be adjusted so that all horizontal lines are evenly spaced. The inability to do so usually indicates a vertical deflecttion problem.

Pincushion distortion is common to a great number of large screen receivers. The outermost vertical and horizontal crosshatch lines are most useful in determining the correct amount of compensation.

A horizontal bar rolling vertically through the crosshatch pattern indicates that 60-Hz hum is entering the receiver circuits. Excessive amounts of hum cause a very noticeable and annoying pattern displacement.

The crosshair pattern provides one vertical and one horizontal line that intersect at screen center. Any visual deviation from screen center may indicate a need for adjustment.

4-3. COLOR TELEVISION ALIGNMENT PROCEDURES

The alignment procedures described in Chapter 3 (Sec. 3-2) are adequate for most black and white receivers. The same procedures can be used for color receivers. However, color receivers have a greater need for proper alignment. (That is, you might be able to get by with a poorly aligned black and white receiver, but poor alignment in a color receiver is totally unacceptable.) Also, color receivers have additional circuits that require alignment.

For these reasons, in this section we shall describe complete color receiver alignment procedures using a B&K 415 sweep/marker generator (discussed in Sec. 2-2.4 and shown in Fig. 2-8). Keep in mind that the same procedures can be used for black and white receivers, if desired.

Before going into detailed alignment procedures, we shall review the composition of the transmitted color signal, the color receiver circuits that require alignment, typical color receiver tuned circuits, and the importance of sweep alignment in color circuits.

4-3.1   The Transmitted Color Television Signal

Figure 4-23 shows the simplified television frequency spectrum of channel 10. The picture carrier is 1.25 MHz above the lower limit of channel 10. Within the channel, frequencies are referenced to the picture carrier frequency.

The I and Q signals (color signals) are centered on the chroma center frequency of 196.83 MHz. The spectrum shows that the Q signal side bands are symmetrical about the chroma center frequency with a distribution of ∕0.5 MHz. The I signal side bands are not symmetrical about the center frequency. Older color receivers have a band pass which will pass the complete bandwidth of the I and Q signals. Modern color receivers use a narrow-band chroma response (∕0.5 MHz from the color subcarrier). This means that portions of the I signal are not used.

Figure 4-23 (b) shows how the relative response changes when the signal is converted to IF by the RF tuner mixer and then passes through the IF stages. The IF frequencies are indicated with the corresponding channel 10 RF frequencies. As discussed in Chapter 3, there is a frequency inversion in the mixer process. For example, the picture carrier is at the low-frequency end in the transmitted signal (input to RF tuner) and at the high-frequency end in the IF stages.

Interference from adjacent channels. As discussed in previous chapters, receiver tuned circuits contain traps to reject unwanted signals from adjacent channels. For example, in Fig. 4-23(a), notice that the sound

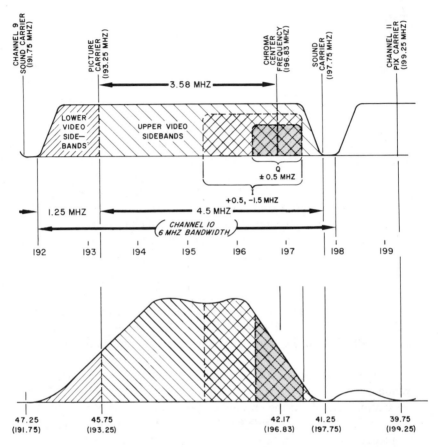

Fig. 4-23. Simplified television signal (channel 10) spectrum *(Courtesy B&K-Precision, Dynascan Corporation)*

carrier of channel 9 (191.75 MHz) is just outside the lower end of the channel 10 band and that the channel 11 picture carrier (199.25 MHz) is 1.25 MHz above the upper band end of channel 10. These two frequencies are called the *adjacent channel sound carrier* and the *adjacent channel picture carrier,* respectively, of channel 10.

Figure 4-24 shows the frequency spectrums of channels 9, 10, and 11. These frequency relationships apply to all channels adjacent to each other. Figure 4-24 shows that the frequency spectrum of interest not only includes the particular channel bandwidth (6 MHz) but also the portions on each side of the band pass (which include adjacent carriers). This is why a sweep width of 10 MHz or more is desired for sweep alignment, even though the channel bandwidth is only 6 MHz. This is also why traps must be included in the tune circuits to remove the undesired signals. (If not rejected, the

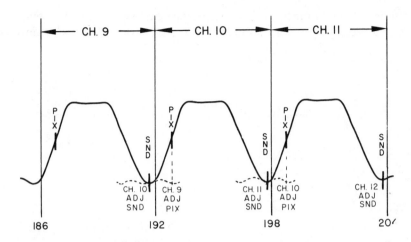

Fig. 4-24. Frequency spectrums of channels 9, 10, and 11 showing relationships of adjacent channel interfering carrier carriers *(Courtesy B&K-Precision, Dynascan Corporation)*

picture carrier of channel 11 could interfere with the sound of channel 10; as could the sound of channel 9 interfere with the picture of channel 10.)

4-3.2   Color Receiver Tuned Circuits

To process the signals properly, the RF, IF, and chroma sections of the receiver must have certain bandwidth characteristics. These are determined by the number of amplifiers and associated tuned circuits. The typical overall IF band-pass curve is shown in Fig. 4-25, with a typical tuner

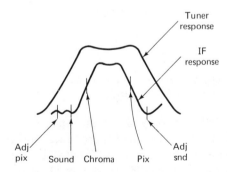

Fig. 4-25. Typical tuner and IF response curves showing relative bandwidths

response curve superimposed. The IF band pass is obviously narrower than the tuner band pass and thus contributes most to band-pass shaping.

Stagger-tuned circuits. To get the bandwidth required in IF amplifiers, it is necessary to use stagger-tuned circuits. This means that the outputs of a series of stages tuned to different frequencies are combined to obtain a desired overall curve. This is shown in Fig. 4-26. Assume that each individual

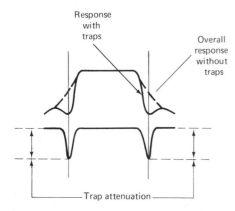

Fig. 4-26. Overall response of stagger-tuned circuits

response curve (dotted lines) represents the output of a single stage in a three-stage amplifier. The overall response would be as indicated by the solid outline. This is the response curve obtained when a sweep generator is used.

Trap frequency circuits. In addition to stagger tuning, traps are added to narrow the "skirts" of the response curve. The effect of traps is shown in Fig. 4-27.

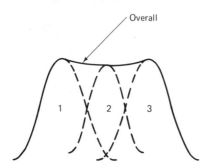

Fig. 4-27. Effects of trap circuits on overall response curves

Overcoupled circuits. Another method of obtaining a flat-top response with tuned circuits is to overcouple. If two tuned circuits are tuned to the same frequency, the overall response of the coils is determined by the amount of the coupling between them. Figure 4-28 shows typical curves obtained by undercoupled, critically coupled, and overcoupled coils. In the overcoupled case, the center dip in response will increase as the coupling is increased to spread the peaks. The maximum amplitudes will also begin to decrease as overcoupling increases.

The overcoupled principle is used in most chroma band-pass amplifier

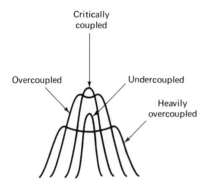

Fig. 4-28. Effects of coupling on overall response of two tuned circuits

transformers. Usually, two tuning slugs are found in these transformers. One slug adjusts the coupling windings, and the other is a tuning adjustment. Sweep alignment of this type of transformer is practically a must.

A compromise between gain and bandwidth is always made in chroma and IF alignment. The amplitude of the response curve can be greatly increased by tuning all adjustments for maximum amplitude, but this will be at a sacrifice in bandwidth, which is equally important in proper circuit operation.

Physical location of tuning slugs. The service literature alignment procedures often specify the locations of the tuning slugs with respect to the mounting board or chassis. The coils are designed so that the proper inductance value is obtained when the tuning slug is *between* the minimum and maximum inductance range of the coil, rather than at minimum or maximum. This means that circuit resonance is obtained at two physical locations of the coil [Fig. 4-29(a)] because at position 1 the slug is surrounded by as many turns of the coil as at position 2.

Now assume that a secondary winding is added at the bottom of the coil form, as shown in Fig. 4-29(b). Resonance is still obtained at either position 1 or 2 of the tuning slug. However, at position 2, the presence of the slug in a portion of the primary and secondary of the assembly increases the coupling between the two windings, as compared to the coupling with the slug at position 1. The amount of coupling affects gain and bandwidth. Thus, to obtain the desired coil or transformer characteristics, the slug must be properly located at resonance.

The effect of slug locations is also important in double-tuned coil assemblies for similar reasons. Figure 4-30 shows a double-tuned transformer, with the equivalent electrical circuit in Fig. 4-30(b). Either tuned circuit of Fig. 4-30(a) can be resonated with the slugs at position 1 or 2. The coil spacing is designed to give the desired response when the tuned circuits

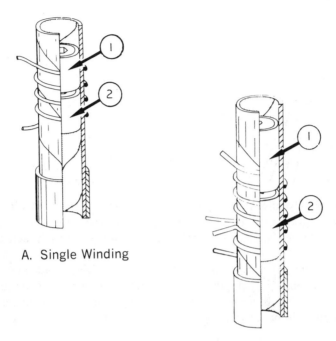

A. Single Winding

Fig. 4-29. Core positions at resonance in tunable coil assemblies *(Courtesy-B&K-Precision, Dynascan Corporation)*

are resonated with the slugs in *only one position,* usually position 1. If the circuits are tuned with either slug in position 2, the coupling will be greater (possibly overcoupled). If the circuits are tuned with both slugs in position 2, greater overcoupling will occur. The range of response curves shown in Fig. 4-28 can be obtained from the double-tuned coil assembly of Fig. 4-30(a) by varying the slug positions.

A special application of *controlled overcoupling* in transformers is the chroma band-pass transformer of Fig. 4-31. In this application, a double-tuned coil assembly is used. Primary tuning and the amount of coupling to the secondary are controlled by the location of a tuning slug with couples the L_1 and L_2 windings. Tuning of the secondary is performed by a slug which varies the inductance of the L_3 portion of the secondary.

4-3.3   The Need for Sweep Alignment

The discussion thus far shows that the picture, sound, and sync information of a television signal are fed through a rather long series of tuned circuits which affect the amplitude on one with respect to the other. If the circuits drift or are misaligned or if the gain of one or more stages changes, the signal properties are affected in several ways. Signal levels may be too

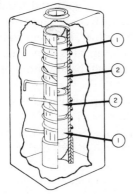

A. Double Tuned Transformer

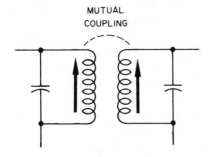

Fig. 4-30. Effects of slug locations in double-tuned circuits (*Courtesy B & K-Precision, Dynascan Corporation)*

low, the bandwidth may become too narrow, the signals may begin to interfere with each other, or if traps are misaligned, the receiver performance may be degraded by interference from undesired signals, such as adjacent channel sound or picture carrier frequencies.

Most of these problems can be overcome by proper alignment. The quickest way to determine the overall condition of the receiver circuits (that require alignment) is to use sweep frequency alignment (with markers), as described in Chapter 3. As the sweep signal is processed through the tuned

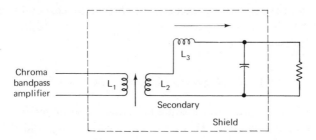

Fig. 4-31. Chroma bandpass amplifier with controlled overcoupling

portions of the receiver, the signal is shaped by the gain and band-pass properties of the various sections (RF, IF, chroma, etc.). Because the signal is channeled from one series of tuned circuits to another, it is important that each section has the proper characteristics. If the signal is demodulated at certain points and the waveform observed, the gain and bandwidth properties *up to that point* can be determined.

Figure 4-32 shows the sweep signal with basic response curves (waveforms) of the RF tuner, IF amplifiers, and chroma band-pass circuits below it. The bandwidths shown are approximately to scale. These outlines are similar to the curves that you would obtain if you monitored corresponding points in the receiver during sweep alignment.

Figure 4-32 includes some reference frequencies to show the importance of proper alignment. Notice that the chroma frequencies are on the *slope* of the IF response curve. This area is the most critical because improper IF alignment in this area will affect the amplitude and shape of the chroma response curve. In turn, this affects color picture quality.

As shown in Fig. 4-23(a), the chroma information is transmitted on a constant-amplitude portion of the broadcast signal. However, as shown in Figs. 4-23(b) and 4-32, the relative amplitudes of the chroma information are modified by passing through the receiver tuned circuits (RF and IF). Notice that the signal information at the upper end of the chroma frequency range (4.08 MHz) is reduced in amplitude with respect to the signal level at the lower end of the chroma frequency range (3.08 MHz).

To compensate for this frequency versus amplitude characteristic of the overall IF response curve, a chroma takeoff coil is used between the IF output and the chroma band-pass amplifier. The chroma takeoff coil is tuned to the upper end of the chroma frequency range, usually about 4.08 MHz, and provides a response as shown in Fig. 4-32, to compensate for the IF slope.

The result of combining the response of the IF curve and the chroma takeoff coil is to produce a flat overall response in the chroma frequency range (3.08 to 4.08 MHz). The resultant signal is then applied to the band-pass amplifier, which is the flat response indicated in Fig. 4-32. Some service literature specifies alignment of the chroma takeoff coil as a separate step. In other literature, the coil is adjusted along with adjustment of the band-pass transformer.

4-3.4   Recommended Sweep Alignment Methods

The service literature of most manufacturers recommends sweep alignment for all tuned circuits. However, the exact method and the order of steps to be performed are not standard. The following notes summarize typical recommendations.

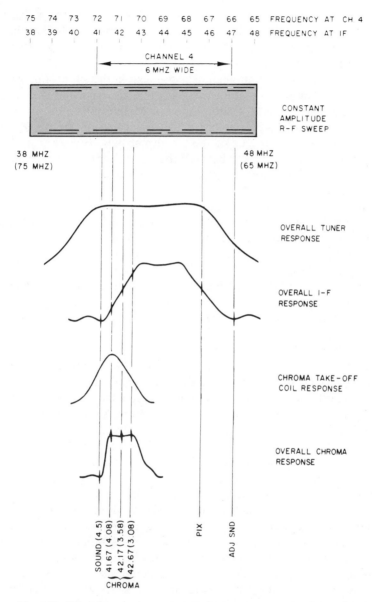

Fig. 4-32. Television response curves obtained using sweep frequency techniques *(Courtesy B&K-Precision, Dynascan Corporation)*

Generally, alignment starts by injecting the sweep signal at the RF input (antenna terminals). You then monitor the IF and chroma outputs for comparison against service literature waveform patterns. If the IF is good but the chroma is not, the problem is between the video detector output and the band-pass amplifier output (input to the color demodulators). If both IF and chroma outputs are abnormal, it is most likely that the IF requires touch-up, particularly if the response is poor on the slope affecting chroma response.

You will seldom find an alignment problem in the RF portion of the tuner (unless there has been tinkering), because the passband is so much greater than that of the IF section (about 6 MHz compared to 4 MHz). However, the mixer output circuit, which is located on the tuner, may require attention. This is part of the tuned matching network (called the "tuner link" or simply "link") between the tuner and the first IF stage. A separate prealignment procedure is given for the link circuits by some manufacturers.

Once you have found the deficient portion of the receiver, you can check alignment of that section. Several methods are recommended. One method is to first connect an RF sweep generator for IF alignment. Then modulate the picture carrier frequency for the channel being used by a video sweep signal. (This is generally called the VSM or video sweep modulation method.) The video sweep modulation is demodulated at the video detector and applied to the chroma band-pass circuits for alignment of these stages.

Other manufacturers recommend an IF sweep frequency injected at the mixer base (or grid) as for IF alignment. The IF picture carrier frequency (typically 45.75 MHz) is then modulated with a video sweep signal (VSM again). As before, the sweep signal is detected at the video detector, and the recovered sweep voltage is used for the chroma circuit alignment.

Still another method is to first video sweep align the chroma circuits directly. (This is generally called the CSS or chroma sweep signal method.) Then the IF is aligned, and VSM is used to modulate the IF picture frequency (45.75 MHz). This signal is used to check the combined effect on the chroma response of IF alignment and chroma alignment. Usually a touch-up of the chroma circuits is done to obtain the desired final overall chroma response. (VSM and CSS are discussed further in Sec. 4-7.)

When performing IF alignment, practically all manufacturers recommend pretuning IF traps by injecting spot frequencies into the IF (usually at specified test points in the tuner). Other procedures outline a prealignment of *all tuned circuits* in the IF before the sweep alignment procedure.

In all cases, always follow the manufacturer's service literature with regard to alignment. If complete alignment of a receiver does not restore normal operation, you must consider that part failure has occurred. In that case, you must proceed with the troubleshooting techniques described in the final sections of this chapter.

4-3.5 Diagnosing the Need for Alignment

You must realize that alignment alone is not the universal cure for poor picture quality. Before attempting to diagnose the need for alignment, you must be sure that the convergence, purity, and focus (and possibly high-voltage regulation) are properly set (Secs. 4-1 and 4-2) and that the receiver has been properly degaussed. You should also eliminate the possibility of interference from other test equipment (caps left off RF signal generator outputs, etc.).

With these problems out of the way, the first step is to connect the test equipment to the receiver, as specified in the test equipment instruction manual. Figure 4-33 shows the basic interconnections between test equipment and the receiver. These interconnections apply to all alignment procedures. Keep in mind that the specific setup shown in Fig. 4-33 applies only to the B&K 415 sweep/marker generator. However, the setup is typical for comparable sweep/marker generators.

1. Read the service literature carefully, paying particular attention to the alignment instructions. Note the location of all signal injection, bias, and signal output points. Note the requirements, if any, for special load or detector circuits. (The B&K 415 is equipped with a direct or basic probe, a demodulator probe, and several load resistances.)

2. Check the bias requirements for overall alignment (input at antenna). Preset the bias supplies to the approximate voltage requirements using an external meter. Connect the bias voltages at the indicated points. An RF AGC and IF AGC bias are usually required. If specified for overall alignment, apply a bias to the color killer. (This bias is usually specified in the chroma alignment section of the service instructions.)

3. Connect the direct cable to the specified video detector test point.

4. Connect the demodulator probe to the chroma band-pass test point. This is usually at the output of the band-pass amplifier. Set the receiver color level and brightness controls to the specified level.

5. Set the IF-RF-VIDEO ATTENUATOR control to minimum.

6. With the impedance selector switch of the RF cable termination set in the 300-Ω position, connect the RF cable to the antenna terminals of the receiver.

7. Set SWEEP WIDTH at maximum. Turn on ADJ PIX (39.75) and ADJ SND (47.25) markers, and use MARKER AMPLITUDE control to set marker amplitude as desired. Using the CENTER FREQUENCY adjustment, center the sweep on the oscilloscope screen as in Fig. 4-34.

8. Set the receiver channel selector to channel 4 and set the FUNCTION switch to CH 4. (Channel 10 may be used if channel 4 is not convenient.)

9. Place PROBES switch in the DIRECT position, and make sure the CHROMA switch is OFF.

10. Slowly increase the generator output by turning the ATTENUATOR control in a clockwise direction until a response pattern of the specified peak-to-peak amplitude is obtained.

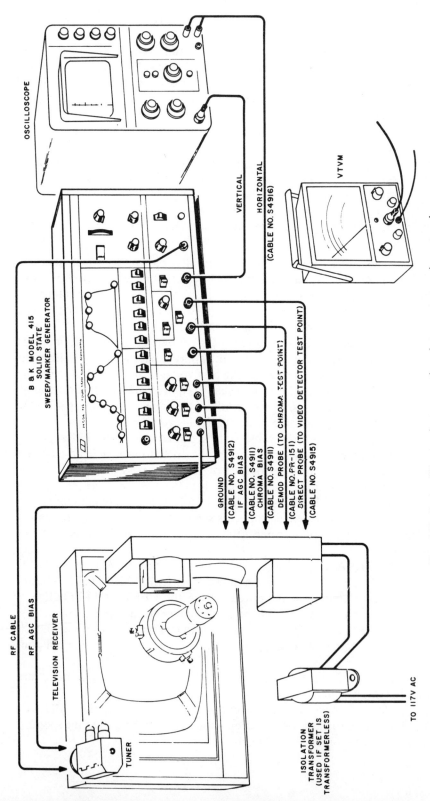

Fig. 4-33. Basic alignment interconnections between test equipment and receiver (*Courtesy B&K-Precision-Dynascan Corporation*)

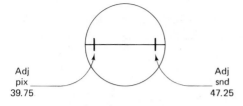

Fig. 4-34. Centering sweep range by use of adjacent picture and sound markers

11. Turn on the SOUND (41.25), CHROMA (42.17), and PIX (45.75) markers. Use the fine tuning as required to center the sound marker in the "sound notch" of the response curve. Refer to Fig. 4-35 for incorrect and correct adjustment of receiver fine tuning.

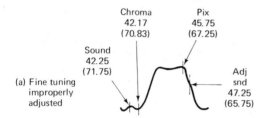

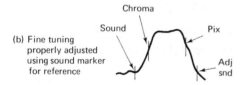

Fig. 4-35. Setting fine tuning using sound notch of response curve (channel 4 marker frequencies shown)

You can get greater resolution by increasing the generator output and oscilloscope sensitivity so that an enlarged portion of the sound trap region is visible, as shown in Fig. 4-36.

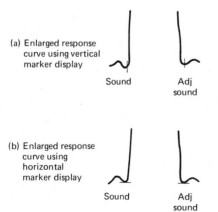

Fig. 4-36. Increasing sound notch resolution to adjust receiver fine tuning control

12. Turn on the remaining CHROMA markers (41.67 MHz and 42.67 MHz), the ADJ PIC (39.75-MHz) marker, and the 44.00- and 45.00-MHz reference markers. The markers should appear approximately as shown in Fig. 4-37. Compare the curve obtained at this point against the service literature.

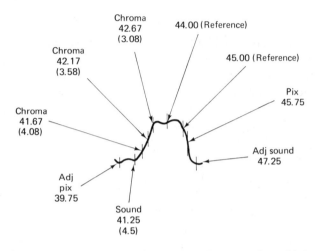

Fig. 4-37. Typical IF response curve with reference markers added

13. Slowly reduce the sweep width while using the CENTER FREQUENCY control to keep the sound and chroma markers as shown in Fig. 4-38.

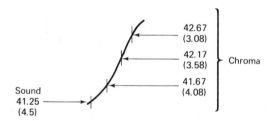

Fig. 4-38. Sweep width reduced to show only sound and chroma portion of IF response

14. Turn the CHROMA switch to ON. Notice that the curve is the same as in Fig. 4-38, except that the direction of sweep is reversed, as shown in Fig. 4-39. (Note that not all generators have this sweep reversal feature.)

The service instructions may specify a signal level adjustment at this point, prior to checking the overall chroma response. Usually, the dc level at the video detector test point is specified when performing chroma alignment. This is done by connecting the meter to the video detector (generally through an isolating resistor) and adjusting the attenuator of the generator to obtain the required dc level at this test point.

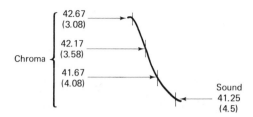

Fig. 4-39. Reversal of sweep

15. Switch the PROBES switch to the DEMOD position. The chroma band-pass response curve should now be observed on the screen with the markers, as shown in Fig. 4-40. Depending on the output polarity of the video detector diode, the polarity of the chroma response obtained may be inverted with respect to the IF response curve. If you want to observe the curve with opposite polarity, place the VERTICAL switch in its alternative position. (Note that not all generators have this polarity reversal feature.)

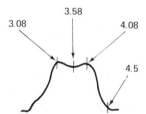

Fig. 4-40. Chroma bandpass response curve

16. If the dc level of the video detector output is not specified, adjust the generator output so that the peak-to-peak value of the chroma response curve is as specified. If no peak-to-peak value is given, observe the curve as the generator control is varied. Note the center of the range over which the response curve amplitude changes without distorting. Set the generator output at the center of this range. Compare the curve obtained with the service literature.

4-3.6 Analyzing the Response Curves

Using the procedures of Sec. 4-3.5, you must now determine if the curves obtained are satisfactory, or if the receiver must be realigned. If alignment is required, to what extent? Is a touch-up required or a complete realignment?

Figure 4-41 shows typical IF and chroma band-pass response curves as they might appear in service literature. Notice that reference marker locations are shown with given tolerances. This means that the response curves obtained may vary within these limits and still give satisfactory performance.

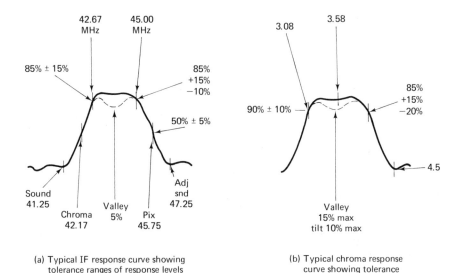

(a) Typical IF response curve showing
tolerance ranges of response levels

(b) Typical chroma response
curve showing tolerance
ranges of response limits

Fig. 4-41. Typical manufacturer's response curves with allowable variations indicated

Figure 4-42 shows some allowable variations based on the limits of Fig. 4-41. You must evaluate the response curves obtained with the allowable tolerances in mind. If the curves fall within the limits indicated, no alignment is required. If you decide to align the receiver, the extent of alignment must be determined. If the response curves are marginal at several or all points but are still recognizable, a touch-up alignment can be performed to correct excessive "tilts" and to restore response levels at various points on the response curves.

The areas to examine (aside from tilt across the curve top) are the areas of the trap frequencies, such as sound and adjacent sound, and the picture

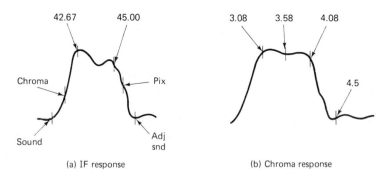

(a) IF response

(b) Chroma response

Fig. 4-42. Response curves within manufacturer's limits

and chroma markers at the 50% reference points. If a trap has been detuned toward the center of the response curve, it will pull the overall response downward. For example, if the sound trap is tuned near the chroma frequency, the curve response at the chroma frequency will be as shown in Fig. 4-43.

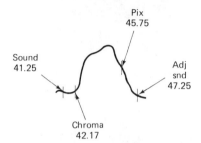

Fig. 4-43. IF response with sound trap tuned to chroma frequency

As discussed in Sec. 4-3.2, for stagger-tuned circuits, the curve will be tilted if the coil (tuned to the approximate center of the response curve) is misaligned. This is shown in Figs. 4-44(b) and 4-44(c). Also, if one of the outside tuned circuits is tuned toward the center of the response curve, the curve will peak toward the center with reduced bandwidth and excessive gain, as shown in Fig. 4-44(d).

By realizing the effect of mistuning traps and other tuned circuits, you can usually localize the mistuned circuit if the approximate alignment frequency of each circuit is given in the service literature. One way of getting this information is to check the alignment procedures thoroughly for *prealignment* instructions. Then cross-reference between marker frequencies and the tuned circuits to be adjusted. If a prealignment of traps and transformers is specified, most or all of the information related to tuned circuit frequencies is included in that section of the service instructions.

4-3.7   Alignment Touch-up Procedures

The following procedures are typical for touch-up of alignment. These touch-up procedures are used when your analysis of the curves indicates that the curves are recognizable but marginal on response limits (such as excessive tilt, abnormal peaking, etc.)

Preliminary touch-up procedure. Perform the following before making any touch-up adjustments.

1. Check if the service literature specifies alignment of the IF section by injecting the sweep signal at the antenna (RF sweep) or at the mixer (IF sweep). The advantage of alignment by injecting at the mixer is that you bypass the tuner fine tuning control. With mixer injection, you

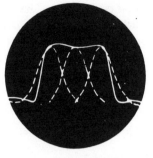

A. Coils Properly Tuned   B. Center Frequency (Tilt) Coil
                                            Tuned to Low End of I-F Response

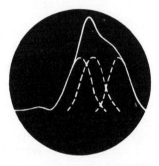

C. Center Frequency Coil Tuned to   D. Low Side Coil Tuned to Center
    High End of I-F Response   of I-F Response Curve

Fig 4-44. Typical response curves showing effects of misalignment of tuned circuits *(Courtesy B&K-Precision, Dynascan Corporation)*

usually bias the RF portion of the tuner to cut off, preventing interference from broadcast signal.

2. Check the service literature for injection points of frequency markers (for prealignment of traps and other tuned circuits). Usually, this is at the mixer test point and is also the sweep frequency injection point when alignment is with an IF sweep.

3. When the mixer test point is specified for marker frequency injection for *prealignment purposes only,* do not use the mixer test point for IF sweep injection. Check the service literature for other injection points (usually RF sweep at the antenna).

4. In checking the location and adjustment of various tuned circuits, note that the location of the tuning slug with respect to the chassis or circuit board is specified. As discussed in Sec. 4-3.2, always observe the instructions in this regard. Also, particularly on older receivers, check for possible broken coil forms or excessively loose tuning slugs. This is indicated by sudden changes in the response curves when the tuning tool is inserted or withdrawn from the tuning slug. This defect must be corrected before completing alignment.

Trap adjustments (touch-up). If trap adjustment is indicated or you suspect trap misalignment, proceed as follows.

   1. Locate the marker injection point. This is usually a test point in the mixer area of the tuner.
   2. Connect the direct probe of the generator to the video detector test point.
   3. Set PROBES switch in DIRECT position.
   4. Apply IF and RF AGC bias voltages as specified for sweep alignment. This will permit you to switch between the MOD MKR and IF positions of the FUNCTION switch as required to see the effect on the IF response of adjusting the traps.
   5. Place the FUNCTION switch in the MOD MKR position. (This modulates the selected marker with a 400-Hz signal.)
   6. Connect the generator cable to the signal injection point of the mixer. Use the 75-Ω termination. Note that the generator has both 300- and 75-Ω output terminations on a single cable.

NOTE: *In selecting marker signals for* **trap alignment,** *use* **only one** *marker at a time. Also, always use the minimum signal level required to give satisfactory tuning indication.*

   7. Select the SOUND (41.25-MHz) marker. Use maximum oscilloscope gain. Adjust the ATTENUATOR, and observe the oscilloscope for a sine-wave pattern (400 Hz). Set the generator output in the middle of the range over which the detected 400-Hz signal varies with the setting of the ATTENUATOR. A typical waveform is shown in Fig. 4-45. The 400 Hz is not synchronized with the sweep and therefore may not be stationary on the pattern.

Fig. 4-45. 400 Hz video detector output when using modulated markers

   8. Locate the sound (41.25-MHz) trap adjustments using the service literature. Adjust the trap for *minimum* indicated 400 Hz on the oscilloscope. The sound adjustment is usually in the last IF transformer and is one of two coils in the same can. Sometimes a potentiometer adjustment is specified in addition to the coil adjustment. When all sound trap (41.25-MHz) adjustments are set for minimum, turn on the SOUND marker.

> NOTE: *In the adjustment of traps in which a coil and potentiometer are involved, the potentiometer should never be set at either extreme of its adjustment range.*

9. Turn on the ADJ SND (47.25-MHz) marker. Use the level adjusting procedures described in step 7.

10. Locate the adjacent sound (47.25-MHz) trap adjustments using the service literature. Adjust the trap for *minimum* indicated 400 Hz on the oscilloscope. The adjacent sound is usually located at the input circuit of the first IF stage. Sometimes a potentiometer adjustment is also required, or a double-tuned trap is used. Alternatively, adjust for minimum indicated 400 Hz, using the minimum generator output required to give a usable scope indication. Turn off the ADJ SND (47.25-MHz) marker.

11. If an adjacent picture (39.75-MHz) trap adjustment is included in the alignment instructions, perform the adjustment as in the preceding steps. The adjacent picture trap is usually located at the first IF stage input circuit. Turn off the ADJ PIX (39.75-MHz) marker upon completion of the adjustment.

12. Set the FUNCTION switch to IF. Turn on the ADJ PIX (39.75-MHz) and ADJ SND (47.25-MHz) markers. Set the SWEEP WIDTH control to maximum, and center the trace on the oscilloscope using the CENTER FREQUENCY control. Adjust the ATTENUATOR until the specified receiver IF response curve amplitude is obtained.

IF response tilt adjustment, mixer injection (touch-up). With the traps properly adjusted, the remaining touch-up should be in correcting excessive tilt in the IF response. If the IF sweep injection is at the mixer, the setup connections described for trap adjustment can be used.

1. Using the service literature, note which IF coil is tuned to the center of the IF range (about 44 MHz). One of these is usually specified as a *tilt adjustment.* Often the last IF coil assembly combines the sound (41.25-MHz) trap adjustment and the tilt coil adjustment. See Fig. 4-44 for the effect of a center-tuned coil on overall response of stagger-tuned circuits.

2. While carefully watching the pattern on the oscilloscope, adjust the coil which affects the tilt so that the *top* of the IF curve is *horizontal.*

3. Turn on the CHROMA (42.17) marker and the PIX (45.75) marker. These markers should fall within the tolerance limits specified in the alignment procedures. If not, proceed to step 4.

4. You can get additional tilt correction by minor adjustment of the IF coils, which affect the height of the high-frequency and low-frequency ends of the IF response curve. These are usually specified in either the *spot frequency alignment* section or the *overall sweep alignment* section.

Chroma circuit adjustment (touch-up). The touch-up in the following steps is limited to the chroma response.

1. With the tilt removed (or at a minimum) in the IF stages, turn on the SOUND (41.25) and CHROMA (41.67, 42.17, and 42.67) markers. Decrease the sweep width, at the same time using the CENTER FREQUENCY control to center the four markers on the oscilloscope.

2. Place the CHROMA switch in the ON position, and recenter the oscilloscope trace if required.

> NOTE: *In performing adjustments in the chroma circuits, particularly the band-pass amplifier transformer, the* **signal level** *at the video detector is usually specified in the service literature. Alignment procedures usually specify that the level of signal injected at the mixer test point be adjusted so that a certain dc voltage is obtained at the video detector output test point. This level is set using a meter to monitor the dc level at the video detector while adjusting the signal level applied to the mixer input. With proper bias levels set in the IF and chroma circuits as specified, no further adjustment of signal levels should be required to produce the proper chroma level in the chroma portions of the receiver. Some procedures require a specified peak-to-peak waveform at the chroma circuit test point, this usually being the output of the band-pass amplifier. If neither a signal level nor a peak-to-peak chroma response is specified, do the following: Observe the chroma response curve and adjust the signal input from the generator so that the amplitude of the chroma response curve is centered in a range over which the amplitude can be varied without distortion or indications of overload.*

3. Set the PROBES switch to DEMOD, and observe the chroma response curve.

4. Perform the indicated chroma circuit adjustments to correct the curve tilt or to bring it within tolerances. The tilt adjustment is usually one of the two slugs in the chroma band-pass transformer.

Tilt adjustments, antenna injection (touch-up). If the specified signal injection point for IF alignment is not the mixer, the signal must be injected at the antenna terminals.

1. Set the FUNCTION switch to CH 4.
2. Set the CHROMA switch to OFF.

3. Place the impedance switch of the RF cable terminating pad in the 300-Ω position.

4. Apply proper RF AGC and IF AGC bias voltages as well as the required bias in the chroma section.

5. Set the PROBES switch to DIRECT. Turn on the ADJ PIX (39.75) and ADJ SND (47.25) markers. Center the sweep on the oscilloscope using the CENTER FREQUENCY control and using maximum sweep width.

6. Increase the generator output until a response curve of the proper amplitude is obtained.

7. Turn on the SOUND (41.25) marker and, using the receiver fine tuning, locate the response curve so that the sound marker falls in the sound notch of the response curve as shown in Fig. 4-35. Increase the oscilloscope gain, and generator output, for additional resolution as shown in Fig. 4-36. Restore the oscilloscope and generator levels.

8. Perform the IF tilt adjustments. That is, adjust the coil which affects the tilt so that the top of the IF curve is horizontal.

9. With the SOUND (41.25) and the three CHROMA (41.67, 42.17, and 42.67) markers on, reduce the sweep width and center the chroma portion of the response curve.

10. Set the CHROMA switch to ON.

11. Set the PROBES switch to DEMOD. Perform the required signal level adjustments as outlined for *chroma circuits adjustment (touch-up)*.

12. Adjust the generator output as required to obtain the specified peak-to-peak chroma response amplitude.

13. Perform the required tilt correction adjustments in the chroma circuits (bring the chroma band-pass tilt within tolerance).

4-3.8   Complete IF and Chroma Alignment

The following procedures are typical for complete alignment of a color receiver. Much of the alignment touch-up procedure information outlined in Sec. 4-3.7 is also applicable in the complete IF and chroma alignment procedures.

Trap alignment using modulated markers. Perform the trap alignment procedures described for touch-up. The standard trap frequencies are the adjacent picture carrier (39.75-MHz), the sound carrier (41.25 MHz), and the adjacent sound carrier (47.25 MHz). However, several manufacturers have an additional trap frequency. For example, some Motorola receivers have a trap frequency at 35.25 MHz, and some Philco receivers have a trap at 38.75 MHz. All traps excepting the sound trap (41.25 MHz) are located at the input to the first IF amplifier. The sound trap circuit is usually located in the last IF transformer or in the output circuit of the last IF stage.

Prealignment of IF band-pass (when specified). Some alignment procedures specify a pretuning of all IF band-pass coils and transformers as well as the traps, as previously outlined. The procedure is identical to the trap adjustments, except that in the case of prealignment, each circuit is tuned for a maximum output as indicated on the oscilloscope using modulated markers. That is, you inject the required marker signal using the MOD MKR (modulated marker) function of the generator and then tune the specified prealignment control for a maximum 400-Hz trace, as shown in Fig. 4-45. Once again, remember to use only one marker at a time. Always turn off the marker when alignment of the particular coil or transformer is complete.

Tuner link adjustment (when specified)

1. Turn off the receiver.

2. Connect any loads to the IF amplifier circuits specified in the service literature. For example, many receivers require that the output circuits of the first and second IF amplifiers be shunted to ground through a capacitor and resistor, as shown in Fig. 4-46(a). These networks swamp (or broaden) the selectivity of the first IF amplifier output. Figure 4-46(b) shows how the model 415 demodulator probe (PR-151) is used with two accessory load networks (called load blocks).

Because the output of the first IF amplifier has been swamped by the loads, the demodulated response across the load resistor is essentially the overall response from the mixer output to the input of the first IF or the *overall response of the tuner link.* Figure 4-47 shows a typical tuner link response curve with significant markers added. In most cases, the link adjustment on the tuner affects the high-frequency (45.75-MHz) side of the response curve, whereas the link adjustment on the IF input affects the low-frequency (42.17-MHz) side.

3. Connect the demodulator probe as shown in Fig. 4-46, and set the PROBES switch to DEMOD.

4. Set the FUNCTION switch to IF. Set SWEEP WIDTH to maximum.

5. Recheck instructions for bias requirements, and apply accordingly. Set the sweep width and markers as described in Sec. 4-3.5, step 7.

6. Turn on the receiver and allow several minutes for warm-up. Adjust the RF-IF-VIDEO ATTENUATOR for a specified signal level observed on the oscilloscope. Turn on the PIX (45.75) and CHROMA (42.17) markers.

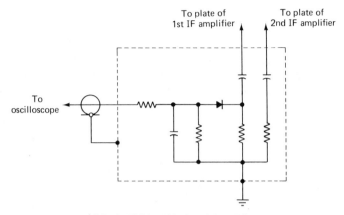

(a) Typical I-F load block and demodulator
assembly used for tuner link adjustment

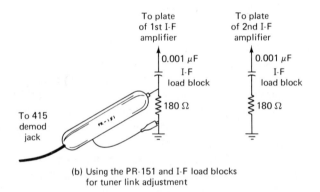

(b) Using the PR-151 and I-F load blocks
for tuner link adjustment

Fig. 4-46. Detector load blocks required for tuner link adjustments (*Courtesy B&K-Precision, Dynascan Corporation*)

    7. Alternately adjust the mixer output tuning adjustment and the IF input adjustment of the tuner link to get the required bandwidth for the tuner link response, as shown in Fig. 4-47.

    8. Upon completion of the tuner link adjustment, turn off the receiver and remove the load networks and demodulator probe.

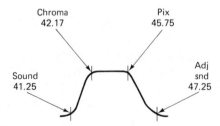

Fig. 4-47. Tuner link response curve with markers added

Sweep alignment of the IF stages

1. Repeat the procedures of Sec. 4-3.5, steps 1 through 7.

2. If signal injection for IF alignment is specified at the antenna terminals, set the channel selector to channel 4 and proceed. However, if the IF sweep injection is at the mixer test point, connect the RF cable to the mixer test point. Then set the channel selector to an unused channel (or as specified). Use the 75-Ω position of the RF cable.

3. Set the PROBES switch to DIRECT, and make sure the CHROMA switch is OFF.

4. Increase the generator output with the ATTENUATOR control until a response pattern of the specified peak-to-peak amplitude is obtained.

5. Turn on the SOUND (41.25), CHROMA (42.17), and PIX (45.75) markers. If antenna signal injection is being used, adjust the receiver fine tuning to center the sound marker in the sound notch, as shown in Fig. 4-35. (If all traps have been prealigned, you should set the sound notch regardless of the overall response of the IF stages.) You can get greater resolution by increasing the generator output and oscilloscope sensitivity so that an enlarged portion of the sound trap region is visible, as shown in Fig. 4-36.

If mixer signal injection is being used, the receiver fine tuning adjustment need not be considered. The IF response curves will be independent of the tuner. If necessary, return the generator output level and oscilloscope sensitivity as necessary to get the required display and peak-to-peak amplitudes.

6. Adjust the first, second, and third IF transformer circuits as outlined in the service literature. In some cases, the procedure will give three reference markers, usually 42.17, 45.75, and 44.00 MHz. (The 44.00 MHz is for sweep center reference). If no order of adjustment is given, keep in mind that (for most receivers) the output circuit of the first IF amplifier adjusts the response of the IF with respect to the 42.17-MHz marker. The output coil of the third IF is tuned to the center of the passband and is adjusted for so-called "tilt" correction. Be sure that the *proper slug* is adjusted in the third IF transformer, because one coil performs the IF tilt correction, and the other coil is the sound trap adjustment (41.25 MHz).

> NOTE: *In some service literature, the term or phrase "set the marker height" is used extensively. This phrase may be misleading since actual marker "height" is not adjustable. (Marker height is not to be confused with marker amplitude, which is adjustable on some generators.) The phrase "set marker height" means that you are adjusting the IF response so that at a particular marker frequency the response is at some percentage* **down from maximum.** *For example, to set the height of the 42.17-MHz marker at 50% amplitude,*

*you adjust the second IF transformer to **alter the curve** so that at 42.17 MHz the response is 50% of maximum.*

7. After completion of the IF alignment, recheck the position of the sound marker, especially if pronounced readjustment of the third IF transformer or coil has been required to restore proper alignment.

8. After you get the desired IF response so that the 42.17- and 45.75-MHz markers are located as specified, turn on the 41.67-and 42.67-MHz markers. These markers, together with the 42.17-MHz marker, define the response portion of the IF curve on which the *chroma information* is carried. The markers should be positioned as shown in the service literature (typically as shown in Figs. 4-38 or 4-39).

Chroma circuit band-pass amplifier alignment using direct video sweep. Upon completion of the IF alignment, recheck the chroma alignment section of the service instructions, and verify that the biases are properly set and that the demodulator probe is properly connected.

1. Set the FUNCTION switch to VIDEO. Set the ATTENUATOR to minimum. Connect the RF cable to the specified injection point (at the input of the band-pass amplifier).

2. If not already accomplished, turn on the 41.25/4.5-, 41.67/4.08-, 42.17/3.58-and 42.67/3.08-MHz markers.

3. Using the SWEEP WIDTH control and CENTER FREQUENCY control to reduce sweep width, center the pattern as required so that the sound and chroma markers occupy approximately 70% of the total oscilloscope sweep.

4. Set the PROBES switch to DEMOD.

5. Adjust the ATTENUATOR control for the specified peak-to-peak amplitude as observed on the oscilloscope.

6. Adjust the band-pass transformer as required to get the desired overall performance. Normally, the band-pass transformer consists of two tuning adjustments. One of the tuning slugs adjusts the band width of the transformer, and the other slug adjusts the center frequency. The bandwidth adjustment is used to position the 3.08- and 4.08-MHz markers on the curve, and the center frequency adjustment is used essentially to correct any tilt.

This direct video sweep alignment is done prior to performing final adjustment of the chroma tuned circuits using video sweep modulation or an equivalent method. Upon completion of the chroma prealignment, overall chroma circuit alignment is performed by injecting either at the mixer test point or at the antenna terminals, as described in the next paragraph.

7. Remove the RF cable from the input to the chroma band-pass amplifier.

Chroma alignment using signal injection at antenna input or at mixer input. Chroma alignment is usually performed by either of these methods, rather than by the direct video sweep injection just described. The obvious advantage is that once all probes and biases have been connected, you may proceed directly from the IF alignment into the chroma alignment without additional connections.

1. Connect the RF cable to the mixer test point or to the antenna terminals, as specified. Set the generator cable impedance switch to 75-Ω for mixer input and 300- for antenna input.

2. Adjust sweep width as necessary to display the complete IF response curve.

3. If not already accomplished, turn on the 41.25, 41.67, 42.17, and 42.67 markers. Set marker amplitude as required.

4. Using the SWEEP WIDTH control and the CENTER FREQUENCY control, reduce the sweep and recenter as necessary so that the sound and chroma portions of the IF response curve are centered on the oscilloscope screen, as shown in Fig. 4-38.

5. If required to match the service literature, set the CHROMA switch to ON. The direction of the sweep should now be reversed as shown in Fig. 4-39.

6. Adjust the signal level. Usually the dc level at the video detector must be set to some specific value. As an alternative, adjust the signal input to get the recommended peak-to-peak response amplitude at the chroma output test point.

7. Set the PROBES switch to DEMOD. The overall chroma response curve should now be present on the oscilloscope. Note the general response of the curve and the position of the markers. If the overall chroma response curve is seriously misaligned so that the general shape is not recognizable, it may be necessary to identify the individual markers by turning each marker on and off. The 3.08-MHz marker should be at the left of the oscilloscope trace (with the CHROMA switch ON) and the sound (4.5-MHz marker) at the right.

NOTE: *An internally generated 3.58-MHz marker may be visible on the response curve, in addition to the generator marker. This is caused by the 3.58-MHz oscillator in the receiver. In some cases, a low-frequency beat may be produced between this marker and the generator marker. The beat can be eliminated by turning off the generator marker.*

4-3.9   Alignment of Sound IF and Audio Detector

The sound IF and audio detector circuits of a color receiver are essentially the same as for a black and white receiver. Thus, the alignment procedures described in Sec. 3-2.5 can be used for color as well as black and white.

4-3.10   Adjustment of Automatic Fine Tuning (AFT) Circuits

Most modern receivers include some form of AFT circuit. The purpose of these circuits is to prevent the local oscillators of the VHF and UHF tuners from drifting in frequency.

Figure 4-48 shows a typical AFT circuit. The 45.75-MHz (picture carrier) output of the last IF amplifier is applied to the input of a tuned AFT amplifier. The AFT amplifier is tuned to exactly 45.75 MHz and delivers its output to the AFT discriminator. The output from the discriminator is applied to a voltage variable capacitor (VVC), which is connected across the resonant circuit of the local oscillator. In some tuners, the base-collector junction of a transistor is sometimes used as the VVC.

If the local oscillator output frequency remains fixed, the outputs of the AFT amplifier and discriminator also remain fixed, and the VVC does not vary in capacitance. Thus, the local oscillator circuits remain unchanged. However, if the local oscillator should drift in frequency, the AFT amplifier and discriminator outputs change, thus changing the VVC capacitance and the local oscillator resonant circuits. The AFT circuits are designed so that the VVC capacitance changes the resonant circuit frequency to offset the initial (undesired) drift. For example, if the local oscillator frequency increases (drifts upward), the resonant circuit frequency is decreased and vice versa. Thus, the local oscillator frequency remains constant (within limits).

Typical alignment procedure. The AFT alignment can be performed upon completion of the IF alignment procedure with no changes in setup connections, other than connecting the direct cable of the generator to the AFT test point (Fig. 4-48) and setting the PROBES switch to DIRECT.

1. The peak-to-peak spacing of the AFT discriminator S curve is usually about 1 MHz, so that at full IF sweep the discriminator curve will appear compressed, as shown in Fig. 4-49(a). Adjust the SWEEP WIDTH and CENTER FREQUENCY as required to increase the size of the discriminator S curve, as shown in Fig. 4-49(b). Only the PIX (45.75-MHz) marker should be used.

2. In the alignment of various AFT circuits, marker frequencies of 45.25 and 46.25 MHz are sometimes specified to locate the peaks of the

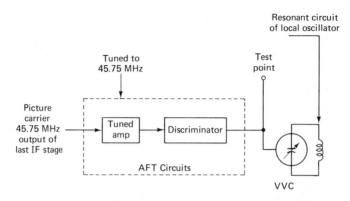

Fig. 4-48. Typical AFT circuit

AFT discriminator curve. Because these frequencies are not standard (used in other alignment procedures), it is not likely that crystal-controlled markers are readily available. Normally it is necessary to use an external variable frequency marker generator to locate the peaks of the AFT S curve.

When using the model 415 generator for AFT alignment, it is only necessary to switch on the 100-kHz markers. This generates a continuous string of markers having 100-kHz separations, and extending over 1 MHz in each direction from the picture carrier marker, as shown in Fig. 4-50. This permits you to set the discriminator adjustments as outlined in any procedure by counting (in increments of 100 kHz) in either direction from the picture carrier.

3. Figure 4-50 shows the method for determining the various frequency points of the AFT discriminator curve. The distance between

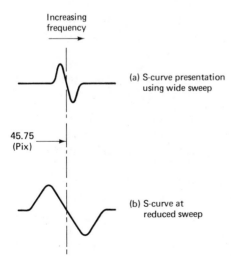

(a) S-curve presentation using wide sweep

(b) S-curve at reduced sweep

Fig. 4-49. AFT discriminator S-curve

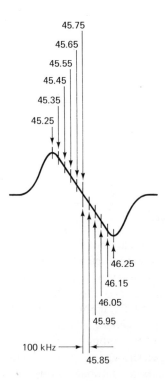

Fig. 4-50. Using 100 kHz markers to determine frequency separation of discriminator peaks

each marker is 100 kHz as indicated, and the method of counting in either direction from the picture carrier is shown. For example, with a discriminator bandwidth of 1 MHz (45.75 MHz ±0.5 MHz), the discriminator adjustment is performed so that the fifth 100-kHz marker on each side of the picture marker falls on each peak of the curve.

4. With the calibration provided by superimposing the 100-kHz markers on the AFT discriminator curve, perform the AFT circuit alignment as described in the service literature.

NOTE: *Most service literature recommends that you disconnect the AFT output to the tuner when performing AFT alignment.*

5. After the discriminator has been aligned, the exact discriminator crossover point is obtained by injecting a 45.75-MHz carrier at the AFT input (last IF output). The dc output voltage of the AFT circuit is then checked at the AFT test point using a meter. This is performed when using the model 415 by setting the FUNCTION switch to MKR and turning on the PIX (45.75) marker (all other markers off). Leave the RF cable connected to the IF sweep injection point. Adjust the AFT circuits as specified to obtain the required dc output voltage.

6. Upon completion of the AFT alignment, perform the AFT check as described in the service literature. Usually this involves connecting the receiver to an external antenna, disabling the AFT, and adjusting the fine tuning manually for best picture and sound with minimum sound interference. Then activate the AFT, and check the "pull-in" effect.

7. An alternative method of checking AFT operation is to adjust the fine tuning in either direction from the best picture setting, with the AFT disabled. Then activate the AFT, and check the effect. The picture should automatically be restored to the best setting obtained by manual adjustment of the receiver fine tuning when the AFT was disabled.

4-4. SIMPLIFIED COLOR TELEVISION ALIGNMENT

In this section, we shall describe simplified color television alignment using a generator such as the RCA WR-514A. Note that these procedures cover essentially the same areas as are covered in Sec. 4-3. However, the generator described here is a much simpler instrument, having fewer features. Also, the procedures described here assume that you have had some experience in color television alignment or that you have read the details and background information found in Sec. 4-3.

These instructions are not intended to replace the service literature. You will need the service literature to prepare the receiver for alignment, locate the appropriate test points, and identify the tunable parts. You may note differences between the service instructions and these procedures. In most cases, the differences occur because the procedures in the service literature may not apply directly to a generator such as the WR-514A. The following procedures may also help you resolve questions concerning discrepancies between the service literature and the receiver being serviced.

4-4.1 Receiver Preparation

1. Carefully examine the service literature for circuit diagrams, alignment instructions, test point locations, etc. Set the tuner on the UHF position, or a particular VHF channel, or in between two VHF channels, as specified for service.

2. Apply RF and IF bias voltages as specified. Keep in mind that bias voltages and connection points often change from one alignment procedure to another.

3. Disable the horizontal deflection by pulling the yoke plug, opening the horizontal output tube cathode jumper, or as recommended in the service literature. In series string vacuum-tube receivers, the output tube may be removed if a resistor of the same resistance as the tube heater is plugged into the socket heater terminals. Some receivers may require a substitute load connected across the B+ supply to prevent an excessive rise in voltage.

4. Disable the vertical deflection, if specified by the service literature. In some receivers, it will only be necessary to set the normal-service raster switch to service.

5. Disable AFT (automatic fine tuning), ACC (automatic color control), and color killer circuits, if required. In some receivers, these circuits are controlled by switches. In other receivers, it is necessary to apply a fixed bias to clamp the circuit out of operation. Set the receiver color control to about midrange. For chroma band-pass alignment, it is often recommended that the receiver 3.58-MHz oscillator be disabled.

6. Do not connect the receiver chassis to a water pipe or other common earth ground. Use an isolation transformer.

7. When connecting test leads to the receiver, be sure to connect their ground clips as closely as possible to the signal clips. Always connect the ground lead first and the signal lead last. Do not allow metal connnectors to come in contact with the chassis.

8. Allow the receiver to warm up for about 10 min, or as specified. As a safety precaution, turn off the receiver momentarily when changing test lead connections.

4-4.2   Test Equipment Precautions

1. One of the usual causes of misalignment is *overloading.* It is good practice to occasionally vary the level and gain controls throughout their range during alignment procedures to make sure that the response does not change when the signal level is varied.

2. Bias voltages applied to the receiver can change when the receiver is turned on. Thus, bias voltages should be checked after turn-on.

3. Some receivers may be "sensitive to touch." That is, the oscilloscope pattern may change shape when the chassis or generator (or connecting cables) is touched. Very often, a simple cure is to connect an insulated wire to the receiver chassis and drape the wire over the generator.

4. In some older receivers, the postinjection markers may not have sufficient amplitude when the oscilloscope vertical input attenuator is set to handle a large signal. Markers can be increased in size by setting the oscilloscope to a higher sensitivity position and then reducing the gain setting of the DEMOD IN LEVEL control of the generator to get a trace of the same amplitude.

4-4.3   Test Equipment Setup

1. On the oscilloscope, calibrate the vertical input signal so that the screen graph indicates the waveform voltages. Set the amplifier to LINE (or 60 Hz). Connect the vertical input to the SCOPE VERTICAL connector on the generator using a WG-427A cable. If the oscilloscope has a microphone-type connector, use the WG-428A connector adapter. Connect a lead from the oscilloscope ground terminal to the receiver chassis.

2. Set the generator CHANNEL selector to 0–50 VF/IF. Set VF/IF TUNING to 45 MHz. Set SWEEP WIDTH to midrange. Use the SCOPE DIRECT function whenever measuring waveform voltages and when setting traps using the audio modulation method. Use the DEMOD AMP position to get postinjection markers on the trace.

3. Sweep voltages in the oscilloscope and generator must be properly phased. This adjustment is usually done during the first step in alignment procedures when a sweep trace is displayed. Depress the BLANKING OFF button, and rotate the phase control on the oscilloscope until the two patterns overlap.

4. In all of the following procedures, the MARKER OUT or the VF/IF OUT will be connected to the mixer input on the tuner using a WG-427A direct cable and a mixer input adapter. Use a WG-433A adapter for solid-state tuners. For vacuum tube tuners, use a WG-429A VF/IF 75-Ω termination adapter with a series capacitor (1500 pF), as specified in some service literature.

5. The RCA accessory kits, specified in these instructions, will work satisfactorily with most receivers. Use WG-402A kits to make up any special probes called for the service literature.

NOTE: *The service literature for some receivers may specify additional markers between 42.67 and 45.75 MHz. However, if the curve has the correct shape and the 42.67- and 45.75-MHz markers are in the proper position, the alignment should be satisfactory.*

4-4.4   Link Alignment and Trap Tuning

As discussed in Sec. 4-3, the link used in many receivers consists of the mixer output coil in the tuner and the first IF input coil in the receiver. The two coils are overcoupled by means of the coaxial cable which links them together. In series with the cable in a dc blocking capacitor which, if variable, is used to adjust the bandwidth of the circuit. Associated with the first IF input coil are one or more adjacent carrier traps. One of the traps may have an associated "Q spoiler," which is a variable resistor (or possibly an extra coil slug) used to improve the effectiveness of the trap.

Link trap tuning. See Fig. 4-51.

1. Connect the generator MARKER OUT to the receiver mixer input. Connect DEMOD IN to the output of the first IF stage using a WG-427A cable and a WG-431-A general-purpose demodulator. Switch to

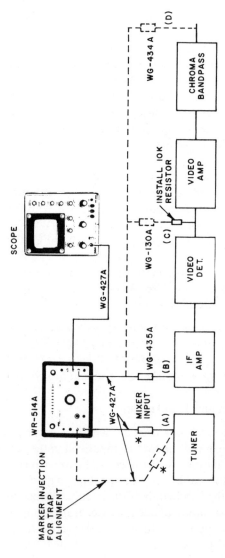

A. Mixer input test point.
B. Output of first IF Amp. Use for link trap adj. and link alignment.
C. Output of video detector. Use for IF trap adj. and IF Amp. alignment.
D. Output of chroma bandpass Amp. In many sets the center lug of the Color Control can be used. Use for chroma bandpass alignment.

*WG-433A for solid-state sets.
 WG-429A with 1500 pF series cap. for tube sets.

Fig. 4-51. Test equipment connections for alignment using WR-514A
(Courtesy RCA)

258

SCOPE DIRECT. Turn on AF MOD (audio modulation). Turn the CHANNEL selector switch to CH. 2 to keep the sweep pattern out of the oscilloscope trace.

2. Set the oscilloscope voltage range switch to 0.05 V, peak to peak.

3. Apply bias voltages as specified. Turn on one of the markers (41.25, 47.25, or 39.75), and bring up the MARKER OUT control to the point where audio modulation appears on the trace. (See Fig. 4-52.) Reduce the bias slightly if necessary to view the audio modulation. Tune the trap to null modulation (or minimum signal). Repeat this procedure for other link traps. If the link contains a sound rejection Q spoiler (variable resistor or coil slug), adjust it to improve the effectiveness of the associated trap. (Do not set the Q-spoiler adjustment to either extreme.)

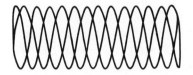

Fig. 4-52. Link trap tuning waveform (pattern is not synchronized)

Link alignment. See Fig. 4-51.

1. Set the oscilloscope voltage range switch to 0.15 V for solid-state receivers or to the 0.5-V range for vacuum-tube receivers.

2. Set the generator to SCOPE DIRECT. Connect VF/IF OUT to the mixer input. Connect DEMOD IN to the output of the first picture IF amplifier (tube plate or transistor collector) using a WG-427A and WG-435A link detector assembly with the red clip to the first picture IF and the yellow clip (loading) attached to the output of the second IF stage. Set the CHANNEL selector to 0–50 VF/IF. Adjust the VF/IF OUT control for a viewable trace, and adjust VF/IF TUNING to center the oscilloscope trace.

3. Apply recommended bias voltages. Adjust the VF/IF OUT attenuator for a trace amplitude as specified for the receiver (typically 0.1 V peak to peak for solid-state and 0.3 V for vacuum-tube). Switch to DEMOD AMP, and adjust DEMOD IN LEVEL for a trace of the same amplitude. Turn on the 42.17- and 45.75-MHz markers. Set the MARKER LEVEL control for markers of the desired amplitude.

4. Tune the mixer output coil and the first IF input coil. One coil will put the 45.75- and 42.17-MHz markers at equal height on the trace, and the other coil will eliminate tilt from the trace top. Adjust the link coupling capacitor (if variable) to put markers at the proper position on the trace. (See Fig. 4-53.)

4-4.5 Picture IF Alignment

There are several different procedures for alignment of the picture IF stages. Typically, the procedures consist of trap tuning and two- and three-

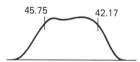

0.1 V Peak-to-peak form most solid-state receivers
0.3 V Peak-to-peak for most vacuum-tube receivers
Markers should be at 70 to 80 per cent.

Fig. 4-53. Link alignment waveform

stage amplifier alignment. The following procedures apply to the majority of receivers.

Picture IF trap tuning. See Fig. 4-51.

1. Connect a 10-kΩ, ½-W isolating resistor to the picture IF detector output (video detector output).

2. Set the oscilloscope voltage range switch to 0.05 V.

3. Connect DEMOD IN using a WG-427A direct cable and WG-430A direct unit to the 10-kΩ resistor. Turn on AF MOD and the 41.25-MHz marker. Adjust the MARKER OUT control for a viewable trace.

4. Apply the recommended bias voltages. Tune 41.25 MHz to null modulation (minimum signal). This pattern will be similar to Fig. 4-52. Adjust the adjacent sound rejection Q spoiler, if present, to improve the effectiveness of the trap. Use the same procedure for the 39.75-MHz trap if the receiver has one. (The WR-514A does not have a 39.75-MHz marker but one can be added easily by installing a 39.75-MHz crystal to the SPARE position.)

Three-stage picture IF alignment. See Fig. 4-51. Three IF stages are common in modern receivers. The output of the first stage uses a single coil which is tuned to 45.75 MHz. The output of the second stage uses a single coil which is tuned to 42.17 MHz. The output of the third stage uses either a single coil (which is tuned in between the two carriers) or a double-tuned transformer which is broadbanded.

1. Connect a 10-kΩ, ½-W resistor to the video detector output. Apply the recommended bias voltages.

2. Set the oscilloscope voltage range switch to 5 V.

3. Switch the generator to SCOPE DIRECT. Connect VF/IF OUT to the mixer input. Connect DEMOD IN to the 10-k Ω resistor using a WG-427A direct cable and a WG-430A. Set the CHANNEL selector to 0.50 VF/IF, and adjust TUNING to center the oscilloscope trace.

4. Adjust the VF/IF OUT control for a 3-V peak-to-peak trace on the oscilloscope. Reset the oscilloscope voltage range switch to 0.5 V. Switch the generator to DEMOD AMP and adjust the MOD IN LEVEL control for a viewable oscilloscope trace. Turn on the 42.17- and 45.75-

MHz markers, and adjust the MARKER LEVEL control for markers of the desired amplitude.

5. Tune the first IF output coil for maximum trace amplitude at 45.75 MHz, and the second IF output coil for maximum trace amplitude at 42.17 MHz. Tune the primary of the third IF output transformer to set the two markers at equal height on the trace. (See Fig. 4-54.) Then tune the secondary to remove tilt from the top of the trace. (If the third IF output uses a single-tuned coil, adjust it to remove any tilt from the trace.) Readjust the first and second IF output coils to put the 45.75- and 42.17-MHz markers at 50% on the trace. Finally, turn on the 41.67- and 42.67-MHz markers, and trim all coils to make the trace conform to Fig. 4-54.

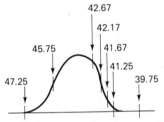

Pix marker (45.75) and color subcarrier marker (42.17) should be at 50 per cent. Location of trap marker is also shown.

Fig. 4-54. Link/IF overall waveform

Two-stage picture IF alignment. See Fig. 4-51.

1. Set the oscilloscope voltage range switch to 0.15 V for solid-state or to 0.5 V for vacuum-tube receivers.

2. Switch the generator to SCOPE DIRECT. Connect VF/IF OUT to the mixer input. Connect DEMOD IN to the output of the second picture IF channel amplifier using a WG-427A cable and a red clip of a WG-435A link detector (the yellow clip is unused). Set the CHANNEL selector to 0–50 VF/IF, and adjust the tuning to center the oscilloscope trace.

3. Apply recommended bias voltages. Adjust the VF/IF OUT control for an 0.1- (solid-state) or 0.3 (vacuum-tube) V trace. Switch the generator to DEMOD AMP, and adjust the DEMOD IN LEVEL control for a viewable oscilloscope trace. Turn on the 45.75- and 42.17-MHz markers, and adjust MARKER LEVEL for markers of desired amplitude.

4. Adjust slugs in the first IF output transformer for a trace of the desired shape. One of the slugs will set the markers at equal height above the base line, and the other slug will remove any tilt from the top of the trace. The markers should be 60 to 70% up on the curve slope, as shown in Fig. 4-55.

5. Connect a 10-kΩ, ½-W resistor to the video detector output. Apply any recommended bias voltages.

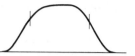

0.1 V Peak-to-peak for most solid-state receivers
0.3 V Peak-to-peak for most vacuum-tube receivers
Markers should be at 60 to 70 per cent.

Fig. 4-55. First IF alignment waveform

6. Switch the generator to SCOPE DIRECT. Connect VF/IF OUT to the mixer input. Connect DEMOD IN to the 10-kΩ resistor using a WG-430A and a WG-427A. Set the CHANNEL selector to 0–50 VF/IF, and adjust TUNING to center the oscilloscope trace.

7. Adjust the VF/IF OUT control for a 3-V trace on the oscilloscope. Then reset the oscilloscope voltage range switch to 0.5 V. Switch the generator to DEMOD AMP, and adjust the DEMOD IN LEVEL control for a viewable oscilloscope trace. Turn on the 42.17- and 45.75-MHz markers, and adjust the MARKER LEVEL control for markers of the desired amplitude.

8. Adjust slugs in the second IF output transformer for a trace of the desired shape. One of the slugs will set the markers at equal height above the base line, and the other slug will remove any tilt from the top of the trace. Finally, turn on the 41.67- and 42.67 MHz markers and fine-tune both transformers so that the curves conform to Fig. 4-54.

4-4.6   4.5-MHz Trap Tuning

The following procedure describes how the 4.5-MHz trap can be set by monitoring at the output of the chroma amplifier. The 4.5-MHz trap can also be adjusted independently of the chroma amplifier, as described in Sec. 4-6.

1. Set the oscilloscope voltage range switch to 0.05 V.

2. Switch the generator to SCOPE DIRECT. Connect MARKER OUT to the mixer input. Connect DEMOD IN using a WG-427A direct cable and a WG-434A detector probe to the output of the chroma amplifier. Turn on the 41.25- and 45.75-MHz markers and AF MOD. Set the CHANNEL selector to CH. 2. Turn up MARKER OUT control for a viewable trace on the oscilloscope.

3. Tune the 4.5-MHz trap to null the modulation (minimum signal).

4-4.7   Chroma Band-pass Amplifier Alignment

1. Connect a meter with a dc probe to the 10-kΩ resistor connected to the output of the video amplifier. Set the receiver color control to about one-third of full on. Turn off the color killer control (or apply a bias to disable the color killer). Apply all recommended bias voltages.

2. Set the oscilloscope voltage range switch to 0.5 V.

3. Set the generator CHANNEL selector to 0–50 VF/IF. Connect VF/IF OUT to the mixer input. Connect DEMOD IN to the output of the chroma band-pass amplifier using the WG-434A chroma/video detector. Switch to SCOPE DIRECT. Turn on the 41.67- and 42.67-MHz markers.

4. Turn on the 45.75-MHz CSS CARRIER, and turn up the LEVEL control until the meter reading increases about 1 V. Readjust VF/IF TUNING to center the trace. Adjust the generator VF/IF OUT control and receiver color control for an 0.25-V oscilloscope trace. Switch the generator to DEMOD AMP, and adjust the DEMOD IN LEVEL control for a viewable oscilloscope trace. Adjust the MARKER LEVEL control for markers of the desired amplitude. Adjust the chroma takeoff coil to position the 4.08-MHz marker properly, and adjust the band-pass transformer for symmetry of the trace.

5. If desired, the chroma band-pass amplifier can also be aligned using a direct video sweep, as described in Sec. 4-7.

4-5. CHECKING AND ALIGNING AFPC CIRCUITS

All color receivers have some form of AFPC (automatic frequency and phase control) circuits. The purpose of these circuits is to lock the 3.58-MHz reference oscillator with the color burst signal. The operation of AFPC circuits is described in the troubleshooting sections at the end of this chapter.

For check and alignment purposes, there are two basic types of AFPC circuits: the *closed-loop* and the *injection locked oscillator.* In this section we shall describe various procedures for checking and aligning both types of AFPC circuits using a color generator such as the RCA WR-515A. AFPC circuits can also be checked and aligned using a vectorscope, as described in Sec. 4-8.

4-5.1   Checking AFPC Without an Oscilloscope

A convenient check of the AFPC circuit, that does not require an oscilloscope, can be made as follows·

1. Connect the generator to the receiver antenna terminals. Press the COLOR and MARK push buttons (10-bar color pattern, as shown in Fig. 4-21). Adjust the generator CHROMA and RF LEVEL controls, along with the receiver fine tuning, brightness, contrast, and other controls to get a suitable color-bar pattern. Advance the brightness control until the spaces between the bars are dimly lighted.

2. Disable the blue and green guns of the picture tube using the gun killer switches.

3. If the AFPC circuit is properly adjusted and all of the color circuits are operating properly, the sixth bar (use the bar "mark" to identify this bar, Fig. 4-21) will have the *same brightness level as the spaces* between the bars, so that the bar will almost disappear. As you turn the tint control back and forth, the sixth bar should become alternately brighter and darker than the spaces between the bars.

4. Disable the red and green guns, viewing only the blue bar pattern. The third and ninth bars should have the same brightness level as the background.

5. Disable the red and blue guns, viewing only the green bar pattern. The seventh bar should have the same brightness level as the background.

4-5.2   Checking AFPC Using an Oscilloscope

As a further check, you can connect an oscilloscope to the red, blue, and green picture tube grids. A pattern similar to Fig. 4-56 should be

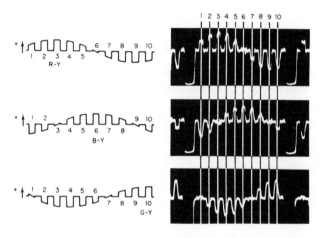

Fig. 4-56. Oscilloscope display at red, blue, and green guns (with rainbow signal at antenna) *(Courtesy RCA)*

obtained. For the red gun, the third bar should have maximum amplitude and the sixth bar should be zero. For the blue gun, the third and ninth bars should be at zero. For the green gun, the first and seventh bars should be at zero.

For positive identification of the bars on the waveform, press the MARK push botton on and off. While the push button is pressed, the third, sixth, and ninth bars will increase in amplitude. (Note that not all color generators have this "mark" feature.)

4-5.3   AFPC Alignment Procedure

Several different demodulation systems are used in various color receivers. The procedures recommended in the service literature for the particular type of receiver should be used. The following procedures are given as a guide for AFPC alignment with either the closed-loop or injection locked oscillator. Figure 4-57 shows a typical set of adjustment points.

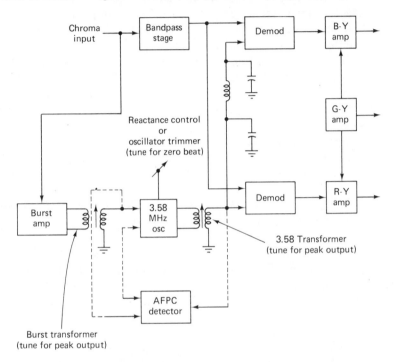

Fig. 4-57. Typical adjustment points for AFPC alignment

Alignment for closed-loop AFPC circuits. The following procedure applies primarily to vacuum-tube receivers, using a reactance tube in the AFPC circuit.

1. Connect the generator RF output to the receiver antenna terminals, or inject the signal to the video stage if desired. Adjust the generator and receiver controls to get a suitable color-bar pattern. Press the MARK push button to place identifying lines on the third, sixth, and ninth bars. Set the generator CHROMA LEVEL control to midrange.

2. Turn off any "Accutint" or other special color circuit, and turn the receiver color control fully on. Set the tint or hue control to midrange. Turn the color killer off (or disable the color killer with a suitable bias). Connect a high-impedance voltmeter to the 3.58-MHz oscillator grid. Tune the 3.58-MHz oscillator transformer to get a maximum negative reading on the meter.

3. Short the grid of the burst amplifier to ground. Connect the voltmeter to the cathode of the phase detector diode. Tune the core of the burst transformer for maximum negative reading on the voltmeter.

4. Remove the short from the amplifier grid. Tune the burst phase transformer for maximum negative voltage.

5. Ground the output of the phase detector (reactance tube grid). Adjust the reactance coil so that the oscillator is in a free-running, zero-beat condition. The correct pattern on the receiver screen will be uniform color bars. However, the bars will change color slowly as the colors appear to drift or float from bar to bar.

6. Remove the ground from the phase detector.

7. Readjust the burst phase transformer to obtain the correct color-bar pattern sequence (third bar red, sixth bar blue, ninth bar blue-green).

Alignment for injection locked oscillator AFPC circuits. The following procedure applies primarily to solid-state receivers.

1. Short out the burst phase transformer, and adjust the oscillator trimmer capacitor, or AFPC potentiometer, for a free-running, "floating" pattern, previously described.

2. Remove the short, and connect an oscilloscope to the burst amplifier output. Adjust the burst amplifier transformer to get a maximum signal indication.

3. Readjust as required, using the oscilloscope on the picture tube control grids (Sec. 4-5.2) or the color-bar cancellation procedure (Sec. 4-5.1).

4-6. SETTING THE 4.5-MHz TRAP

In Section 4-4.6 we described a procedure for setting the 4.5-MHz sound trap by monitoring at the output of the chroma amplifier. The chroma circuits pass only those frequencies in the 3- to 4-MHz region and are inefficient at 4.5 MHz. Thus, considerable amplification is needed for the adjustment, and this produces a noisy oscilloscope pattern. It is possible to use a demodulator probe, such as the RCA WG-449A, to set the 4.5-MHz trap during IF alignment at readily accessible test points.

The following is a step-by-step procedure describing 4.5-MHz trap adjustment using the RCA WG-449A probe and the WR- 514A sweep/

marker generator as a signal source. Other generators that have a suitable 45.75- and 41.25-MHz signal, with audio modulation, can be used in a similar manner.

The procedure is intended as a continuation of IF alignment. The receiver should be prepared for alignment as described in the service literature. Figure 4-58 shows a schematic of the probe, the test connections, and typical oscilloscope patterns.

1. Connect the output from the generator MARKER OUT connector

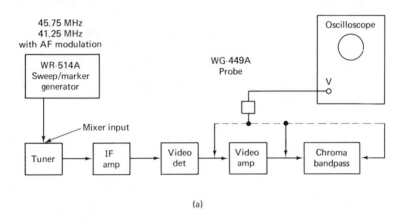

(a)

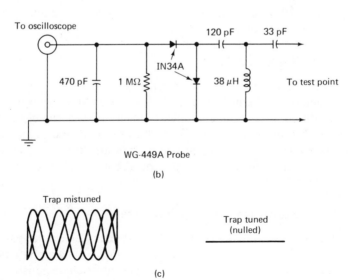

(b)

(c)

Fig. 4-58. Test connections and probe for setting 4.5 MHz trap in color receivers *(Courtesy RCA)*

to the receiver mixer input through an appropriate pad. Connect the WG-449A probe to a convenient point at the output of the video detector.

2. The probe can be connected at other test points after the video detector, such as the output of the video detector or the output of the chroma band-pass amplifier. One suitable test point that is convenient on most receivers it the top (high side) of the receiver color control. Use a shielded cable to connect the probe to the oscilloscope input.

3. Set the generator 45.75, 41.25, and AF MOD switches to ON. Adjust the oscilloscope gain so that a pattern of suitable height is obtained with a minimum amount of signal from the generator. Set the generator to channel 2.

4. Adjust the receiver 4.5-MHz trap to null the pattern. The before and after patterns should be as shown in Fig. 4-58(c).

4-7. SERVICING VIDEO AND CHROMA AMPLIFIERS WITH VIDEO SWEEP

As discussed in Chapter 3 (Sec. 3-2.6), a video sweep can be used to check operation of a receiver from the antenna to the picture tube. It is also possible to use the same technique to service the video and chroma amplifiers in a color receiver. In this section we shall describe how to use a sweep generator such as the RCA WR-514A and an RCA WG-450 marker probe to monitor the video and chroma circuits. This procedure should be used as a supplement to other alignment/troubleshooting alignment test connections (biases, AFT and color killer on or off, etc.).

4-7.1 The Marker Probe

A schematic diagram of the marker probe is shown in Fig. 4-59. The probe contains "suckout" traps at 3.08 and 4.08 MHz, with the 4.08-MHz trap identified by a "touch contact" which removes the dip in the curve when touched by your finger. The probe uses no dc blocking, so when the probe is connected into a circuit where dc of any kind (B+, bias voltages, etc.) is present, connect a capacitor of about $0.47\mu F$ between the probe and the circuit test point.

4-7.2 Example of Servicing Procedure

Assume that there is a color problem in the receiver and that an overall check from the mixer input to the output of the chroma amplifiers indicates a loss of response, or a discrepancy in the response characteristics (but there is a video signal). At this point, the chroma and video amplifiers are suspect, with the chroma amplifiers being the most likely problem.

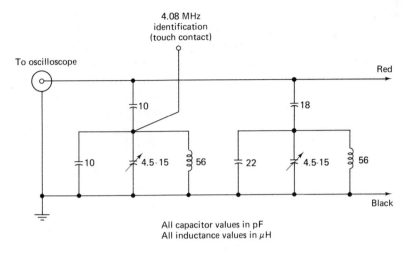

Fig. 4-59. WG-450A marker probe schematic *(Courtesy RCA)*

With a video sweep available, you can inject a signal into the last chroma output stage and observe the output with a wide-band oscilloscope (or a narrow-band oscilloscope and a chroma/video detector probe). Typical connections are shown in Fig. 4-60. If all is well in the last chroma stage (before the demodulators) you should get a response curve similar to that of Fig. 4-60. (A similar curve may be given in the service literature.)

Next, inject the sweep signal into the input of the first chroma amplifier stage (if there is more than one chroma amplifier). Again, you should get a pattern similar to Fig. 4-60.

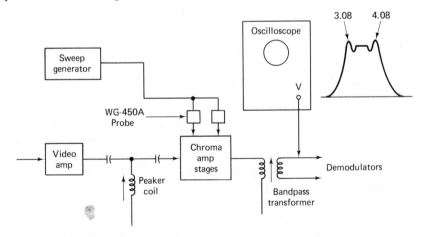

Fig. 4-60. Test connections for video sweep check of chroma amplifier stages *(Courtesy RCA)*

Now inject the signal into the input of the peaker coil (if any) or the color takeoff stage, as shown in Fig. 4-61(a). Now you should get a response curve similar to Fig. 4-61(b), with some possible variations indicated in the service literature. This curve is the complement of the color slope of the IF band-pass curve, and the two are shown in Fig. 4-61(c). Note that 42.67 and 4.08 are of equal height up on the curve, that the two curves cross over at 3.58 and 42.17, and that 3.08 and 41.67 are also equal in height.

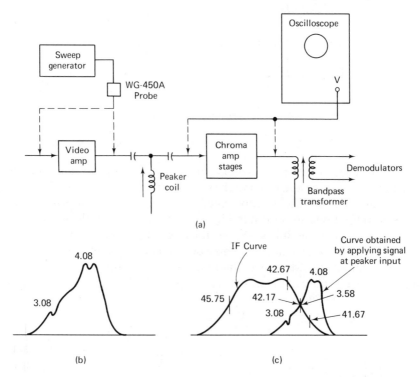

Fig. 4-61. Test connections for video sweep check with input at peaker coil (or at color take-off stage) *(Courtesy RCA)*

Finally, inject a sweep into the video amplifier as shown in Fig. 4-62. This should produce a curve that is flat to about 3 MHz, as shown in Fig. 4-62.

This procedure should pinpoint which section (video or chroma amplifier) is at fault. The problem can then be corrected by alignment and/or troubleshooting, as described in other sections of this chapter.

In some receivers, the chroma amplifier may become unstable when the video signal is connected. If this occurs, connect an 0.05-μF capacitor and an 82-Ω resistor between the probe and the circuit test point. It may also be

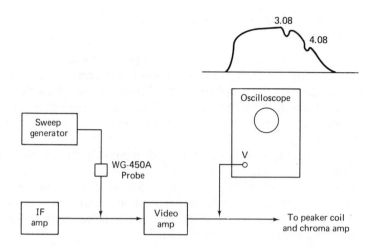

Fig. 4-62. Test connections for video sweep check with input into the video amplifier *(Courtesy RCA)*

necessary to connect an 82-Ω resistor from the probe to ground, as shown in Fig. 4-63. Also, if the 3.58-MHz oscillator is not disabled, a 3.58-MHz "marker" may appear on the pattern.

4-8. USING THE VECTORSCOPE

The basic principles of the vectorscope are covered in Chapter 2. As discussed, the vectorscope is used in television broadcast work as well as to troubleshoot and adjust the color demodulators of receivers. The vectorscope also permits you to judge the general condition of all color circuits (chroma band pass, burst amplifier, reference oscillator, phase detectors, etc.). In this section we shall cover basic vectorscope techniques that apply to practical television service. Further use of the vectorscope in troubleshooting is described in the troubleshooting sections at the end of this chapter.

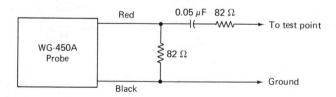

Fig. 4-63. Test connections for added stability when using WG-450A *(Courtesy RCA)*

4-8.1 Connecting a Conventional Oscilloscope as a Vectorscope

1. Connect the oscilloscope vertical input to the driven element of the red gun, usually the grid. [See Fig. 4-64(a).] If the cathode is the driven element, then connect the probe tip to the cathode. (The driven element is the element to which the output signal of the color amplifier is applied.) Use a low-capacitance probe. Connect the probe ground lead to the receiver chassis.

2. Connect the oscilloscope horizontal input to the driven element of the blue gun. Connect the probe ground lead to the receiver chassis.

3. Connect a color-bar generator RF output to the receiver antenna terminals. Reset the channel selector, fine tuning, and other controls to obtain a color-bar pattern. Turn off any special color controls, such as "Accutint," automatic color control, etc.

4. Reset the oscilloscope controls as necessary to obtain a circular rosette pattern on the oscilloscope, as shown in Fig. 4-64(c). Note that the pattern has 10 petals or vectors, one for each color bar. The third petal (counting clockwise from the upper left area of the trace) represents the red bar. With proper adjustment of the hue or tint control, the third bar should be located at or near the top of the pattern (that is, near a 12 o'clock position).

5. Figure 4-64(b) shows a typical vector overlay available for many

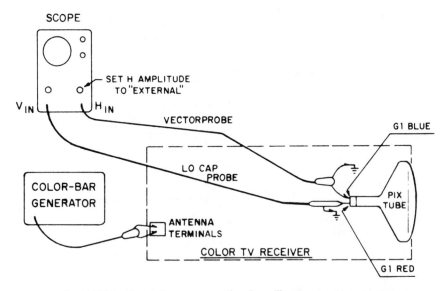

Fig. 4-64(a). Connecting a conventional oscilloscope as a vectorscope (*Courtesy RCA*)

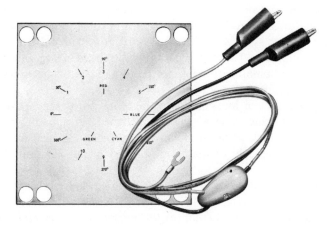

Fig. 4-64(b). RCA WG-409A "Vectorprobe" Oscilloscope Accessory.

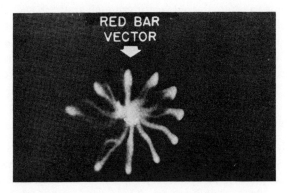

Fig. 4-64(c). Typical "rosette" pattern. Pattern shape will vary somewhat with different receivers and color bar generators.

commercial oscilloscopes. The overlay provides a reference for each of the color petals or vectors. Figure 4-64(b) also shows a set of probes designed specifically to convert a conventional oscilloscope into a vectorscope.

4-8.2   Basic Vectorscope Alignment Procedure

1. Turn the tint or hue control through its range, and note the effect on the vector pattern. The pattern will turn but should *not change in size.* If necessary, adjust the 3.58-MHz oscillator coil until the pattern is the same size throughout the range of the hue or tint control.

2. Set the tint or hue control to the center of its range. Adjust the reactance coil (or oscillator trimmer capacitor in some receivers) so that the color signal stays in sync as the chroma control on the color-bar generator is turned to minimum. The vector pattern will reduce in size, but it should not rotate. Rapid spinning of the vector pattern indicates misadjustment of the reactance coil (or oscillator trimmer).

3. Set the tint or hue control to the center of its range. Adjust the burst phase transformer to move the third vector (corresponding to the red bar) to the 12 o'clock position. Readjust the reactance coil (or trimmer capacitor) simultaneously to maintain color sync (no rotation of vector pattern).

4. Rotate the tint control fully in each direction from its center range position, and note the effect on the vector pattern. If necessary, readjust the burst phase transformer so that the third vector moves an *equal distance* each side of its normal vertical position as the tint control is varied in each direction from midrange. The tint control should have sufficient range to rotate the pattern so that the third vector moves to each side at least as far as the normal position of the adjacent bars (vectors 2 and 4). In many receivers, it will be possible to rotate the red vector as much as 45° in each direction.

5. Repeat steps 2, 3, and 4 until best results are obtained. If proper adjustment cannot be made, check to be sure that the cores are set to their proper peak, which may be either above or below the coil (refer to Sec. 4-3.2). Also refer to the service literature for information on color circuit alignment.

4-8.3   Interpreting Vectorscope Patterns

The display of Fig. 4-65 is that produced by color receivers using a 90° demodulator system (which is typical of the Zenith color system). In a typical vectorscope (or on a vectorscope overlay), the reference burst signal is considered as 0° (or no phase shift). This is represented by the 0, or burst mark, on the vectorscope screen or overlay. The red bar of the standard 10-bar pattern is the third bar and is 90° from the burst. This is at the top

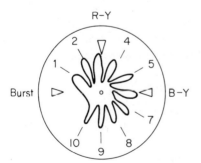

Fig. 4-65. Vectorscope display produced by color receivers using a 90° demodulator system

of the vectorscope screen or overlay. The red bar is marked R-Y. The numbers 1 and 2 represent the first and second bars of the pattern and are all 30° apart on the screen. The B-Y mark (blue) is 90° away from the R-Y mark and 180° from the burst.

4-8.4 Basic Demodulator Adjustments with Vectorscope Patterns

The vectorscope pattern is very useful for adjusting the phase angle between the red and blue demodulators for a 90°, 105°, or 116° demodulator system. (Note that the demodulators of some receivers are not adjustable.)

To align a 90° demodulator system (Zenith type, Fig. 4-65), adjust the demodulators until the third bar rests on the R-Y mark and the sixth bar is on the B-Y mark.

If the receiver calls for a demodulator angle of 105° (typical of the General Electric type, Fig. 4-66), then the third bar should rest on the R-Y mark, with the sixth bar halfway between the B-Y and No. 7. Note that the pattern of Fig. 4-66 is not so round as that of Fig. 4-65. This is due to the increased angle of demodulation and is normal (for that type of demodulator).

As the angle is increased to 116° (such as the old Motorola-type demodulation, Fig. 4-67), the pattern appears more squared, as shown. The sixth bar is past the No. 7 mark.

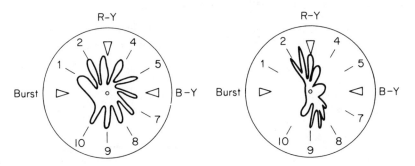

Fig. 4-66. Vectorscope display produced by color receivers using a 105° demodulator system

Fig. 4-67. Vectorscope display produced by color receivers using a 116° demodulator system

As can be seen, the overall shape of vector patterns will describe the type of demodulation used as well as the existing state of demodulator adjustment. Always consult the service literature concerning recommended demodulator alignment procedures (and corresponding vector patterns).

4-8.5   Alignment of an Off-frequency 3.58-MHz Oscillator

This procedure is an alternative to that described in Sec. 4-8.2 and should be used when the 3.58-MHz oscillator is way off frequency.

1. Disable the correction signal to the 3.58-MHz oscillator as described in the service literature. Usually, the correction signal can be disabled by a bias applied at some test point in the circuit.

2. With the correction signal disabled, the oscillator will be free-running, and the pattern will appear to rotate or possibly will appear as a blurred circle, depending on how far the oscillator is off frequency. (A free-running oscillator can also be verified by a "barber-pole" effect on the receiver picture tube. The color bars will be in a broken diagonal pattern.)

3. Adjust the 3.58-MHz oscillator until the vectorscope pattern stands still or is as close as possible to a motionless condition. Then restore the correction signal to the oscillator.

4-8.6   Band-pass Amplifier Alignment with Vectorscope Patterns

It is possible to touch up band-pass amplifier alignment using a vector-scope. However, complete alignment of any stage in a receiver should be done using sweep/marker techniques, as described in other sections of this chapter. A poorly aligned band-pass amplifier will produce a display similar to that of Fig. 4-68.

4-8.7   Troubleshooting with Vectorscope Patterns

A loss of the R-Y signal will cause no vertical deflection and the pattern will appear as shown in Fig. 4-69. The B-Y signal will deflect the beam

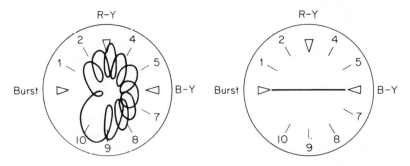

Fig. 4-68. Vectorscope display produced by color receiver with poorly aligned bandpass amplifier

Fig. 4-69. Vectorscope display produced by color receiver with loss of R-Y signal, no vertical deflection

along the horizontal axis, producing a bright line. This indicates that the trouble lies in the R-Y demodulator, matrix, or difference amplifier, depending on the circuit used in the receiver. If the R-Y signal is weak, some deflection will be noted, and an extremely distorted pattern will result, again pointing to the R-Y circuits.

A loss of B-Y signal will result in no horizontal deflection and will appear as shown in Fig. 4-70. The R-Y signal will cause the beam to deflect vertically. This indicates that the problem is in the B-Y difference amplifier, matrix , or demodulator circuits. Again, if the B-Y is weak, some deflection will be noted, and a distorted display will appear.

If there is a complete loss of color, the pattern will appear somewhat like that in Fig. 4-71.

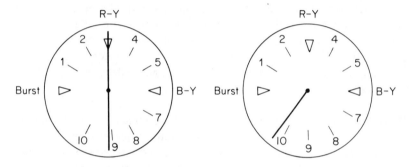

Fig. 4-70. Vectorscope display produced by color receiver with loss of R-Y signal; no horizontal deflection

Fig. 4-71. Vectorscope display produced by color receiver where there is a complete loss of color

Keep in mind that not all circuits produce identical vectorscope patterns. For example, as demodulation angle is increased (such as with General Electric and Motorola), the pattern will appear more "square" than with 90° demodulation systems (such as Zenith), even though all of the receivers are operating properly.

The patterns discussed here are for reference only and must be considered as typical. Always consult the vectorscope instruction manual and all service literature.

4-9. BASIC COLOR CIRCUIT TROUBLESHOOTING

In this section we shall describe those troubleshooting techniques that apply to all of the color circuits. In the remaining sections of this chapter we shall cover specific color circuit groups (sync, matrix, demodulator, etc.).

Before going into the troubleshooting, we shall discuss some typical color circuits. Keep in mind that these circuits represent only a fraction of the many types of color receivers available today.

4-9.1   Closed-loop Color Circuit

Figure 4-72 shows a closed-loop color circuit which is typical of older, vacuum-tube receivers. The major sections are the following:

Band-pass amplifier. The band-pass amplifier is used to separate and amplify, from the video signal, the band of frequencies from 3 to 4 MHz. All of the color information used in modern receivers is within this band.

3.58-MHz reference oscillator. The reference oscillator is used to restore the color carrier in the demodulator circuits. The oscillator is kept in phase with the original carrier at the transmitter by color bursts which appear immediately after the horizontal sync pulse (on the horizontal "back porch"). The oscillator provides two signals (separated by a phase shift) for the demodulators.

Demodulators. The demodulators beat the reference oscillator signals against the chroma signal from the band-pass amplifier to produce color information signals for the picture tube guns. In some receivers, the color-gun grids are driven directly from the demodulator outputs. In other receivers, the color signals are amplified by R-Y, G-Y, and B-Y amplifiers (sometimes known as the matrix) before being applied to the grids.

Burst amplifier. The burst amplifier separates and amplifies the burst pulse from the video signal to control the 3.58-MHz reference oscillator.

Color killer. The color killer produces a bias in the absence of a color signal to cut off the band-pass amplifier. The color killer is controlled by the phase detector.

Phase detector. The phase detector compares the burst signal from the burst amplifier with the reference oscillator signal. The phase detector controls the phase of the reference oscillator signal (and controls the color killer, in this particular circuit).

Signal flow in the Fig. 4-72 circuit is as follows. The video signal is applied from the chroma takeoff coil to the band-pass amplifier, burst amplifier, and blanker. The color phase information is amplified by the band-pass amplifier and applied to the demodulators. The color saturation control (also called *color level* or simply *color*) sets the level of the color signal to the demodulators.

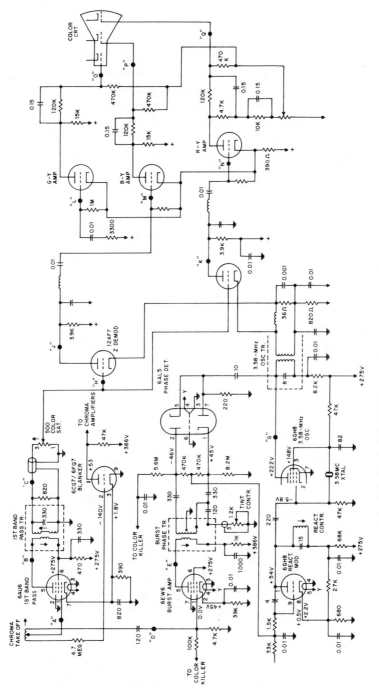

Fig. 4-72. Typical closed-loop color circuit *(Courtesy B&K-Precision, Dynascan Corporation)*

The burst signal is separated and amplified by the burst amplifier, and is applied to the phase detector through the adjustable burst phase transformer. The phase detector also receives a signal from the 3.58-MHz oscillator. If the 3.58-MHz signal is in phase with the burst, the phase detector produces a zero output (or fixed output), and the reactance modulator remains unchanged. If there is a phase difference between the burst and the reference oscillator signals, the phase detector output changes, as does the reactance modulator. Any changes in the reactance modulator causes the 3.58-MHz oscillator to shift in phase as necessary so that the oscillator reference signal is in phase with the burst. The tint (or hue) control shifts the phase detector output (and thus the reference oscillator phase) over a narrow range to set the desired tint in the picture tube display.

The color killer (not shown) receives burst information and produces a bias to control circuits ahead of the band-pass amplifier. If the color burst is absent (black and white transmission, failure of receiver circuits, etc.), the bias cuts off the circuits and prevents color information from being applied to the picture tube.

The blanker tube receives signals (burst and horizontal blanking pulses) and functions to cut off the chroma amplifiers during the blanking interval (when both the horizontal blanking pulse and the color burst are present). Thus, the picture tube should receive no color information during the blanking period.

4-9.2   Injection Locked Oscillator Color Circuit

Figure 4-73 shows an injection locked oscillator circuit which is typical of the more recent, solid-state receivers. Basic operation of the circuit is as follows.

The chroma signal (burst and phase information) is amplified by the first band-pass amplifier Q_1 and is applied through the color intensity (same as color saturation or level) control to the second band-pass amplifier Q_2 and to the burst amplifier Q_3-Q_4. The signal from Q_4 locks the 3.58-MHz oscillator in phase with the burst. The output of Q_5 is amplified by Q_6 and is applied to the demodulators. The tint control shifts the burst signal phase over a narrow range to get the desired picture color tint.

The output of the burst amplifier is applied through the color killer Q_9-Q_{10} and color amplifier Q_8 to the emitter of Q_2. If the burst is absent, Q_2 is biased off. If the burst is present, Q_2 is biased on, and the signal passes to the chroma demodulators. Q_2 is also biased off by the burst killer Q_7 when the horizontal pulse is present. Thus, no color passes during the horizontal sync pulse blanking interval.

Note that the level of the killer amplifier Q_8 can be set by a killer potentiometer. This accommodates burst levels of different amplitude and permits the color killer operating level to be set manually. With this

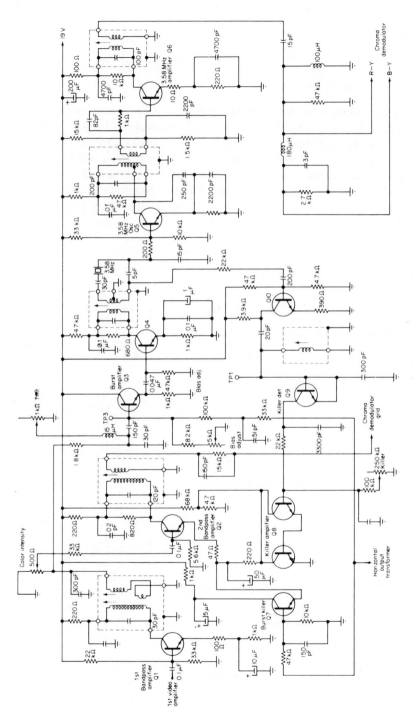

Fig. 4-73. Typical injection locked oscillator color circuit (*Courtesy B&K-Precision, Dynascan Corporation*)

potentiometer, the color killer function can be turned on or off during troubleshooting if required.

4-9.3   Demodulator Circuits

Figure 4-74 shows a matrix-demodulator circuit which is typical of the more recent, solid-state receivers. The circuit receives three inputs. The X and Z demodulators receive chroma signals (color phase information) from the band-pass amplifiers and a fixed reference signal from the 3.58-MHz oscillator. These signals are amplified and applied to the red, blue, and green picture tube guns through the three matrix transistors. The relative phase of the signals at each gun is dependent on the phase relationship between the fixed reference signal (3.58-MHz oscillator) and the color phase signals.

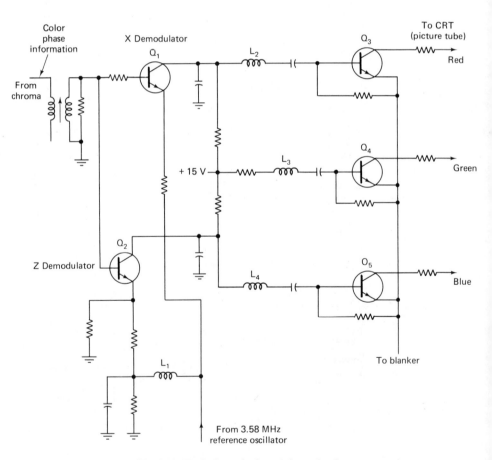

Fig. 4-74. Typical matrix-demodulator circuit.

Note that the circuits are not adjustable. Thus, any failure must be corrected by parts replacement. Also note that the output color signals from the matrix to the picture tube are cut off by blanking pulses (during the horizontal blanking interval).

4-9.4 Preliminary Color Circuit Check

As a first step in troubleshooting, check operation of the color circuits using the procedure of Sec. 4-5 or 4-8. All of these procedures require a color-bar generator. The procedures of Sec. 4-5 can be done with an oscilloscope or by using the receiver picture tube to indicate proper response to the color-bar signals. The procedures of Sec. 4-8 require a vectorscope or a conventional oscilloscope connected as a vectorscope. The choice of procedures is yours.

No matter what procedure is used, if the receiver performs properly, leave it alone. Your problem is one of a bad antenna, operator trouble, etc. If operation is almost normal, try correcting the problems using the alignment procedures of Secs. 4-5 and 4-8. If you cannot bring the color circuits into proper operation with these alignment procedures, go on with the troubleshooting as described in the remainder of this section and in the following sections of this chapter.

4-9.5 Localizing Color Troubles

Color troubles can be grouped into four classifications: no color, wrong color, weak color, or no color sync. The following procedures describe the basic approach to color trouble localization, using either type of color generator (color-bar generator or analyst).

    1. Connect the generator output to the receiver antenna input.

    2. Adjust the generator controls to produce a normal color output.

    3. Set the receiver tint or hue control to midrange.

    4. In turn, connect an oscilloscope to each of the points shown in Fig. 4-75. Observe the following notes for each particular symptom.

No color. A no-color symptom checkout should begin with an observation of the picture detector output. If there is no color output from the picture detector, no color can be expected from the chroma circuits. Likewise, with a normal color signal to start with, a no-color symptom can be localized to the chroma section that follows the video amplifier and/or detector.

Start at either chroma demodulator stage (point A) and work back to the video detector. If the input to the chroma demodulator is normal at point A, then check for the correct signal at point B. This signal is generated

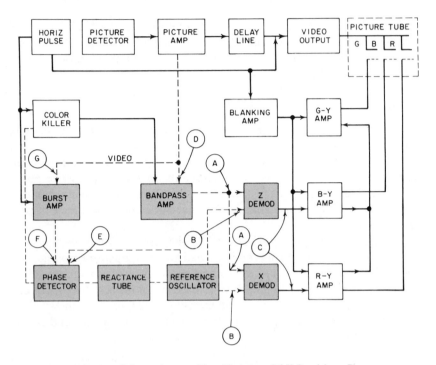

Fig. 4-75. Localizing color troubles *(Courtesy B&K-Precision, Dynascan Corporation)*

by the reference oscillator, which is controlled by the phase detector and reactance tube. The signals at points A and B are both at 3.58 MHz. The signal amplitude should be indicated in the service literature. If not, it is reasonable to assume that the amplitude is correct if both amplitudes (A and B) are the same.

The signal at A is passed through the band-pass amplifier, which, in turn, is controlled by a bias from the color killer. If the bias is applied incorrectly (say due to some circuit failure), the color killer will cut the band-pass amplifier off, and there will be no signal at A. Always check the bias and/or color killer operation first in the event of a "no-color" symptom.

Wrong color. The demodulators are the most likely points to start in localizing a wrong-color symptom. The procedure is essentially the same as that for a no-color symptom, except that the demodulator outputs must also be checked. These are shown in Fig. 4-75 as point C.

If the demodulator inputs are correct but there is no output from one demodulator, the trouble is localized immediately. If the demodulator outputs are present but not phased properly, the trouble is most likely improper alignment of the demodulator stages (if the demodulators are adjustable). If the demodulator output are correct, the trouble can then be localized to the B-Y, R-Y, or G-Y amplifiers (also known as matrix amplifiers) or to the picture tube itself.

Weak color. Again, the demodulator inputs are the most likely places to start looking. If the information is available, check the demodulator input amplitudes against those in the service literature. Note that the input from the reference oscillator is fixed, whereas the input from the band-pass amplifier can be varied by the color control. If it is necessary to advance the color control beyond its normal setting to produce the correct waveform amplitudes or if the signals are still weak with the color control wide open, check the drive to the band-pass amplifier (at point D in Fig. 4-75). Again, remember that the band-pass amplifier bias is controlled by the color killer, which, in turn, is controlled by the burst signal. It is also possible that subnormal amplitudes in the band-pass circuits can be caused by improper alignment.

No color sync. Color sync is controlled primarily by the reference oscillator (3.58 MHz), the reactance tube, and the phase detector. A phase difference between the 3.58-MHz oscillator and the generator burst will cause loss of color synchronization. That is, color bars will not be in proper sequence, or the colors will be erratic and confetti-like at each color bar.

The phase detector receives inputs from both the burst amplifier and the reference oscillator (points E and F in Fig. 4-75). If the amplitude of either input to the phase detector is subnormal in comparison to that given in the service literature, the cause can be localized rather quickly. However, both inputs can be present but out of sync without any evidence in the waveform pattern. Likewise, the burst signal can be contaminated by chroma signals and produce a loss of sync. This will usually show up in the waveform at the burst amplifier input and output (points F and G in Fig. 4-75). A likely cause of chroma signals passing through the burst amplifier is incorrect bias from the color killer.

4-9.6   Localizing Color Troubles with an Analyst Generator

In the remainder of this section we shall illustrate how an analyst generator such as the B&K Model 1077 can be used to localize troubles in the color circuits.

No color (black and white normal). See Figs. 4-76 and 4-77. This symptom can be caused by a failure in the color IF amplifiers (band-pass amplifiers) or the color killer. You can locate the specific stage as follows.

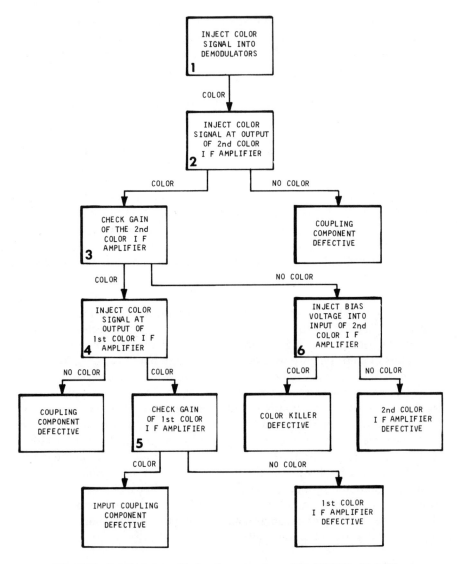

Fig. 4-76. Condensed troubleshooting procedure, NO COLOR, BLACK AND WHITE NORMAL *(Courtesy B&K-Precision, Dynascan Corporation)*

1. Inject a maximum amplitude COLOR signal from the generator at the color demodulator input. It is very improbable that all demodulators have failed simultaneously, so you can assume that the trouble lies before the demodulators. If synchronized color is displayed on the picture tube, the 3.58-MHz oscillator is operating. Go to step 2.

2. Inject a maximum amplitude COLOR signal at the input of the

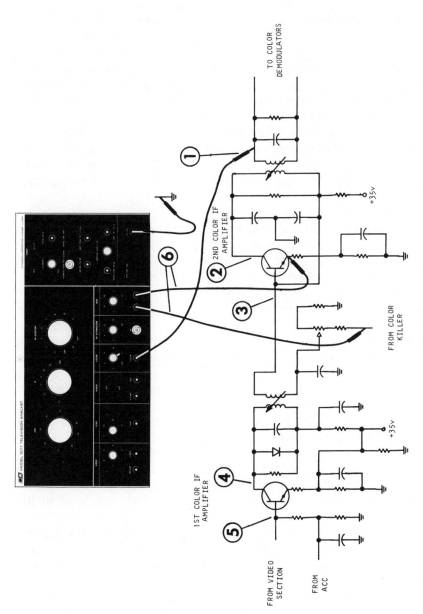

Fig. 4-77. Color IF and color killer schematic diagram *(Courtesy B&K-Precision, Dynascan Corporation)*

second color IF amplifier. If no color is displayed, check the coupling components between the second color IF amplifier and the color demodulators. Use signal injection to isolate the defect to as small an area as possible.

If color is displayed, the coupling components are good. Go to step 3.

3. Check the gain of the second color IF amplifier by moving the signal injection point to the input of the stage and observing the difference in color intensity on the picture tube. If the amplifier is operating properly, the intensity of the color display should increase radically when the signal injection point is changed to the input of the stage.

The relative amount of gain may be checked more accurately by reducing the COLOR control setting until color is barely seen with the color signal injected at the output of the stage, then moving the injection to the input of the stage, and again reducing the COLOR control setting until color is barely seen.

If gain is normal, go to step 4.

If gain is low or no color display is provided with signal injected at the stage input, the trouble could be in the second IF amplifier or the color killer. Go to step 6 for isolation procedures.

4. Inject a medium-amplitude COLOR signal at the output of the first color IF amplifier (also known as the first band-pass amplifier in many receivers). If no color is displayed, the coupling components between the first and second color IF amplifiers are defective.

If color is displayed, the coupling components are good. Go to step 5.

5. Check the gain of the first color IF amplifier. Refer to step 3 for procedures.

If gain is normal, check the coupling components between the video amplifier section and the first color IF amplifier.

If gain is low or color is not displayed, the first color IF amplifier is inoperative. Locate the component that has disabled the stage.

6. The procedures of this step will isolate the no-color symptom to the second IF amplifier or the color killer. The bias output of the color killer normally keeps the second IF amplifier cut off until a color signal is received. If the color killer is defective, the amplifier will stay cut off all the time.

Connect the RF output of the generator to the receiver antenna terminals, and apply a color signal. Also, connect the BIAS power supply to the input of the second IF amplifier, and adjust the BIAS control until the amplifier is not cut off.

If the color display is restored, the color killer is the defective stage.

If no color can be displayed, the second color IF amplifier is the defective stage.

One color absent. See Figs. 4-78 and 4-79. With this symptom, all

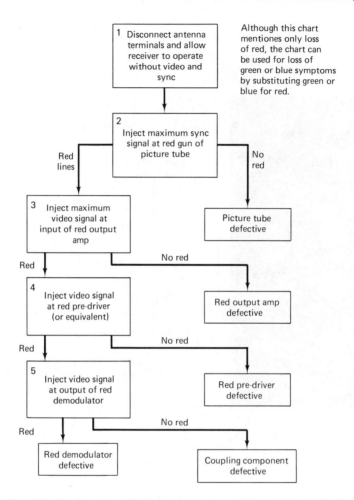

Fig. 4-78. Condensed troubleshooting procedure, ONE COLOR ABSENT
(Courtesy B&K-Precision, Dynascan Corporation)

color will be missing on one of the primary color bars (red bar 3, blue bar 6, green bar 10) when the tint or hue control is adjusted to obtain correct color on one of the other primary color bars. All other color bars will probably be of incorrect color because color mixing cannot be performed when a primary color is missing.

The symptom can be caused by an inoperative demodulator, color (matrix) amplifier, or color gun in the picture tube (depending on design of the color circuits). The localization procedures are essentially the same for red, blue, and green color circuits. For simplicity, the following procedure used the red color cirucit only, but a blue or green circuit can be substituted.

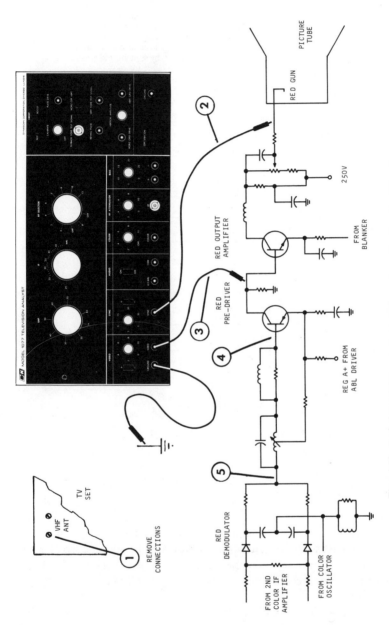

Fig. 4-79. Red demodulator and amplifier schematic diagram (*Courtesy B&K-Precision, Dynascan Corporation*)

REMOVE CONNECTIONS

VHF ANT

TV SET

RED DEMODULATOR

FROM 2ND COLOR IF AMPLIFIER

FROM COLOR OSCILLATOR

RED PRE-DRIVER

RED OUTPUT AMPLIFIER

REG A+ FROM ABL DRIVER

FROM BLANKER

250V

RED GUN

PICTURE TUBE

MODEL 1077 TELEVISION ANALYST

1. Remove the antenna terminal connections (if any), and disable the first or second video amplifier (remove the tube, short across the base and emitter of the transistor, etc.). Under these conditions, the receiver is without video or sync.

2. Inject a maximum amplitude SYNC signal (+) at the red gun of the picture tube. You are not concerned with a test pattern or picture. Instead, you want to check the ability to produce red on the picture tube screen. It may be necessary to adjust the receiver horizontal hold to produce visible patterns. The typical pattern will be diagonal bars or lines.

If red lines are displayed, the picture tube red gun is good. Go to step 3.

If no red color is displayed, the red gun is defective.

3. Inject a maximum amplitude VIDEO signal at the input of the red output (matrix) amplifier. Try both the positive and negative polarities of VIDEO.

If red is displayed, the red output stage is good. Go to step 4.

If no red is displayed, inject the SYNC signal at various points in the output stage to isolate the trouble.

4. If there is a red predriver or equivalent stage, such as shown in Fig. 4-79, inject a VIDEO signal at the input to that stage.

If red is displayed, the stage is good. Go to step 5.

If no red is displayed, the stage is defective.

5. Inject a VIDEO signal at the red demodulator output. If red is displayed, the trouble is in the demodulator.

If red is not displayed, check the coupling components between the demodulator and red amplifier.

No color sync. See Fig. 4-80. With this symptom, the color bars are not in the proper sequence, or colors are erratic and confetti-like at each color bar. The symptom is caused by loss of synchronization to the 3.58-MHz oscillator. For color sync circuits such as Fig. 4-72, the problem can be in the burst amplifier, phase detector, reactance modulator, or possibly the oscillator. For circuits such as Fig. 4-73, the problem is most likely in the burst amplifier, although it could be the first band-pass amplifier or oscillator. Color sync troubleshooting is discussed further in Sec. 4-10.

The following procedure applies when an analyst generator is used to localize trouble in a circuit similar to that of Fig. 4-80.

1. Inject a bias voltage at the input of the reactance modulator, and vary the bias control setting. If the reactance modulator stage is good, the hue of the color bars should vary as the bias control is adjusted. It should also be possible to adjust for proper color display with the bias voltage. If varying the BIAS control has no effect, the reactance modulator or oscillator stage is defective.

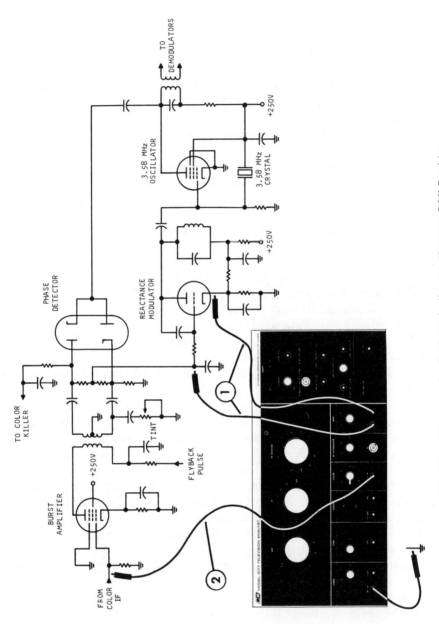

Fig. 4-80. Color sync circuit schematic diagram *(Courtesy B&K-Precision, Dynascan Corporation)*

2. Inject a COLOR signal into the burst amplifier. If the burst amplifier and phase detector are good, injection of the color signal (which includes the required burst) should restore color sync. If either is defective, the symptom will remain unchanged.

4-10. TESTING COLOR SYNC CHARACTERISTICS

This section is devoted entirely to the testing of color sync circuits. Before making any of these tests, the basic color tests of Sec. 4-5 or 4-8 should be performed. Even though the basic color tests reveal the presence of a sync signal, sync trouble can be experienced if the sync signal is low in amplitude or if the amplitude is fluctuating (or if the sync is mistimed).

4-10.1   Checking Burst Amplifier Circuit

The obvious starting point for checking the sync circuits is at the burst amplifier. If the burst amplifier output is normal but the reference oscillator is not locked in sync, the trouble is in the oscillator or its phase control circuits. However, if the burst amplifier output is not normal, the circuits after it cannot be expected to operate properly.

1. Connect a color generator to the receiver antenna terminals. Adjust the generator and receiver controls for normal color reception (rainbow display).

2. Connect an oscilloscope to the output of the burst amplifier using a low-capacitance probe. In a circuit similar to Fig. 4-72, check at test point E. In the Fig. 4-73 circuit, check at the collector of Q_4.

3. Check the amplitude of the burst signals against the service literature. If the burst amplifier output is not correct but the input is correct, the trouble is obviously in the burst amplifier. Note that some receivers produce a symmetrical burst signal at the burst amplifier output, whereas others produce a high positive (or high negative) signal. Also, the burst amplitude varies greatly from receiver to receiver. Figure 4-81 shows an expanded view of a typical burst signal taken from the burst amplifier output.

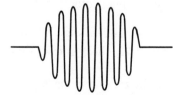

Fig. 4-81. Expanded view of a typical burst signal taken from the burst amplifier output

4-10.2   Checking Color Sync Point with Variable Burst Input

If the generator burst signal (or color output) can be varied in amplitude, it is possible to check the point at which color sync is lost or takes over. This can be compared with the service literature to determine if the overall sync circuits are operating on or near the borderline. Note that the color output of the generators described in Chapter 2 can be varied.

1. Connect the color generator to the receiver antenna terminals. Adjust the generator and receiver controls for normal color reception (rainbow display).

2. If the receiver is provided with AGC or color killer threshold controls, make certain that these are properly adjusted.

3. Gradually reduce the burst amplitude (color output) while watching the color display on the picture tube.

4. When the color display loses sync, check the amplitude of the burst signal with an oscilloscope. If the service literature specifies correct amplitude at the burst amplifier output, measure at that point. If not, measure at the output of the video detector. Most service literature will specify some signal level of amplitude at the video detector output. Either way, compare the amplitude with that of the service literature.

5. Most color receivers will operate satisfactorily even if the color burst signal amplitude is within about 25% of the correct value. This is due to operation of the AGC and AFPC circuits. If color sync is lost when the burst amplitude is reduced slightly, this usually indicates a defective reference oscillator or its control circuits.

4-10.3   Checking Burst Amplifier Input Signal

If the burst amplifier signal is absent or abnormal, the drive signal to the burst amplifier should be checked (before making voltage-resistance measurements at the amplifier circuits). Keep in mind that the drive signal consists of two parts: the gating pulse from the horizontal circuits (flyback transformer) and the 3.58-MHz chroma signal. In this case, the chroma signal is being supplied by the color generator.

1. Connect the color generator to the receiver antenna terminals Adjust the generator and receiver controls for normal color reception (rainbow display).

2. Connect an oscilloscope to the input of the burst amplifier using a low-capacitance probe. In a circuit similar to Fig. 4-72, check at test point D. In the Fig. 4-73 circuit, check at TP$_3$.

3. Check input signals to the burst amplifier against the service literature. If the gating pulse is present but there is no burst or chroma, check the input back to the video detector through the chroma takeoff, first band-pass amplifier, or whatever. If the video or burst is present but the gating pulse is gone, check the input back to the horizontal (flyback) circuit.

4-10.4   Checking Burst Gate Timing

If the gating pulse from the horizontal circuits to the burst amplifier is improperly timed, there will be loss of color sync, or the sync will be incorrect. Likewise, if the gating pulse is too wide, it is possible for the first color information (such as the first color bar from a color generator) to pass the burst amplifier to the reference oscillator and/or its control circuits. Either way, the color sync will be incorrect.

Figure 4-82 shows the theoretical waveforms of the burst signal sequence.

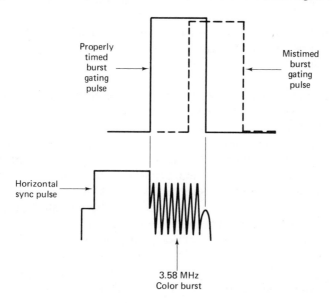

Fig. 4-82. Theoretical waveforms of the burst signal sequence

Note that the burst signal *follows* the horizontal sync pulse. Figure 4-82 also shows the correct, and typically incorrect, timing of the burst signal. With correct timing, the burst amplifier is opened during the entire burst by the horizontal sync but is closed immediately before the color information signal. With incorrect timing (late), the burst amplifier is closed during part of the burst signal and is open during part of the color signal that follows the burst. Either of these conditions can result in improper sync.

The burst gate pulse (or burst keying pulse, as it may be called) to the burst amplifier is usually supplied by a winding on the flyback transformer.

Since the burst signal is delayed from the horizontal sync pulse during transmission, the burst gate pulse must also be delayed in the receiver. In many receivers, this is done by an RC circuit at the flyback transformer, as shown in Fig. 4-83.

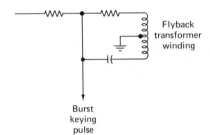

Burst
keying
pulse

Fig. 4-83. Typical gate-pulse time-delay (RC) circuit

No matter what system is used, it is sometimes difficult to distinguish the exact burst gating interval. However, the following simple test will usually show if there is a timing problem.

1. Connect a color generator to the receiver antenna terminals. Adjust the generator and receiver controls for normal color reception (rainbow display).

2. Vary the generator color output (amplitude) from zero to maximum (usually, maximum is about 200% saturation, for most color generators).

3. The color saturation should change as the color output amplitude is varied, but there should be no appreciable change in tint or hue.

4. If the tint or hue changes appreciably, this indicates that the burst gate timing is incorrect or that the pulse is too wide. Of course, a similar condition can be caused by defective automatic color control or AFPC circuits, since a small mistiming of the burst gate can be overcome by properly operating color control circuits. However, if the burst timing is way off, the automatic color control circuits cannot correct the condition.

4-10.5   Checking AGC Effect on Color Sync

In addition to automatic control (ACC or AFPC) of the color circuits, most color receivers have AGC (automatic gain control) in the black and white circuits. A defective AGC can cause clipping or depression of the color signal at normal 100% saturation. This can be checked as follows.

1. Connect a color generator to the receiver antenna terminals. Adjust the generator and receiver controls for normal color reception (rainbow display).

2. Set the generator color output to zero, and measure the burst amplifier output signal amplitude as described in Sec. 4-10.1.

3. Increase the generator color output to 100% saturation, and measure the burst amplifier output signal.

4. If there is a *drastic change* in burst amplifier output signal amplitude when the color output is varied, apply a fixed bias to the AGC line, and repeat the test. If the burst signal amplitude remains constant with the fixed bias, it is possible that the AGC circuits are defective. It is also possible that there may be regeneration in the IF amplifiers. In either case, these circuits are in the black and white portion of the receiver and should be treated as discussed in Chapters 3 and 6.

4-10.6   Checking Input Signal to the Phase Detector

In color circuits similar to Fig. 4-72, once the burst amplifier is established as operating normally, the phase detector should be checked for proper input and output. (Note that there is no phase detector, or comparable circuit, in the color circuits of Fig. 4-73. In the Fig. 4-73 circuit, the oscillator will lock in if the burst signal is present, unless the oscillator itself is defective.)

The ideal way to check the detector is to compare input waveforms and output voltage against the service literature. In the absence of such data, it is possible to check the phase detector of most receivers as follows.

1. Connect a color generator to the receiver antenna terminals. Adjust the generator and receiver controls for normal color reception (rainbow display).

2. Connect an oscilloscope through a low-capacitance probe to the phase detector as shown in Fig. 4-84. Move the oscilloscope probe to each phase detector input in turn.

3. Check the waveform at the phase detector input as displayed on the oscilloscope. Figure 4-85 shows a possible phase detector input waveform. However, the exact waveform and amplitude will vary. The important point to remember is that you should find equal (or almost equal) amplitudes at each of the phase detector inputs.

4. It should also be remembered that the phase detector receives an input from the reference oscillator. This signal should be present with or without a burst signal. To check for the reference oscillator signal, move the oscilloscope probe to the phase detector input from the oscillator. Check on both sides of the coupling capacitor between the phase detector and oscillator. If the service literature is available, check the oscillator output voltage for correct value. A low-voltage output from the oscillator can cause loss of color sync as well as weak colors and possibly distorted colors.

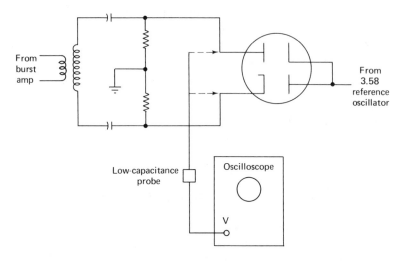

Fig. 4-84. Checking input signal to phase detector

5. If all inputs to the phase detector appear to be correct, check for correct output. This is a variable dc voltage applied to the reactance tube. Connect a meter to the phase detector output line, and measure the voltage to the reactance tube with a color signal applied from the color generator (connected to the antenna input terminals). Then remove the color signal by switching the generator off. There should be a change in the meter reading, indicating that the phase detector output changes when the color burst is removed.

4-10.7   Checking for Open Decoupling Circuits and Ripples

Open decoupling (bypass) capacitors in the sync circuits can cause a loss of color sync, as can excessive ripple. These conditions can be checked with a high-gain oscilloscope and low-capacitance probe in the usual manner (refer to Chapter 3). Although a color generator signal is not essential to locate these problems, it is usually easier to find decoupling faults if there is a signal present.

1. Connect a color generator to the receiver antenna terminals. Adjust the generator and receiver controls for normal color reception (rainbow display).

2. Connect the oscilloscope probe to the hot side of each decoupling

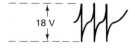

Fig. 4-85. Typical input signal to phase detector

capacitor in the sync circuits, particularly the reactance tube and the reference oscillator. There should be substantially no alternating current present at any of these points (it should be bypassed through the capacitor to ground).

3. If there is substantial alternating current at all of the decoupling capacitors, this indicates ripple from the power supply. Check the power supply circuits in the normal manner.

4. If there is ripple at only one decoupling capacitor, check that capacitor for a possible open.

4-11. TESTING DEMODULATION CHARACTERISTICS

This section is devoted entirely to testing of the demodulation circuits. Before making any of these tests, the basic color tests of Sec. 4-5 or 4-8 should be performed. The following tests are a supplement to those of Secs. 4-5 and 4-8.

4-11.1 Notes on Demodulation Troubleshooting

Often, the colors that appear on the screen of a defective receiver provide a clue as to the defect. For example, a defect in one color demodulator will create colors lying along the other demodulator axis. If the receiver demodulates on the R-Y and B-Y axes and the R-Y demodulator is faulty, then the screen will have a greenish-yellow or bluish cast, with red appearing gray, yellow appearing green, and magenta appearing blue.

If the B-Y demodulator is faulty, the screen will take on a reddish or cyan cast, with blue appearing gray, cyan appearing green, and magenta appearing red.

If the G-Y signal (G-Y demodulators, as such, are not used) is faulty, green will appear gray, and cyan will appear blue.

If the receiver demodulates on the I and Q axes (as in the case of older receivers) and the Q signal is faulty, the effect will be similar to that of a loss of the B-Y axis, except for some slight differences in the hues. When the I signal is inoperative, then the loss will be similar to the loss of R-Y, except for the slight differences in hues.

Most modern color receivers demodulate on or near the R-Y/B-Y axes. Many older receivers use I and Q axes. This design practice has been discontinued. However, some older color generators have provisions for I and Q test signals. In any case, the two axes are in quadrature (90° apart).

Some manufacturers use neither of these axes, and the ones they do use are not in quadrature. Many use a pair of axes with one within several degrees of the R-Y axis, while its partner lies in the B-Y sector at some angle other than 90° from the first axis. One manufacturer uses an X axis 10.9° off the R-Y, with an accompanying Z axis 62.5° toward the B-Y axis.

No matter what system is used, the outputs of the color demodulators invariably work into a matrix (sometimes called color amplifiers or color-gun amplifiers) whose values are so arranged as to produce R-Y, G-Y, and B-Y signals. Thus, all transmitted colors will fall into their correct positions.

Figure 4-86 shows four typical demodulator and matrix circuits in block form. The XZ system is the most common. In this system, the R-Y, B-Y, and G-Y signals first appear in the outputs of the amplifier stages following the demodulators. (The circuit of Fig. 4-74 is typical of the XZ demodulator system.)

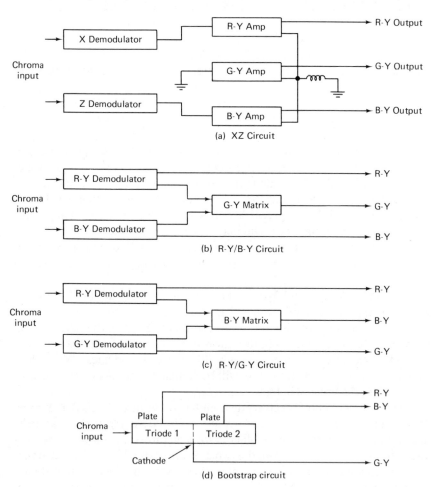

Fig. 4-86. Typical chroma demodulator and matrix circuits

In the R-Y/B-Y and R-Y/G-Y system (b and c of Fig. 4-86), the circuits are essentially the same, except that the B-Y/G-Y functions are interchanged as demodulator or matrix. In the bootstrap circuit [Fig. 4-86(d)], twin diodes are used to develop all three signals. In this older system, the R-Y and B-Y outputs appear at the triode plates, with the G-Y output appearing at the common cathode terminal.

Universal chroma troubleshooting chart. The chart shown in Fig. 4-87 is a handy tool for rapid color problem analysis. Assume that the predominant color on the screen is yellow. Place your finger on the portion of the circle marked "yellow," and pass a straight line from this point through the center point (marked "white") to the other side of the circle. This shows that the blue gun is faulty, and the circuit should be investigated from the demodulators through the matrix and amplifiers to the blue gun. By the same method, if the picture has a predominant magenta cast, then the problem lies in the green gun, and if the picture is cyanish, the red gun circuit is suspect. Of course, it is assumed that the receiver fine tuning, hue, and any color controls are properly set when using the chart of Fig. 4-87.

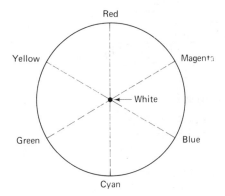

Fig.   4-87. Universal   chroma troubleshooting chart

4-11.2   Checking Demodulators with Rainbow Generators

The basic procedure for checking demodulators using a rainbow generator (without a vectorscope) is described in Sec. 4-5. The following procedure is a supplement to those of Sec. 4-5.

1. Connect a rainbow generator to the receiver antenna terminals. Adjust the generator and receiver controls for normal color reception. Set the receiver tint or hue control to midrange.

2. In turn, connect the oscilloscope to each of the demodulator outputs, and check for presence of the pattern. Figure 4-88 shows the typical keyed sine-wave display that should appear at each of the outputs

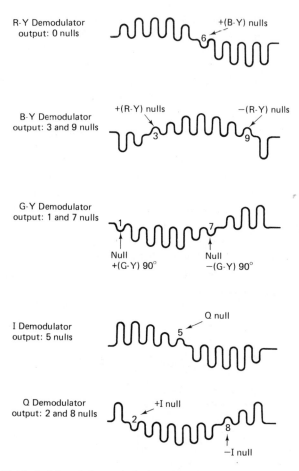

R-Y Demodulator
output: 0 nulls

+(B-Y) nulls

B-Y Demodulator
output: 3 and 9 nulls

+(R-Y) nulls   −(R-Y) nulls

G-Y Demodulator
output: 1 and 7 nulls

Null
+(G-Y) 90°

Null
−(G-Y) 90°

I Demodulator
output: 5 nulls

Q null

Q Demodulator
output: 2 and 8 nulls

+I null

−I null

Fig. 4-88. Typical keyed sinewave display with rainbow generator at input *(Courtesy B&K-Precision, Dynascan Corporation)*

when a standard 10-bar keyed or gated rainbow signal is applied. The phases shown are at the picture tube gun inputs (usually the grids) and will be 180° reversed at the amplifier inputs (such as at the bases of transistors Q_3, Q_4, and Q_5 of Fig. 4-74).

3. Check for proper phasing by counting the number of color-bar signals (after the burst or sync) at which null occurs. For example, the R-Y demodulator should show a null on the sixth color bar signal. This is the point at which the B-Y signal is maximum positive. Likewise, the B-Y signal will show nulls at the third and ninth color-bar signals. These are the points where the R-Y signal is maximum.

4. Note that Fig. 4-88 shows the waveforms produced by a keyed

rainbow generator on receivers that demodulate on the I and Q axes. Such receiver circuits are not common. However, it is possible that you may run into an older receiver with IQ demodulators.

   5. If any of the nulls occur at incorrect points, adjust the demodulator phasing controls as necessary to produce correct nulls. Keep in mind that all receivers may not have adjustable demodulator circuits. For example, the circuits of Fig. 4-74 are not adjustable. It is possible to correct very minor phasing problems in the demodulators by slight adjustment of the 3.58-MHz oscillator phasing. However, *never attempt* to correct any major problem in the demodulators or matrix by adjustment of the 3.58-MHz oscillator phasing. Correct the problem in the demodulator circuits first. Also, make certain to *recheck all three outputs* (R-Y, B-Y, and G-Y) after adjustment of any demodulator or phasing controls. This is necessary since the controls may interact.

4-11.3   Checking Demodulator Output Amplitudes with Rainbow Generator

   Although the exact amplitudes of demodulators will differ with each particular generator and receiver, and *relative* amplitudes should remain approximately the same when a rainbow signal is applied. When troubleshooting a receiver where the correct amplitudes are given in the service literature, always use those amplitudes for comparison. However, when the correct amplitudes are not known, a check of the relative amplitudes may prove quite helpful. For example, if the relative amplitudes are way off, the demodulator circuits are suspect.

   1. Connect a rainbow generator to the receiver antenna terminals. Adjust the generator and receiver controls for normal color reception. Set the receiver tint or hue control to midrange.

   2. In turn, connect the oscilloscope to the R-Y, B-Y, and G-Y demodulators outputs, and check amplitudes of the patterns. (If XZ demodulation is used, check the amplitudes at the R-Y, B-Y, and G-Y amplifier outputs.)

   3. Figure 4-89 shows the patterns and relative amplitudes that should appear at each of the outputs when a standard 10-bar rainbow signal is applied.

   4. As shown in Fig. 4-89, the B-Y output is usually maximum, no matter what demodulator circuit is used. The B-Y amplitude can be considered as 100%. Using this as a basis, the R-Y output should be approximately the same (slightly less in some cases), while the G-Y output should be approximately 45%.

Fig. 4-89. Relative output amplitudes of demodulators *(Courtesy B&K-Precision, Dynascan Corporation)*

4-11.4   Notes on Matrix Troubleshooting

In many receivers it is difficult to differentiate between the demodulator circuits and the matrix circuits. The following notes apply to those circuits which follow the demodulators and precede the picture tube (such as the circuits of Q_3, Q_4, and Q_5 in Fig. 4-74). The notes also include the Y amplifier, also known as the video amplifier or luminance amplifier.

The Y amplifier and demodulator circuits are closely associated with the matrix. The demodulator outputs are fed to the matrix or form a portion of the matrix. The Y amplifier circuits are related in that the output of both matrix and Y amplifiers are applied to the picture tube guns (at the cathodes, or grids, or both, as shown in Fig. 1-9). The following troubleshooting notes are included to help establish the relationship of demodulator, Y-amplifier, and matrix circuits.

Low-capacitance probes. Low-capacitance probes should be used whenever possible to minimize circuit loading. Such loading can cause waveform distortion and is a particular problem with color circuits. For example, the capacitance of a probe will affect a peaking coil so that the high-frequency response is weakened.

The input capacitance of a probe is determined by the length of the probe cable, the capacitance per foot, and the wire size. Such capacitance is also inversely proportional to the attenuation factor. The greater the attenuation, the lower the capacity. Since the permissible attenuation factor is limited by the maximum gain available in the oscilloscope vertical amplifier, it may be necessary to compromise between attenuation and input capacitance.

Checking Y signals. The output of the Y amplifier can best be measured at the picture tube input (usually the cathodes), since this point provides a final, overall check of the Y signal. The exact waveform and amplitude will, of course, vary from receiver to receiver and should always be checked against the service literature. Also, the exact waveshape will depend on the signal source.

One of the major problems in the Y signal is that it often picks up stray signals (hum, noise, etc.) or excessive 3.58-MHz signals. The Y circuits usually include traps to eliminate the 3.58-MHz signal at the picture tube cathodes. The presence of the 3.58-MHz signal in the Y-signal circuits requires the use of a wide-band oscilloscope (about 5 MHz or wider is preferable).

Some technicians feel that it is sufficient to check the Y signal using the broadcast station signal. This can be misleading. If the signal is weak, it may not be able to override the noise and hum which are normally present in the receiver. This will cause *apparent* excessive interference. Such interference will disappear when a signal of sufficient amplitude is applied.

In general, noise on the Y-signal amplifier will cause the signal waveform to appear ragged. Hum voltage, on the other hand, will cause the waveform to thicken. A faulty bypass capacitor in the circuit can cause undesired ripple to weave through the signal pattern.

Checking demodulator-matrix signals. One point to remember in checking any demodulator-matrix circuit is that a completely dead circuit can be caused by the color killer. For example, misadjustment of the color killer threshold control can bias off most of the demodulator-matrix circuitry.

Another point to remember is that peaking coils in the demodulator output circuits (such as L_2, L_3, and L_4 in Fig. 4-74) serve a filtering function as well as maintaining the frequency response at the high end of the color response. If a wide-band oscilloscope is used at both the input and output of these peaking coils, the input (demodulator output) may show considerable 3.58-MHz signal, but the coil output should show little or no 3.58-MHz residual signal.

Waveforms at picture tubes. When checking waveforms at the picture tube grids (matrix, demodulator, or Y) you may notice that the waveforms at the grids also appear on the cathodes. This may lead you to the belief that the demodulator or matrix signal has leaked into the Y-signal circuitry. While this is possible, it is more likely that the picture tube is producing a cathode-follower effect. As in the case of other tubes, whatever signal appears on the grids will also appear on the cathodes, if there is a cathode dropping resistor of sufficient value to develop a measurable voltage.

If the grid signal does appear on the cathode of the picture tube, try disabling the Y amplifier. If the signal remains, it is a result of cathode-follower action in the picture tube and is probably normal for the receiver. If it is not practical to disable the Y amplifier or remove the Y signal, carefully measure the relative amplitudes of the grid and cathode signals. The cathode signal should be considerably lower in amplitude since there is no gain in a cathode follower.

5

Television Service Hints and Tips

In this chapter, we shall discuss hints, tips, and other procedures that apply to all phases of television service (black and white, color, solid-state, etc.). As examples, we shall discuss how to test antennas and leads, how to interpret service literature and test patterns, and how to check deflection yokes, transformers, and coils associated with the picture tube circuits, and we shall tabulate frequencies commonly used in television receivers.

5-1. TESTING TELEVISION ANTENNA SYSTEMS

Sweep generators, as well as conventional RF signal generators, can be used to test television antennas, transmission lines (lead-ins), baluns, couplers, TVI filters, lightning arresters, tuning stubs, Q bars, etc. These uses for generators may not be too well known by the average television technician, although similar tests have been put into practice by industrial and laboratory technicians for many years.

The sweep generator is a far better instrument for these tests than the RF generator, since the sweep generator provides a simultaneous test of many frequencies. The sweep generator must be capable of providing signals at the fundamental frequencies under test (not at harmonics). Likewise, the sweep generator must have a flat (constant-amplitude) output over the frequency range used for test. Some older (and less expensive) sweep generators operate on harmonics to cover all TV channels. Such generators are not suitable for the following antenna system tests. Likewise, the output of some sweep generators is not flat, as discussed in Chapter 2. If there is any doubt, check the sweep generator for output uniformity, as discussed in Chapter 2.

It is recommended that you try all antenna system tests described here on your own antenna and lead-in. The procedures are interesting and instructive and will help you to have a better understanding of antenna and transmission line characteristics. With experience, you can quickly make meaningful checks for defects in any antenna installation.

5-1.1 Checking Antenna Transmission Line (Lead-In) Loss

Television antenna transmission lines (300-Ω twinlead or 75-Ω coax lead-ins) can deteriorate with age, exposure to weather, or accidental damage. The signal loss can be quite drastic, creating a serious problem with reception. Snow and noise caused by a bad transmission line can be just as annoying as that caused by a defective antenna. To complicate the problem, it is sometimes impossible to get at the lead for visual inspection. Using a sweep generator, you can check questionable line installations for excessive loss, including shorts or opens.

The sweep check is most effective for transmission lines with lengths from 50 to 100 ft. Fortunately, most installations fall into this category. Lines less than 50 ft will not produce suitable indications (enough standing waves) on the lower VHF television channels. Lines longer than 100 ft may have so much loss that the test becomes inconclusive.

Procedure (300 twinlead). Use the following procedure for a 300-twinlead of any type.

1. Connect the equipment as shown in Fig. 5-1. It is important that

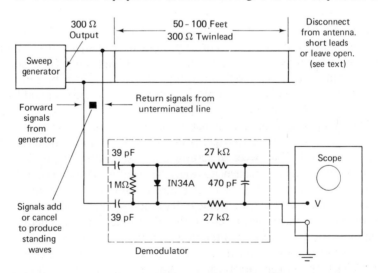

Fig. 5-1. Checking antenna transmission line (lead-in) loss, 300-Ω twinlead

you either short the leads at the antenna connection or completely disconnect both leads from the antenna, leaving the leads open. To short the leads, simply twist the ends of the wire together. Disconnect any distribution boxes, splitters, or amplifiers from the line under test.

   2. Set the generator to channel 11. Set the scope controls, and generator RF attenuator, as necessary to obtain a pattern of suitable height.

   3. Set the generator sweep width control to about 10 MHz or to some width that will provide a pattern similar to that shown in Fig. 5-2.

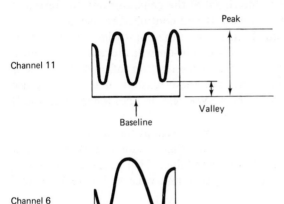

Fig. 5-2. Typical standing wave patterns for 300-Ω unterminated twinlead

   4. Note that the pattern shown in Fig. 5-2 is formed as follows: The signal from the generator travels through the transmission line to an open or short circuit (no termination). If the line were connected to an antenna (and were perfectly matched to the antenna), all of the signal would be absorbed by the antenna. However, since the line is not terminated, the signal returns back down the line. During the return trip, the signal encounters forward-going signals (from the generator) of different phase that will vary with the length of the line. The forward-going and returning signals add or cancel at the demodulator (Fig. 5-1) and produce patterns similar to those of Fig. 5-2.

   5. To evaluate the line for *relative loss*, observe the distance from the "valleys" of the Fig. 5-2 pattern to the base line, and compare this with the total distance from the base line to the pattern "peaks." The ratio of peak amplitude to valley amplitude is known as voltage standing wave ratio, or VSWR, and is expressed as

$$\text{VSWR} = \frac{\text{Peak amplitude}}{\text{Valley amplitude}}$$

   6. On an unterminated line, a higher ratio is the desired condition. This is the opposite of a terminated line (where the antenna is connected). In a perfect (theoretical) system where the antenna is perfectly matched to the line, the antenna would absorb all of the signal, there would be no return signal, there would be no standing waves (the pattern would be flat), and the VSWR ratio would be 1:1. However, in an unterminated line as shown in Fig. 5-1, there will be standing waves. As a guideline, in a good 300-Ω line, 50 to 100 ft long, the distance from the base line to the valley should be approximately 20%, or less, of the distance to a peak. As an example, if the peaks are 10, the valleys should be 2, or less.

    7. As line length is increased, the loss will become greater, since the loss is the difference between the signal at the generator and the reflected signal producing the standing waves. If you continued to increase the line length, the loss would eventually increase to the point where the loss itself would appear as a termination. This condition will prevent useful unterminated VSWR checks. (As a matter of interest, an infinitely long line is theoretically an ideal termination.) For these reasons, it is generally not practical to make this unterminated VSWR test on lines longer than 100 ft (at VHF television frequencies).

    8. An unterminated line (open or shorted at the antenna) that is in *very poor* condition may have so much loss that it exhibits a pattern similar to that of the pattern obtained from a good line. That is, there will be little difference in amplitude between the peaks and valleys. Figure 5-3 shows

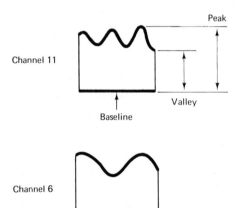

Fig. 5-3. Standing wave patterns for 300-Ω unterminated twinlead in poor condition

typical patterns obtained from poor line with considerable loss. Any unterminated line that shows patterns similar to those of Fig. 5-3 should be inspected for damage and replaced if necessary.

    9. Check the line on all television channels. Note that the pattern will normally vary from channel to channel, even though the VSWR may (or may not) remain constant. Typically, there will be more loss on higher channels. The typical differences between channels 6 and 11 are illustrated in Figs. 5-2 and 5-3.

Procedure (75-Ω coax line). Use the following procedure for a 75-Ω coax line of any type.

1. Connect the equipment as shown in Fig. 5-4. Again, either short

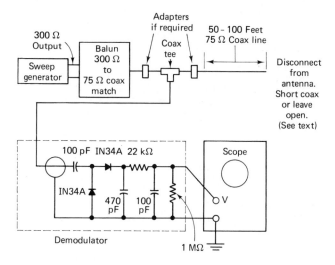

Fig. 5-4. Checking antenna transmission line (lead-in loss 75-**Ω** coax

the coax at the antenna connection or disconnect the coax from the antenna. In most installations, it is easier to simply disconnect the coax at the antenna. Disconnect any distribution boxes, splitters, etc., from the line under test.

2. Use the same procedure for a 75-Ω coax as that described for a 300-Ω twinlead. In general, for equivalent lengths, the loss will be greater for coaxial lines than for twinlead lines. The losses shown by coax lines will vary with the type of line involved. This is illustrated in Fig. 5-5, which shows typical traces produced by three representative types of coaxial line.

5-1.2 Checking Antenna Lead-In for Shorts or Opens

The sweep frequency test described for checking line loss can also be used to check lead-ins for shorts or opens. (Note that this test will not be effective if the short or open is close to the generator end of the line.)

1. Use the same connections as shown in Fig. 5-1 or 5-4, except connect a carbon (noninductive, not wire-wound) resistor across the lead at the antenna end, as shown in Fig. 5-6.

2. Compare the patterns produced when the terminating resistor is connected with those without the resistor (line unterminated). There should

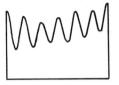

Typical trace obtained with 50-foot
length of small diameter video
frequency type of 75-ohm cable

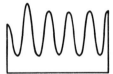

Typical trace obtained with 50-foot
length of RG-59/U cable

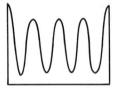

Typical trace obtained with 50-foot
length of "foam" type cable similar
to RG-59/U

Fig. 5-5. Typical standing wave patterns produced by three representative
types of coaxial line

be a drastic change when the resistor is connected. Typically, the peaks and
valleys (representing the standing waves) will disappear and the pattern will
be relatively flat when the resistor is connected. If the resistance is matched
exactly to the line impedance, the pattern will be perfectly flat (no standing
waves) since the resistor is absorbing all of the energy, and no signal is being
returned down the line.

3. In making this test, the main concern is that there is a drastic
change in pattern after the resistor is connected. If there is little or no
change (with and without the resistor), the line is either open or shorted at
some point.

4. Do not be concerned if the pattern indicates some standing waves
after the resistor is connected. This is probably due to a mismatch between
resistance value and line impedance. Or there may be some reactance in the
line (or the resistor leads) which is producing a mismatch at some frequencies.
To get a good termination, indicated by a fairly flat pattern, it may be
necessary to move the connections in or out on the resistor leads as shown
in Fig. 5-6. This compensates for the effect of reactance.

5. Do not use the antenna to make this test for shorts and opens in
the line. Unlike the carbon resistor, any antenna will have some reactance
(and impedance) which changes with frequency. Thus, the antenna
impedance will match the line impedance only at one precise frequency. At

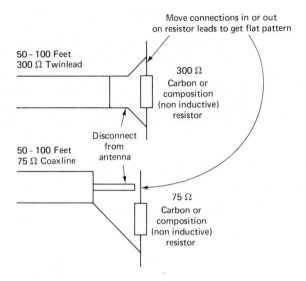

Fig. 5-6. Checking antenna lead-in for shorts or opens

all other frequencies, the antenna will not absorb all of the energy, and some signal will be reflected down the line to produce standing waves. You may not be able to tell these standing waves from those produced by shorts or opens. Figure 5-7 shows typical patterns produced when a 300-Ω line is properly terminated with a 300-Ω carbon resistance and the generator is swept across channels 6 and 11.

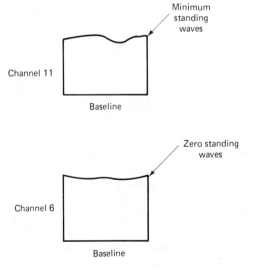

Fig. 5-7. Typical patterns produced by good 300-Ω line terminated with 300-Ω carbon resistance

5-1.3   Checking Lead-In to Antenna Match

The basic sweep frequency test can be used to check for a proper match between antenna and lead-in. If this test is made with a known good antenna and a drastic mismatch is indicated, the lead-in is probably defective. Likewise, if the lead-in is known to be good, and you get a mismatch indication during the test, the antenna is probably bad. At least, any indication of severe mismatch is an indication that the antenna and/or lead-in are not performing properly at the frequency where the mismatch is indicated.

1. Connect the equipment as shown in Fig. 5-8. Note that this is essentially the same as for the previous sweep frequency tests, except that the lead-in is connected to the antenna in the normal manner. If convenient, you can make this impedance match test first, before checking the lead-in for loss, opens, or shorts.

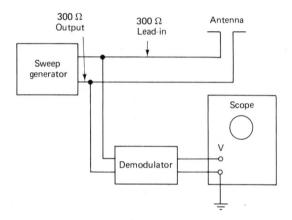

Fig. 5-8. Checking lead-in to antenna match

2. Check the scope pattern on all television channels. Set the sweep generator to produce a sweep of about 10 MHz across each channel.

3. If the antenna and lead-in are matched, the pattern will show very little standing waves on all channels. Ideally, the pattern will be almost flat on one channel (the frequency at which the antenna is matched to the lead-in) and relatively flat on all other channels.

4. The main concern in making this test is that there is no drastic change in standing wave patterns when you switch from channel to channel. If all channels show essentially the same pattern, you have a good match (and the antenna and lead-in are probably good). If one or more channels shows a large standing wave indication, with other channels showing low standing waves, there is a mismatch.

5. It is recommended that you make this test on your own antenna system, and on several known good systems, before attempting to judge any installation. As discussed, all antenna systems will show some standing waves (indicating some mismatch) at some frequencies. However, with experience, you should be able to judge what is a normal, acceptable level of mismatch for a typical system.

5-1.4   Checking Lead-In Impedance

The basic sweep frequency test can be used to measure the impedance of antenna lead-in wire or cable (either twinlead or coax).

1. Connect the equipment as shown in Fig. 5-9. Note that this is essentially the same as for the previous sweep frequency tests, except that the lead-in is terminated in a variable resistance (carbon or composition, noninductive, not wire wound).

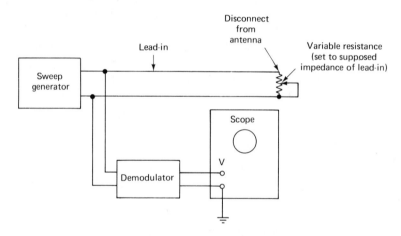

Fig. 5-9. Checking lead-in impedance

2. Set the variable resistance to the supposed value of the lead-in impedance (typically 300Ω for twinlead, 75Ω for coax; however, some coax may be 50).

3. Check the scope pattern on all television channels. Set the sweep generator to produce a sweep of about 10 MHz across each channel.

4. If the lead-in and resistance are matched, the pattern should be flat (no standing waves). If not, adjust the variable resistance until the pattern is flat, or at a minimum.

5. Disconnect the variable resistance (without disturbing its setting) and measure the resistance with an ohmmeter. This value is equal to the lead-in characteristic impedance (at the frequency, or band of frequencies, used for the test).

6. Note that it may be necessary to readjust the variable resistance to get a match at each television channel. If the resistance readings (or characteristic impedance of the lead-in) remain approximately the same (within 10%) from channel 2 through 13, this can be considered a normal condition. If the resistance indications are drastically different from channel to channel, this can indicate a defective lead-in. However, it is also possible that your test connections are introducing some reactance. Again, experience in making these tests on similar lead-ins will make you the best judge.

7. Another problem to watch is when you get a perfectly flat indication on all channels. Theoretically, this indicates a perfect match between lead-in and variable resistance. From a practical standpoint, it usually indicates a defective lead-in, improper test connections, or defective test equipment. To remove doubt, try adjusting the variable resistance to some value other than the supposed lead-in impedance. If the standing wave indications increase as the variable resistance is changed away from the supposed lead-in impedance but decrease when the resistance is returned to the correct impedance, it is reasonable to assume that the lead-in, test connections, and test equipment are operating normally. If there is no change in the pattern when the resistance is changed drastically (say from $300\,\Omega$ to $700\,\Omega$), check your connections, equipment, and lead-in.

5-1.5 Checking TVI Filters

The basic sweep frequency test can be used to measure the impedance of TVI (television interference) filters. The generator can also be used to check the efficiency of such filters.

Procedure (input impedance). Connect the equipment as shown in Fig. 5-10. Note that this is essentially the same as for the previous sweep frequency tests, except that the filter is connected to the generator through a sample of lead-in wire and to a resistance equal to the lead-in impedance. The resistance can be fixed or variable, as described in Sec. 5-1.4, whichever is convenient. However, the resistor must be noninductive (not wire wound).

1. Check the scope pattern on all television channels. Set the sweep generator to produce a sweep of about 10 MHz across each channel. Do not set the generator to a frequency that the filter is supposed to suppress.

2. If the filter is matched to the lead-in, the pattern should be flat (no standing waves). For example, if the line impedance is a true $300\,\Omega$ and a match is indicated by a flat trace, then the filter impedance is also $300\,\Omega$.

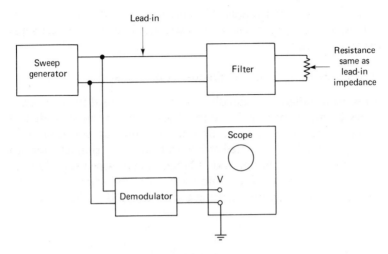

Fig. 5-10. Checking TVI filters (for input impedance)

3. If the trace is not perfectly flat and you are in doubt as to a match, disconnect the filter and connect the resistance to the lead-in. If the trace is flatted with only the resistance connected, there is some mismatch between lead-in and filter. If the trace is essentially the same with the filter in and out, the filter is properly matched to the lead-in.

Procedure (filter efficiency). Connect the equipment as shown in Fig. 5-11.

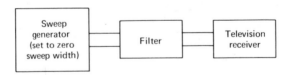

Fig. 5-11. Checking TVI filters (for efficiency)

1. Adjust the generator sweep width to zero. Set the generator to the frequency that the filter is supposed to suppress. If possible, operate the sweep generator to produce a modulated output.

2. Observe the interference pattern on the television receiver (1) with the filter out of the circuit and (2) with the filter connected.

3. The interference should be eliminated or greatly reduced when the filter is connected (if the filter is operating properly). If the filter is tunable, adjust the filter for maximum attenuation of the generator signal.

4. It should be noted that if the interference frequency is exactly on (or very near) the channel frequency (either sound or picture), the filter will also attenuate the carrier frequency. Filters are of little value in combating interference of this type. Generally, filters are designed to reject signals in the IF frequency range.

5-1.6 Checking Resonant Frequency of Stubs

In some installations, television antennas are tuned to cover a specific frequency by means of a tuning stub. These stubs are usually made up of two trimmer capacitors and a section of 300-Ω line, as shown in Fig. 5-12. In most cases, the line is shorted at one end, with the trimmer capacitors at the opposite end. The shorted stub is better than the open-ended stub, since there is less radiation from the end away from the trimmers. Such stubs can be checked and/or adjusted using a generator and oscilloscope.

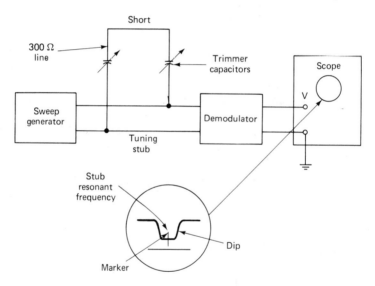

Fig. 5-12. Checking resonant frequency of stubs

1. Connect the equipment as shown in Fig. 5-12. Set the generator to the channel or frequency at which the stub is supposed to resonate. Adjust the sweep width to cover the entire channel (or frequency range).

2. Adjust the trimmer capacitors on the stub so that the dip in the oscilloscope trace is at the trace center. The stub should now be tuned to the approximate frequency range.

3. If greater accuracy is necessary, use the sweep generator markers (or an external marker generator). Superimpose the marker on the trace tip at the maximum dip point as shown in Fig. 5-12. Read the stub resonant frequency as indicated by the marker generator dial. If desired, set the marker to some precise frequency, and then adjust the stub trimmer capacitors until the marker is superimposed on the trace tip at the maximum dip point.

   4. It should be noted that the sharpness of the dip is dependent on the length of the stub line versus the trimmer values. Large trimmer values and short lines give broader response, whereas small trimmer values result in sharp resonant peaks (or dips).

5-1.7 Adjustment of Matching Q Bars

   Some television antennas are tuned by means of Q bars. These Q bars are connected at the point where the lead-in attaches to the antenna and are usually cut to length so that the antenna will be flat across a wide range of frequencies (such as across the entire VHF television band of channels). A sweep generator can be used to determine if the Q bars are cut to the correct length and are properly spaced.

   1. Connect the equipment as shown in Fig. 5-13.

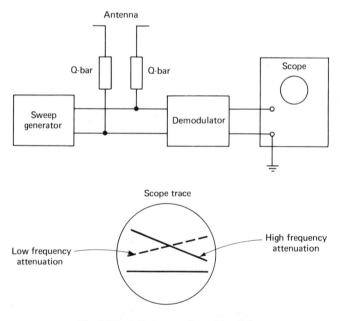

Fig. 5-13. Adjustment of matching Q-bars

   2. In turn, adjust the sweep generator to the low and high channels used in that area (or to channels 2 and 13). Operate the sweep generator at full sweep width (at least 10 MHz).

   3. The oscilloscope pattern should show a flat trace on both channels, indicating that the Q bars are cut to the correct length and are properly matched.

4. On either or both channels, if the oscilloscope trace slopes to the right [Fig. 5-13(b)] indicating that the higher frequencies are being attenuated, the Q bars are spaced too close together. If the opposite condition occurs and the low frequencies are attenuated, the Q bars are spaced too far apart.

5. If the oscilloscope trace is flat on one channel but not on others, or if there is a drastic change in trace amplitude from channel to channel, the Q bars may be cut too long or short, even though the bars are properly spaced.

5-1.8 Checking A Balun

A balun or any similar device that matches a single-ended unit to a double-ended unit (such as a 50- to 75-Ω output to a 300-Ω line) can be checked with a sweep generator.

1. Connect the equipment as shown in Fig. 5-14.

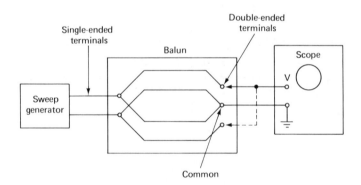

Fig. 5-14. Checking a balun

2. Adjust the sweep generator to the frequency at which the matching unit will be used. Operate the sweep generator at full width (at least 10 MHz).

3. Check the pattern across both double-ended terminals. Typically, there are three double-ended terminals (one common and two line terminals). Measure across the common and each line terminal, in turn, as shown. If there is any significant difference in the pattern across either set of terminals, the balun or other matching device is unbalanced.

4. To be thorough, make this test at all the frequencies with which the balun will be used.

5-1.9   Checking a Lightning Arrester

Various types of lightning arresters are used in television antenna installations. These arresters are supposed to bypass the lightning to ground, while passing the desired signals between antenna and receiver. If the arrester is not properly matched to the particular line, or is defective, the desired signals can be attenuated. A sweep generator can be used to check this condition.

1. Connect the equipment as shown in Fig. 5-15.

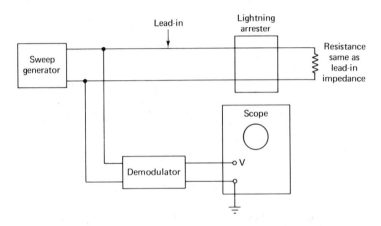

Fig. 5-15. Checking a lightning arrester

2. In turn, adjust the sweep generator to each of the frequencies with which the arrester is to be used. Operate the sweep generator at full width (at least 10 MHz).

3. First check the oscilloscope pattern with the arrester out of the line. Then insert the arrester and check for any change in the pattern. There should be no noticeable attenuation, or no substantial change in standing wave patterns, when the arrester is inserted. Also, the attenuation (if any) and standing wave patterns should be substantially the same for all channels or frequencies.

4. If there is any noticeable attenuation, increase in standing waves, or a pronounced slope or dip in the trace (when the arrester is inserted), the arrester is defective or is not properly matched to the line.

5-1.10   Checking Couplers and Splitters

There are a number of couplers that function to couple two (or more) television receivers to one antenna. These couplers can be checked for correct impedance match, loss, and flatness by means of a sweep generator.

Procedure (impedance check). Connect the equipment as shown in Fig. 5-16. Note that this is essentially the same as for the impedance test of TVI filters (Sec. 5-1.5), except that the coupler is connected to the generator through a sample of lead-in wire and to resistances that are equal to the

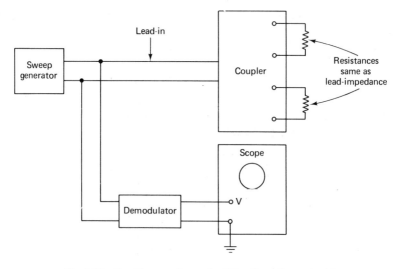

Fig. 5-16. Checking couplers and splitters (impedance check)

lead-in impedance. The resistances can be fixed or variable, whichever is convenient. However, the resistances must be noninductive (not wire wound).

1. In turn, adjust the sweep generator to the low and high channels used in that area (or to channels 2 and 13). Operate the sweep generator at full sweep width (at least 10 MHz).

2. If the coupler is matched to the lead-in, the pattern should be flat (no standing waves). For example, if the line impedance is a true $300\,\Omega$ and a match is indicated by a flat trace, then the coupler impedance is also $300\,\Omega$.

3. If the trace is not perfectly flat and you are in doubt as to a match, disconnect the coupler, and connect one resistance to the lead-in. If the trace is flat with the coupler out of the circuit, there is some mismatch between lead-in and coupler (or the coupler is defective). If the trace is essentially the same with the coupler in and out, the coupler is properly matched to the lead-in.

Procedure (flat response and attenuation). Connect the equipment as shown in Fig. 5-17.

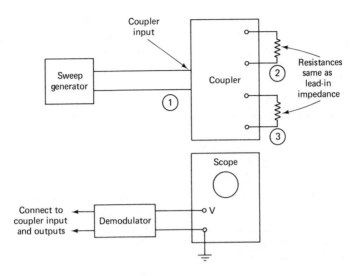

Fig. 5-17. Checking couplers and splitters (flat response and attenuation)

1. In turn, adjust the sweep generator to each of the channels with which the coupler is used. Operate the sweep generator at full width (at least 10 MHz).

2. Temporarily connect the oscilloscope probe to the coupler input (sweep generator output). This will establish a vertical reference deflection on the oscilloscope trace.

3. Connect the oscilloscope probe to each coupler output in turn. Note any difference in oscilloscope pattern at each of the outputs.

4. Note that all couplers will show some loss or attenuation. (The oscilloscope deflection at the output will be lower than at the input.) However, both (or all) outputs should show the same loss. If not, the coupler is defective.

5. It is also possible that some couplers will show considerable loss on a few channels but not on all channels. This may be due to coupler design, rather than to a defect. Further, some couplers may show a consistent loss over all channels but may not show a flat trace (indicating standing waves) on some channels. This indicates that the coupler response is poor on that particular channel. As a guideline, generally, those couplers with the greatest loss on all channels have the most even response (lack of standing waves).

5-2. TESTING TRANSFORMERS, YOKES, AND COILS

Transformers, yokes, and coils are often difficult components to test. Of course, if one of these components has a winding that is completely open or shorted or that has a very high resistance, this will show up as an abnormal voltage and/or resistance indication during troubleshooting. However, a winding that has a few shorted turns or is leaking to another winding may produce voltage and resistance indications that appear close to normal and will pass unnoticed.

An oscilloscope can be used to perform a *ringing* test on such components. In this procedure, a pulse obtained from the oscilloscope is applied to the component under test. The condition of the component can then be evaluated by the amount of *damping* observed in the waveform.

The best way to obtain reliable results from a ringing test is to compare the ringing waveform obtained from the part being tested with the waveform from a duplicate part that is known to be good. However, there are many times when a duplicate part is not available for comparison, and you must judge the part being tested by a study of the waveform. To gain experience in evaluating the ringing waveform, it is helpful to try the procedure several times, both with good parts and with parts that you have purposely shorted with various resistances.

5-2.1   Basic Ringing Test Procedure

Connections for the basic ringing test are shown in Fig. 5-18. The basic ringing test procedure is as follows.

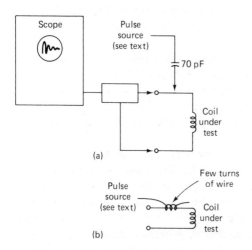

Fig. 5-18. Basic connections for ringing test of coils, transformers and yokes

1. Remove power from the part or circuit to be tested. Do not apply power to the circuit at any time during the test procedure.

2. Disconnect the part to be tested from the related circuit. Although many parts can be tested in circuit, it is usually necessary to remove circuit connections, especially in solid-state circuits where the diodes and transistors have a loading effect on the part.

3. Connect the oscilloscope to the part terminals as shown. Use a low-capacitance probe.

4. Connect a pulse source to the part through a capacitor as shown in Fig. 5-18(a) or through a few turns of wire as shown in Fig. 5-18(b). All oscilloscopes have circuits that produce pulses suitable for use in ringing tests. (These pulses are the same ones used to trigger the oscilloscope horizontal sweep.) Unfortunately, the pulses are not always accessible without modification of the oscilloscope. However, most oscilloscopes designed specifically for television service have an external terminal (either front panel or rear panel) that provides for electrical connection to the pulse source.

5. Operate the oscilloscope controls to produce a ringing waveform as shown in Fig. 5-19(a) or 5-19(b). The waveform of Fig. 5-19(a) is a typical

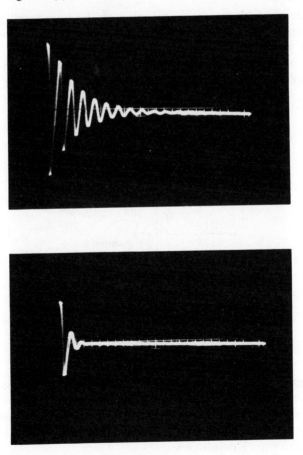

Fig. 5-19. Typical waveforms found in ringing tests *(Courtesy RCA)*

ringing pattern for a normal coil or transformer winding. The waveform of Fig. 5-19(b) is representative of patterns obtained from defective coils and transformers (with shorted turns or leakage in the windings). Keep in mind that these waveforms are typical. Experience is necessary to judge the waveforms when no duplicate parts are available for comparison.

   6. To help you make the judgment, connect a resistance across the part terminals (or make a loop consisting of a few turns of solder and pass the loop around the part). Note any change in the ringing pattern. There should be a drastic change in the pattern when the resistor (or solder coil is added. If not, the part is probably defective.

5-2.2   Ringing test for high-voltage (flyback) transformer

   To test the high-voltage transformer in the horizontal system of a receiver, remove the horizontal output and high-voltage rectifier tubes, or disconnect the equivalent leads in solid-state receiver circuits. Using the procedure described in Sec. 5-2.1, obtain a ringing waveform across the transformer primary. Typical test connections are shown in Fig. 5-20.

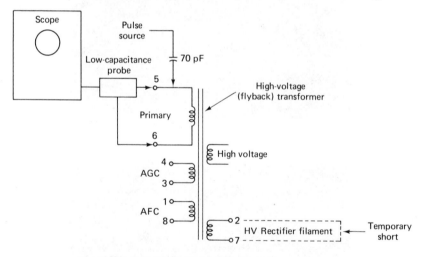

Fig. 5-20. Ringing test for high-voltage (flyback transformer)

   With the ringing pattern established, connect a short across the transformer filament winding. In the transformer of Fig. 5-20, connect the short across terminals 2 and 7. If a significant change in the waveform is noted, the transformer is probably good. Little or no waveform change indicates that the transformer is defective.

5-2.3 Ringing Test for Picture Tube Yoke

To test the horizontal and vertical windings of the deflection yoke, disconnect the yoke from the circuit. Using the procedure described in Sec. 5-2.1, obtain a ringing waveform across each winding, in turn. Typical test connections are shown in Fig. 5-21.

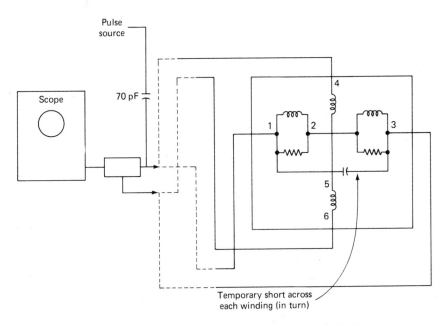

Fig. 5-21. Ringing test for picture tube yoke (series windings)

Note that the horizontal winding of the yoke in Fig. 5-21 consists of two sections. Alternately short each section. If the winding is good, the effect on the waveform should be the same as each section is shorted.

The vertical winding of the yoke also consists of two sections, with a damping resistor connected across each section. Disconnect these damping resistors, as well as the capacitor that is in parallel across both vertical sections. Then alternately short each section. If the winding is good, the effect on the waveform should be the same as each section is shorted.

In some yokes, the two sections of each winding are connected in parallel, as shown in Fig. 5-22. In such cases, disconnect the sections at one end. Connect the oscilloscope across both sections, as shown in Fig. 5-22(b), and then short each section, in turn. Again, the effect on the waveform should be the same as each section is shorted.

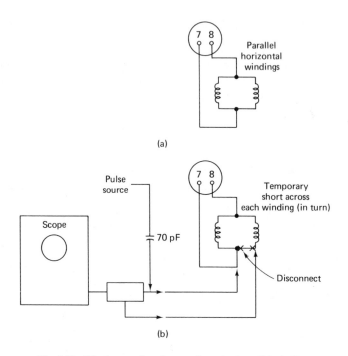

(a)

(b)

Fig. 5-22. Ringing test for picture tube yoke (parallel windings)

5-2.4   Ringing Test for Picture Tube Coils

Use the procedure described in Sec. 5-2.1 to test the width, linearity, focus, etc., of coils of a picture tube if they are suspected of having shorted turns or leakage. The solder loop usually produces the best results. Again, if the solder loop causes a significant change in the waveform, the coil under test is probably good. However, experience is the best judge when making any ringing test.

5-3. TELEVISION CHANNEL FREQUENCIES

The picture and sound carrier frequencies for all television channels (VHF and UHF) are given in Fig. 5-23.

5-4. INTERPRETING TELEVISION SERVICE LITERATURE

Although there is no standardization on service literature available for all television receivers, there is a general pattern for the literature. Usually, television receiver service literature is in the form of data sheets, rather than instruction books or manuals. As a minimum, the data sheets contain a

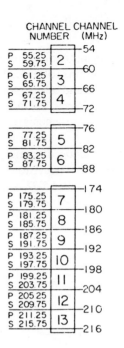

Channel	P (MHz)	S (MHz)	Band (MHz)
2	55.25	59.75	54–60
3	61.25	65.75	60–66
4	67.25	71.75	66–72
5	77.25	81.75	76–82
6	83.25	87.75	82–88
7	175.25	179.75	174–180
8	181.25	185.75	180–186
9	187.25	191.75	186–192
10	193.25	197.75	192–198
11	199.25	203.75	198–204
12	205.25	209.75	204–210
13	211.25	215.75	210–216

Channel	P (MHz)	S (MHz)	Band (MHz)
14	471.25	475.75	470–476
15	477.25	481.75	476–482
16	483.25	487.75	482–488
17	489.25	493.75	488–494
18	495.25	499.75	494–500
19	501.25	505.75	500–506
20	507.25	511.75	506–512
21	513.25	517.75	512–518
22	519.25	523.75	518–524
23	525.25	529.75	524–530
24	531.25	535.75	530–536
25	537.25	541.75	536–542
26	543.25	547.75	542–548
27	549.25	553.75	548–554
28	555.25	559.75	554–560
29	561.25	565.75	560–566
30	567.25	571.75	566–572
31	573.25	577.75	572–578

Channel	P (MHz)	S (MHz)	Band (MHz)
32	579.25	583.75	578–584
33	585.25	589.75	584–590
34	591.25	595.75	590–596
35	597.25	601.75	596–602
36	603.25	607.75	602–608
37	609.25	613.75	608–614
38	615.25	619.75	614–620
39	621.25	625.75	620–626
40	627.25	631.75	626–632
41	633.25	637.75	632–638
42	639.25	643.75	638–644
43	645.25	649.75	644–650
44	651.25	655.75	650–656
45	657.25	661.75	656–662
46	663.25	667.75	662–668
47	669.25	673.75	668–674
48	675.25	679.75	674–680
49	681.25	685.75	680–686

Channel	P (MHz)	S (MHz)	Band (MHz)
50	687.25	691.75	686–692
51	693.25	697.75	692–698
52	699.25	703.75	698–704
53	705.25	709.75	704–710
54	711.25	715.75	710–716
55	717.25	721.75	716–722
56	723.25	727.75	722–728
57	729.25	733.75	728–734
58	735.25	739.75	734–740
59	741.25	745.75	740–746
60	747.25	751.75	746–752
61	753.25	757.75	752–758
62	759.25	763.75	758–764
63	765.25	769.75	764–770
64	771.25	775.75	770–776
65	777.25	781.75	776–782
66	783.25	787.75	782–788
67	789.25	793.75	788–794

Channel	P (MHz)	S (MHz)	Band (MHz)
68	795.25	799.75	794–800
69	801.25	805.75	800–806
70	807.25	811.75	806–812
71	813.25	817.75	812–818
72	819.25	823.75	818–824
73	825.25	829.75	824–830
74	831.25	835.75	830–836
75	837.25	841.75	836–842
76	843.25	847.75	842–848
77	849.25	853.75	848–854
78	855.25	859.75	854–860
79	861.25	865.75	860–866
80	867.25	871.75	866–872
81	873.25	877.75	872–878
82	879.25	883.75	878–884
83	885.25	889.75	884–890

Fig. 5-23. Television channel frequencies *(Courtesy B&K-Precision, Dynascan Corporation)*

schematic diagram which shows all component values, component location diagrams (often called transistor placement diagrams), and complete alignment and adjustment instructions. Also, most data sheets describe receiver characteristics, including any special features.

The schematic diagrams show all component values, normal operating voltages, coil resistances, waveforms, and peak-to-peak voltages. The schematic diagrams given in Chapter 6 of this book are typical of those found in television receiver data sheets.

Figure 5-24 is typical of the placement diagrams. These diagrams show

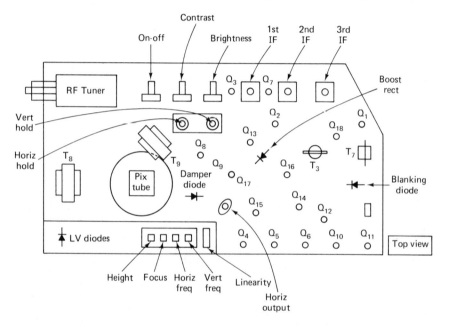

Fig. 5-24. Typical component location diagram (transistor placement diagram) found on television receiver data sheet

the physical relationships of important parts (generally all transistors and major diodes), all test points, and all alignment/adjustment controls. Often the placement diagrams are supplemented by photographs with callouts that show the actual appearance of each part on the chassis. The diagrams may also be supplemented with a list of suitable replacement parts, disassembly instructions, and receiver specifications.

Figure 5-25 is typical of the alignment and adjustment instructions. Usually, these instructions are quite detailed. You should always follow the specific instructions found in the receiver data sheet, rather than the generalized instructions found in Chapters 3 and 4 of this book. However, a study of the information in Chapters 3 and 4 should make it possible for you to quickly interpret and perform any receiver data sheet procedures.

Use an isolation transformer and maintain voltage at 117 volts. Allow a 20-minute warm-up period for the receiver and test equipment.
Suggested Alignment Tools: L_1 through L_{18} .... GENERAL CEMENT #9087 .... WALSCO #2528
WALSCO #2584

VIDEO IF ALIGNMENT

Connect the synchronized sweep voltage from the sweep generator to the horizontal input of the oscilloscope for horizontal deflection.
Use only enough generator output to provide a usable indication. Note: Response may vary slightly from those shown.
Connect a variable bias supply to the IF AGC line (point ◇) and adjust to obtain a response curve which shows no indication of overload.
Disable Oscillator section of Mixer-Osc. Set the Channel Selector to any non-interfering channel.

	Indicator	Sweep/marker generator coupling	Sweep generator frequency	Marker generator frequency	Adjust	Remarks
1.	Connect DC probe of a EVM through a 47 K resistor to point ◇. Common to ground.	Connect high side through a 100 pF capacitor to point ◇ on Tuner. Low side to ground.		22.25 MHz 28.25 MHz	L_1 L_2	Adjust for MINIMUM.
2.	Connect DC probe of a EVM through a 47 K resistor to point ◇. Common to ground.	Connect high side through a 100 pF capacitor to point ◇ on Tuner. Low side to ground.		22.5 MHz 26.3 MHz 23.6 MHz 24.3 MHz	L_3 L_4 L_5 Mixer collector coil	Adjust for maximum.
3.	Connect vertical input of a scope to point ◇. Low side to ground.	Connect high side through a 100 pF capacitor to point ◇ on Tuner. Low side to ground.	25 MHz (10 MHz Sweep)	22.25 MHz 23.5 MHz 25.25 MHz 26.75 MHz 28.25 MHz		Check for maximum gain and symmetry of response with markers as shown in Figure 1. In order to obtain a proper response, it may be necessary to SLIGHTLY retouch L_1 through L_4 and Mixer Collector Coil.
4.	Connect DC probe of a EVM to point ◇. Common to ground.	Connect high side through a 100 pF capacitor to point ◇ on Tuner. Low side to ground.	44 MHz (10 MHz Sweep)	26.75 MHz	L_8	Adjust for maximum.

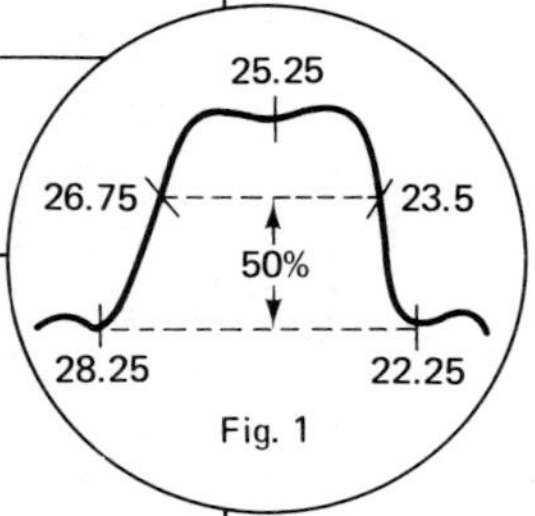

Fig. 1

Fig. 5-25. Typical alignment instructions found on television receiver data sheets

Well-written data sheets will show such vital information as at what specific test point to connect indicators (meters or scopes) for signal tracing, where to inject signals, exact frequencies to be used, the specific receiver controls to adjust, the recommended tool for adjustment, as well as all general test conditions to be observed during adjustment.

For example, in row 2 of the example instructions (Fig. 5-25), you are told to connect a signal-tracing meter to test point J, inject a signal from a sweep/marker generator at test point D, set the marker generator to produce signals at specific frequencies (25.5, 26.3, 23.6, and 24.3 MHz), and adjust receiver alignment controls L_1, L_2, L_3, and L_4 for a maximum indication on the meter. In the notes prior to these specific instructions, you are told how to operate the sweep/marker generator, at what test point to clamp the AGC line, as well as what alignment tools are suggested.

5-5. INTERPRETING TELEVISION TEST PATTERNS

Most television stations broadcast test patterns at some time (usually at the beginning and end of the broadcast day). Test pattern generators, such as described in Chapters 2 and 3, are also becoming quite popular. The instruction manuals supplied with the test pattern generators describe how their particular patterns can be used to adjust the receiver as well as how to determine receiver characteristics.

Figures 5-26(a) and 5-26(b) are typical of the test patterns produced by generators. The following notes describe how these test patterns are to be interpreted. Most test patterns (both from generators and broadcast stations) will have characteristics similar (but not identical) to those shown in Figs. 5-26(a) and 5-26(b). If you are using a pattern generator, follow the instructions supplied with that particular model. If you are without any instructions, use the following notes as a starting point to interpret test patterns available to you.

1. The pattern circles are provided for accurate checking and adjustments of picture height, width, linearity, and centering. All pattern circles should be round, and the center of the pattern should be adjusted so that it is at the physical center of the receiver screen. In the pattern of Fig. 5-26(a), the instructions recommend that the circles not overlap the sides of the picture tube by more than ¾. The instructions for the pattern of Fig. 5-26(b) recommend that you set the top and bottom of the circle to the top and bottom of the receiver screen.

2. The pattern of Fig. 5-26(a) has vertical and horizontal grid lines [not found in Fig. 5-26(b)] used to check the standard aspect ratio of 4:3. (Color as well as black and white television pictures are broadcast with an aspect ratio of 4:3, width to height.) The correct ratio is indicated if the

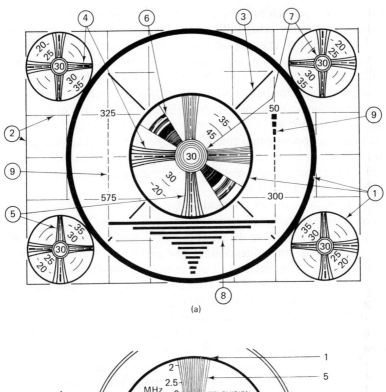

(a)

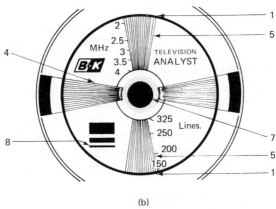

(b)

Fig. 5-26. Typical television test patterns

grid lines appear as squares. If the grid lines appear rectangular but straight, the ratio is not correct. If either the vertical or horizontal lines are not straight, even though they form a square, there is some distortion (such as the pincushion effect or keystoning). Thus, the grid lines can be used to detect any form of vertical or horizontal nonlinearity in the picture (that might not appear when checking against pattern circles).

3. The pattern of Fig. 5-26(a) also has diagonal lines not found in the pattern of Fig. 5-26(b). These diagonal lines provide for a check of interlacing. Poor interlacing (also known as line splitting or line pairing) will make the diagonal lines appear to be ragged. As discussed in Chapter 6, poor interlacing is usually associated with the vertical sweep circuits and is an indication of undesired signals entering the vertical system.

4. Both patterns have horizontal wedges that can be used to check vertical resolution (in terms of horizontal lines). The pattern of Fig. 5-26(a) is provided with numbers to evaluate the vertical resolution. The point at which the wedges become indistinct (lines begin to merge) is an indication of vertical resolution (in terms of a specific number of horizontal lines). For example, if the lines begin to merge at 37 on the screen, add a zero to find a vertical resolution of 370 lines.

5. Both patterns have vertical wedges that can be used to check horizontal resolution (in terms of vertical lines), and receiver bandwidth. In the pattern of Fig. 5-26(a), note the number on the screen at which the vertical wedges become indistinct (lines merge), and add a zero. This gives the horizontal resolution in vertical lines. For example, if the lines merge at 32, the horizontal resolution is 320. Divide this number by 80 to obtain the receiver bandwidth in megahertz. Using the example of 320 lines, the bandwidth is 4 MHz (320/80 = 4). In the pattern of Fig. 5-26(b), use the top vertical wedge to find the frequency response of the receiver. Simply note the point at which the lines merge on the top wedge, and read the frequency (in megahertz) from the corresponding number. For example, if the lines merge only at the very bottom of the top vertical wedge (near the screen center), the bandwidth is 4 MHz. Use the bottom vertical wedge of the Fig. 5-26(b) pattern to find horizontal resolution. Note the point at which the lines merge on the bottom wedge, and read the corresponding number.

6. The pattern of Fig. 5-26(a) is provided with diagonal wedges that indicate variations in contrast. The diagonal wedges have four shading tones. If all four shades are present and distinct, the receiver video amplifier has a linear response (and the video portion of the picture tube is good). The diagonal wedges can also be used to adjust brightness, contrast, and AGC controls.

7. The pattern of Fig. 5-26(a) is provided with a number of small center circles (often called bull's-eyes). These circles can be used to check and adjust the picture tube beam focus. The center circle of the Fig. 5-26(a) pattern can also be used to check focus.

8. Both patterns are provided with horizontal bars to indicate low-frequency response of the receiver as well as phase-shift, ringing, and overshoot conditions. If all of the bars are reproduced clearly, the low-frequency response of the receiver is good. If any of the bars are not clearly reproduced, the receiver has poor low-frequency response. This may be due to improper alignment or a defect in the video amplifier circuits. Such problems are usually evidenced by black or white trailing edges on the bars (and possibly on the vertical wedges). If the trailing edge is black, this usually indicates low-frequency phase shift. A white trailing edge usually shows ringing or overshoot in the video amplifier.

9. The pattern of Fig. 5-26(a) is provided with resolutions lines on each side of the center circle. These lines indicate horizontal resolution, in steps of 25, ranging from 50 to 575 lines. The lines are used to check the video amplifier for ringing and overshoot. If ringing occurs at some video frequency, the resolution line near that frequency will be repeated several times at evenly spaced intervals. To convert the resolution line indications into the frequency at which ringing occurs, divide the number of lines by 80. For example, if the 575 resolution line is repeated, the video amplifier is ringing at about 7.2 MHz ($575/80 = 7.2$).

5-6. ADJUSTING THE TELEVISION PICTURE WITH AN ANALYST TEST PATTERN

It is difficult to adjust centering, size, and linearity of the television picture without the aid of a test pattern such as an analyst generator provides. Using the picture that is normally transmitted by a station, no image remains fixed long enough to permit complete adjustment. Also, any image not containing a fairly large circular pattern will not provide enough visible markings to establish true linearity. If this method is attempted, the non-linearity will be apparent when a circular object finally does appear on the screen.

The test pattern produced by the B&K 1077 Television Analyst is the ideal signal source for making centering, size, and linearity adjustments. The pattern is unchanged over the duration of the adjustment sequence, has a large outer circle as a standard for making size adjustments, and has a number of smaller circles which should all be perfectly round when the linearity of the deflection is properly adjusted. Use the following procedure for all black and white receivers as well as for the starting point for adjustment of color receivers.

1. Measure from opposite diagonal corners of the picture tube screen, and mark the physical center of the screen with a grease pencil or similar marker, as shown in Fig. 5-27.

2. Connect the analyst to the receiver antenna terminals. Operate the analyst controls (as described in the analyst instruction manual) and the receiver controls as necessary for the best test pattern display.

3. Adjust the ion trap or centering adjustment of the receiver so the center circle of the test pattern coincides with the physical center of the

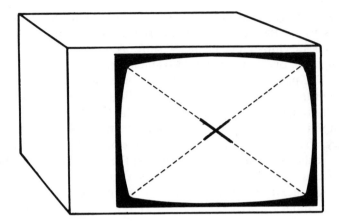

Fig. 5-27. Marking physical center of screen *(Courtesy B&K-Precision, Dynascan Corporation)*

screen. It should be noted that if the test pattern is badly distorted, the linearity adjustments (or troubleshooting and repair if necessary) should be performed before the centering adjustment is completed.

    4. If the receiver has a width adjustment, adjust it so the raster just fills the screen.

    5. Adjust the height adjustment so the outer circle of the test pattern just fills the screen vertically.

    6. Adjust the vertical linearity and horizontal linearity adjustments for a round outer circle and round inner circles on the test pattern display. The display should appear as shown in Fig. 5-28 when properly adjusted.

    7. Wipe the grease pencil centering marks from the face of the picture tube.

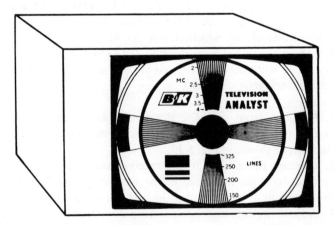

Fig. 5-28. Proper centering, size, and linearity *(Courtesy B&K-Precision, Dynascan Corporation)*

6

Troubleshooting Solid-State
TV Circuits

This chapter is concerned with practical troubleshooting techniques for solid-state TV receivers. It is assumed that the reader is already familiar with basic troubleshooting procedures. Such basic information is discussed in Chapter 1 through 5 and will not be repeated here. However, since it is necessary to understand the operation of a circuit to do a good trouble-shooting job, a brief discussion of operational theory for each circuit is given.

A consistent format is used throughout the chapter for each circuit. First, typical solid-state TV circuit diagrams are given, along with the basic theory of operation. Second, the recommended troubleshooting approach for that particular type of circuit is discussed. Then a group of typical troubles is described, along with the most likely causes of such troubles.

6-1. LOW-VOLTAGE POWER SUPPLY

Figure 6-1 is the schematic diagram of a typical low-voltage power supply for solid-state TV receivers. The circuit consists essentially of a full-wave bridge rectifier CR_1, followed by a solid-state (Zener diode CR_2 and transistor Q_1-Q_3) regulator circuit. As with many portable solid-state TV receivers, the set can also be operated with self-contained rechargeable batteries. Thus, the power supply circuit also provides for recharging the batteries. The set can also be operated from an auto battery (12 V) by means of an adapter plug connected to the auto's cigarette lighter output. This arrangement is typical of small portable TV receivers.

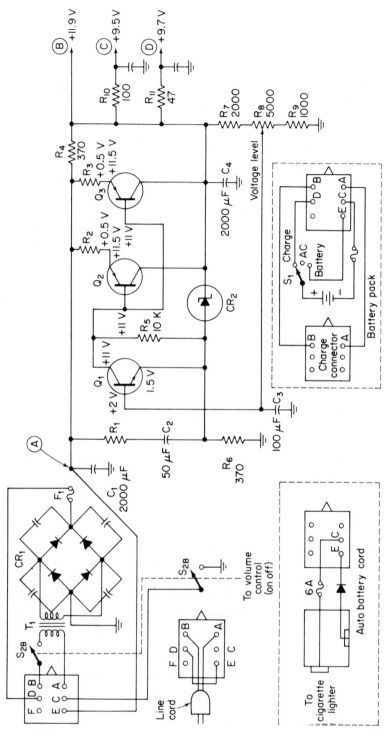

Fig. 6-1 Low-voltage power supply

6-1.1   Basic circuit theory

The basic function of the circuit is to provide dc at approximately 10 to 12 V. This voltage is sufficient to operate all circuits of the set except the high voltage required by the picture tube. Typically, high voltages are supplied by the flyback circuit (horizontal output and high voltage) as described in Sec. 6-2. Note that in the circuit of Fig. 6-1 the negative side of the power supply is grounded. Positive grounding is used equally as often.

The line voltage is dropped to about 12 V by transformer T_1 and is rectified by CR_1. The output from CR_1 is regulated by Q_1, Q_2, and Q_3 as well as by Zener CR_2 and is distributed to three separate circuit branches, each at a slightly different voltage level.

Operation of the regulator is standard. The emitter of Q_1 is held constant by Zener diode CR_2, whereas the base of Q_1 depends on output voltage. Any variations in output voltage (resulting from changes in input voltage or variations in load) change the base voltage in relation to emitter voltage. These changes appear as a variation in Q_1 connector voltage and, consequently, Q_2-Q_3 base voltage.

Transistors Q_2 and Q_3 are connected in series with the output in CR_1 and the load. Thus, Q_2 and Q_3 act as variable resistors to offset any changes in output voltage. For example, if the output voltage increases, the base of Q_1 swings more positive, causing a drop in Q_1 collector voltage. This causes the bases of Q_2 and Q_3 to swing negative and results in more collector current flow (since Q_2 and Q_3 are PNP). Thus, the increase in output voltage is offset. The level of the output voltage is set by R_8.

When the set is to be operated on battery power, switch S_1 is set to *battery*, the ac power plug is removed, and the battery pack cord is plugged in. When the battery is to be recharged, switch S_1 is set to *charge*, and the ac power cord is plugged into a special connector on the battery pack. Although there is no standardization, rechargeable batteries provide about 4 to 6 h (hours) of operation, and require about 8 to 12 h of recharging time. When the set is to be operated from an auto's cigarette lighter, the ac power plug is removed, and the auto battery power cord is plugged in.

6-1.2   Recommended troubleshooting approach

If the symptoms indicate a possible defect in the low-voltage supply, the obvious approach is to measure the dc voltage. If there are many branches (three output branches are shown in Fig. 6-1), measure the voltage on each branch. If any of the branches are open (say, an open R_{10} or R_{11}), the voltage on the other branches may or may not be affected. However, if any of the branches are shorted, the remaining branches will probably be affected (the output voltage will be lowered).

Current measurements. If a clamp-type current probe is available, measure the current in each branch. Most TV service shops are not equipped with current probes. Likewise, many solid-state sets use *printed circuit boards* where the output leads are etched wiring rather than a wire or group of wires and a probe cannot be used. The use of printed circuit boards also eliminates the possibility of disconnecting each output branch, in turn, until a short or other defect is found. This practice is common on vacuum-tube sets or those sets without printed wiring.

Resistance measurements. If one or more output voltages appear to be abnormal and it is not practical to measure the corresponding current, remove the power and measure resistance in each branch. Compare actual resistances against those in the service literature, if available. If you have no idea as to the correct resistance, look for obvious low resistance (a complete short or a resistance of only a few ohms).

Isolation transformer. Always use an isolation transformer for the ac power line. In many solid-state circuits, one side of the ac line is connected to the chassis, even though the set may have a power transformer. As a convenience, have a 12-V dc supply to substitute for the low-voltage power supply circuit of the set. A 12-V battery eliminator makes a good source if a conventional supply is not available. Ideally, the substitute power supply should be adjustable (in output voltage) and have a voltmeter and ammeter to monitor power applied to the set.

Battery operation. If the set can be operated on batteries, switch to battery operation and see if the trouble is cleared. If so, the problem is definitely localized to the low-voltage power supply circuit. Likewise, if the set can be operated from an auto battery or similar arrangement, switch to that mode and check operation.

Zener replacement. If it becomes necessary to replace a Zener diode in the regulator circuit, always use an *exact replacement.* Some technicians will replace a Zener with a slightly different voltage value and then attempt to compensate by adjusting the regulator circuit. This may or *may not* work.

Oscilloscopes. Do not overlook the use of an oscilloscope in troubleshooting the power supply circuits. Even though you are dealing with dc voltages, there is always some ripple present. The ripple frequency and the waveform produced by the ripple can help in localizing possible troubles. Also, an oscilloscope with a dc/ac vertical input switch will work as a dc voltmeter.

6-1.3   Typical Troubles

In the following paragraphs we shall discuss symptoms that could be caused by defects in the low-voltage power circuits.

No sound and no picture raster. When there is no raster on the picture tube screen and no sound whatsoever, it is likely that the low-voltage power circuits are totally inoperative or are producing a very low voltage (less than 20 to 30%). If the circuits are producing a voltage about 50% of normal (say 6 V in a 12-V system), there will be some sound, even though the raster may be absent. (This symptom is discussed further in a later paragraph.) The most likely defects are filter capacitor leakage, a defect in the regulator (regulator completely cut off), or a short in the output line.

Start the troubleshooting process by making voltage measurements at test points A, B, C, and D or their equivalent.

If the voltage is absent at all test points, check all fuses and switches as well as CR_1 and T_1. If the voltage at B, C, and D is absent or very low but the voltage at A is high, the regulator circuit is probably at fault. If all the voltages are low, look for a short in one or more of the output lines.

To check the regulator, first make sure that all three transistors are forward-biased. Typically, the base of Q_1 will be about 2 V, with an emitter voltage of 1.5 to 1.8 V. The bases of Q_2 and Q_3 will be about 11 V, with the emitters at 11.5 V. In any event, all three transistors must be forward-biased for the regulator to operate.

If any of the transistors are not forward-biased, remove power and check the transistors in circuit (with an in-circuit tester) or out of circuit (ohmmeter method) as described in Chapter 1. If the transistors appear to be good but not forward-biased, check all of the associated resistors and capacitors with an ohmmeter on a point-to-point basis.

Note that it is possible for transistor Q_1 to be cut off while Q_2 and Q_3 are forward-biased. However, such a condition will quickly point to a fault in Q_1 or its associated parts.

A common fault in such regulator circuits is a base-emitter short in the current-carrying transistors Q_2 or Q_3. Since these transistors are in parallel, it may be difficult to tell which transistor is at fault. If necessary, disconnect each transistor and check it separately. If either Q_2 or Q_3 has a base-emitter short, the regulator will be cut off, and both the base and emitter will be high.

No sound, no picture raster, and transformer buzzing. When there is no sound or picture, accompanied by transformer buzzing, there is probably an excessive current being drawn in the power supply circuit. The most likely defects are a short circuit ahead of the regulator (test point A, for example), shorted rectifiers, or shorted turns in the power transformer.

Again, start the troubleshooting process by making voltage measurements at test points A, B, C, and D. Keep in mind that prolonged excessive current will cause one or more fuses to blow. Thus, with these symptoms, the assumption must be excessive current that is below the rating of the fuses. Such a condition will normally result in a very low voltage, but voltage will not be completely absent.

Unlike the previous symptoms, the regulator is probably operating properly within its capabilities. Thus, concentrate on short circuits, particularly the rectifiers, and possibly shorted transformer turns. Both T_1 and CR_1 can be checked by substitution or by an ohmmeter test, whichever is convenient.

If this does not localize the problem, check for short circuits in each branch of the output. Also look for overheated parts, such as the transformer or burned printed circuit wiring.

Distorted sound and no raster. These symptoms are similar to those previously described, except that there is some sound (usually with inter-carrier buzz). Generally, this indicates that the power supply output voltage is below normal but not absent or near zero. These symptoms are produced when the power supply output is about 40 to 60% of normal. The most likely defects are the filter capacitor, rectifiers, and power transformer.

Start the troubleshooting process by making voltage measurements at test points A, B, C, and D or their equivalents. Next, measure the amplitude and frequency of the waveforms at the same test points, particularly at the bridge output (test point A) and the regulated dc output (test point B). An analysis of the waveforms in a solid-state supply can often pinpoint troubles immediately. The following are some examples.

The normal waveform at the bridge output (test point A) is a sawtooth (almost) at twice the line frequency (usually 120 Hz), as shown in Fig. 6-2(a). If the filter and regulator are operating properly, the 120-Hz signal will be suppressed at the final output. However, there may be some line frequency (60-Hz) ripple at the regulator output (all three branches).

If there is a strong 120-Hz waveform at the regulator output, this indicates that the filter and/or regulator are not suppressing the 120-Hz bridge output. The most likely cause is an open filter capacitor C_1 or a regulator circuit defect. Capacitor C_1 could be leaking. However, excessive leakage or a short would blow the fuse.

Capacitor C_1 can be checked for an open by connecting a known good capacitor in parallel. Make sure to use the right capacity (and right polarity if C_1 is an electrolytic). *Avoid connecting* the test capacitor leads across C_1 with the *power applied.* The voltage surge could damage transistors

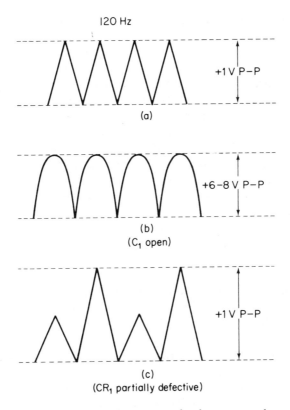

Fig. 6-2 Typical solid-state, regulated power supply waveforms

throughout the set. Remove the power, connect the test capacitor, and then reapply the power. Keep in mind that a test capacitor will not show leakage in C_1. Substitution is the only true test.

The waveform at test point A will also indicate the condition of capacitor C_1. If the waveform is a 120-Hz half sine wave similar to Fig. 6-2(b), rather than the sawtooth wave of Fig. 6-2(a), capacitor C_1 is probably open. This is confirmed further if the amplitude of the waveform at test point A increases from a typical 1 V (or less) to several volts, as shown in Fig. 6-2(b).

The condition of rectifier CR_1 is also indicated by the waveform at A. If the waveform is 60 Hz (line frequency), then one-half of rectifier CR_1 is defective (most likely shorted or open). If the waveform is not symmetrical [Fig. 6-2(c)], one diode of CR_1 is probably defective (most likely leaking or with a poor front-to-back ratio).

Note that if any of the diodes in CR_1 are shorted, transformer T_1 will probably run very hot. Thus, a hot T_1 does not necessarily indicate a bad T_1. However, if the waveform at test point A is good (indicating a good

CR_1 and C_1) but T_1 is hot, the most likely problem is a defective T_1 (probably shorted turns).

Note that all four diodes in CR_1 are paralleled with capacitors to protect the diodes in case of sudden voltage changes that might exceed the breakdown voltage. If one of these capacitors is shorted, this can give the appearance of a shorted diode. If CR_1 is a sealed package with self-contained capacitors, the entire package must be replaced. If the capacitors can be replaced separately from the diodes, make sure that *both the capacitor and diode* are good before replacing either. For example, if the capacitor is open, the corresponding diode may be damaged. If the diode is replaced, the trouble will be repeated unless the capacitor is also replaced.

Keep in mind that an increase in load (say, due to a short or partial short) will cause an increase in ripple amplitude, even with a good filter and regulator, because a larger load (lower load resistance) causes faster discharge of the filter capacitor between peaks of the sine-wave pulses from the rectifier.

Picture pulling and excessive vertical height. Thus far, we have discussed symptoms and troubles that result from a low power supply output due to defective parts or shorts. It is possible that the power supply can produce a high output voltage, resulting in picture pulling (the raster is stretched vertically and is bent). Hum bars across the picture screen are usually present with this condition.

Assuming that the trouble is definitely in the low-voltage supply, the most likely cause is in the regulator circuit. A defect in the rectifier, filter, or transformer is likely to result in a *low-voltage output*, producing the symptoms described previously.

It is possible for the regulator circuit to be cut off, causing the output lines to increase in voltage. The logical approach is to start the troubleshooting process by making voltage and waveform measurements at all test points.

If the voltage at test point B is nearly the same as at test point A (within about 0.5 V), the regulator is probably cut off. The voltage at test points C and D will also be high. The condition can be confirmed further if the waveform at test points B, C, and D is at 120-Hz (indicating that the regulator is not suppressing the 120-Hz bridge output).

With the trouble definitely pinned down to the regulator, test all of the transistors (Q_1 through Q_3) and the Zener diode (CR_2) as previously discussed.

Low brightness with insufficient height. When the low-voltage power supply output drops below normal but is not really low, one of the first symptoms is low picture brightness. Picture height and width are also reduced, but height is generally reduced more.

There are several possible causes for a small reduction in power supply output voltage. One of the bridge rectifier diodes (or the related protective capacitor) can be shorted. Or the diode can be open, but this is less likely. Capacitor C_1 can be leaking. There may be a partial short circuit in one of the output lines, thus dropping the voltage at all outputs by a small amount.

Start by checking the voltages at test points B, C, and D. Try to correct a low voltage condition at all outputs by adjustment of the voltage level control (R_8 of Fig. 6-1). It is possible that the low-voltage output is due to component aging, and a simple adjustment of R_8 does not restore the correct output voltage, or if it is necessary to set R_8 to one extreme in order to have the correct voltage, look for trouble in the rectifier CR_1 or the filter C_1. Use the procedures previously discussed for check of CR_1 and C_1. For example, a waveform similar to Fig. 6-2(b) at test point A indicates an open C_1. The waveform of Fig. 6-2(c) indicates a partial defect in CR_1. Keep in mind that a cold solder joint or a break in printed-circuit wiring can simulate a defective (open) C_1.

Insufficient height and width, weak sound, and snowy picture. These are also symptoms of below-normal voltage from the low-voltage power supply. Again, start by checking voltages at test points B, C, and D, and try adjustment of R_8. Also measure the waveform and voltage at test point A. If the indications are correct at A but the voltages at B, C, and D are low (and cannot be adjusted), the regulator is at fault. This means that you must dig into the circuit, looking for such problems as short circuits (from solder splashes) and breaks in printed circuit wiring, as well as defective transistors, resistors, and capacitors. Let us consider a few examples.

When troubleshooting any circuit such as the regulator of Fig. 6-1, voltage and resistance measurements are generally more useful than oscilloscope checks. Likewise, capacitor failures are more likely than resistor failures.

Assume that R_7 is open. This will show up in several ways. First, R_8 will have little or no control of the regulator circuit (little effect on output voltages). The base voltage of Q_1 will be abnormal, as will the resistance at test point B. However, the waveforms throughout the regulator circuit could remain normal (or so close to normal as to be unnoticed).

Now assume that C_3 is shorted. This will cut off Q_1 and provide a partial short across the output lines. Such a defect is generally easy to locate since the voltage at the base of Q_1 (normally 2 V) will be zero. Likewise the resistance from the base of Q_1 to ground will be zero.

When tracking troubles by means of voltage-resistance measurements, it is often necessary to check several components for one abnormal reading. For example, assume that the emitter of Q_1 shows a high reading (say 3 V instead of 1.5 V). There are three logical suspects. First, C_2 could be leaking. This will apply a large voltage to the emitter of Q_1 through R_1. Next on the list is a defective R_6 where the resistance has increased (say due to a partial break in the composition resistance material). Of course, Zener diode CR_2 is supposed to overcome minor changes in R_6 resistance. However, a large change in R_6 could prevent CR_2 from maintaining the desired 1.5 V at the emitter of Q_1. Finally, CR_2 could be at fault, even though Zener diodes are generally rugged.

6-2. HIGH-VOLTAGE SUPPLY AND HORIZONTAL OUTPUT

There is very little standardization in the high-voltage and horizontal output circuits of solid-state TV receivers. There are two basic circuit types: the *hybrid* circuit, which uses a vacuum tube (usually subminiature) for the high-voltage rectifier, and the *all-solid-state,* which uses solid-state rectifiers. At one time, most sets were of the hybrid type. Later, the hybrid circuit was used mostly for large-screen TV. Today, the trend is toward all-solid-state circuits. However, since there are many hybrid sets in use, both circuits are discussed in this section.

6-2.1 Basic Circuit Theory

Figure 6-3 is a typical hybrid circuit; Fig. 6-4 is a typical all-solid-state circuit. In both cases the main functions of the circuits are to provide a high voltage for the picture tube second anode and to provide a horizontal deflection yoke. In practically all cases, the circuits also supply a boost voltage for the picture tube focus and accelerating grids and possibly a voltage source for the video output transistor. [In many solid-state sets the video output transistor operates at a higher voltage (40 to 70 V) than the other transistors (typically 12 V).]

In some cases, the horizontal output circuits also supply an AFC signal to control the frequency of the horizontal oscillator and an AGC signal to provide *keyed* AGC. These functions are discussed in later sections.

Although there is little standardization, most horizontal output circuits have certain characteristics in common that must be considered during the troubleshooting process. The circuit receives pulses from the horizontal driver (at 15,750 Hz) synchronized with the picture transmission. The horizontal output transistor is normally biased at or near zero so that one edge of the pulse (negative in this case since the transistor is PNP) will drive the transistor into heavy conduction (near saturation); the opposite swing of the pulse will cut off the horizontal output transistor. In effect, the horizontal output transistor is operated in the switching mode.

The collector current of the horizontal output transistor is applied through a winding on the flyback transformer, resulting in a pulse waveform

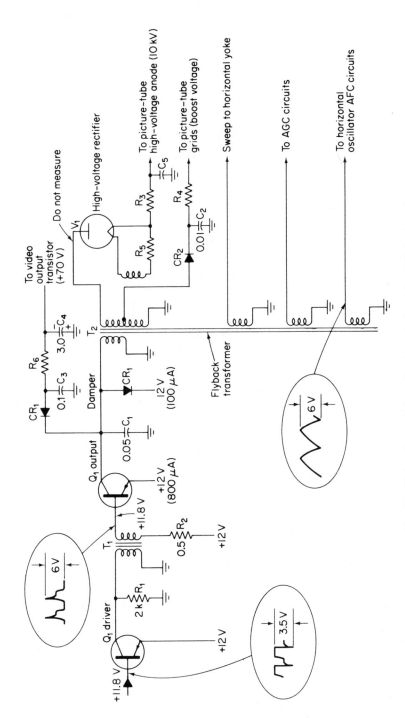

Fig. 6-3. High-voltage supply and horizontal output circuit (hybrid type)

347

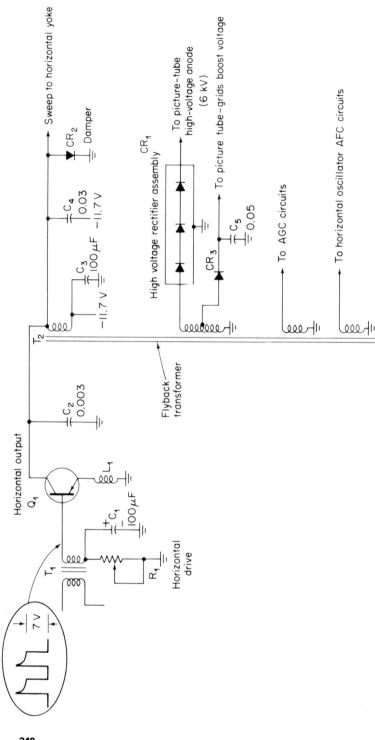

Fig. 6-4. High-voltage supply and horizontal output circuit (all solid-state type)

Sweep to horizontal yoke

CR₂
Damper

To picture-tube
high-voltage anode
(6 kV)

CR₁

C₄
0.03
−11.7 V

To picture tube–grids boost voltage

C₃
100 μF
−11.7 V

High voltage rectifier assembly

CR₃

C₅
0.05

To AGC circuits

To horizontal oscillator AFC circuits

T₂

Flyback
transformer

C₂
0.003

Horizontal output
Q₁

L₁

+ C₁
−
100 μF

T₁

R₁

Horizontal
drive

7 V

at all of the other windings. The high voltage is rectified and applied to the picture tube anode; the boost output is rectified and applied to the picture tube focus and accelerator grid anode. On some sets a separate winding is provided for the horizontal deflection yoke. On other sets the horizontal output transistor current is passed through the deflection yoke. Either way, the horizontal output pulses produce the horizontal sweep.

The AFC and AGC windings (not found in all sets) are applied to the horizontal oscillator AFC circuits and the tuner/IF stage AGC circuits, respectively. In some sets the output of one winding is rectified and applied to the video output transistor as a collector voltage.

Because of the great variety in horizontal output circuits, it is difficult to arrive at a typical theory of operation. However, the important point to consider in practical troubleshooting is that the damping diode starts to conduct when the transistor is cut off. The diode continues to conduct until the horizontal output transistor conducts.

During the horizontal forward scan (when the picture is displayed), the diode conducts and the transistor is cut off from the start of the sweep at about the midpoint. Then the diode is cut off, and the transistor conducts for the remaining half of the sweep. This sequence is shown in Fig. 6-5.

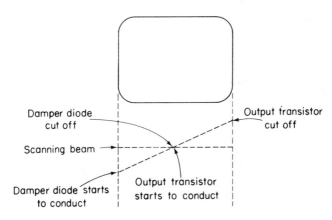

Fig. 6-5. Relationship of damper diode and output transistor conduction periods to picture display horizontal sweep

The scan sequence is important in troubleshooting, since any problems in the right-hand side of the picture are probably the result of defects in the transistor (or related components), whereas trouble in the left-hand side is probably the result of a defective damping diode.

When the transistor is cut off, the picture tube is blanked, and the current flows rapidly through the horizontal yoke in the opposite direction, pulling the electron beam back to the left side of the screen (known as horizontal retrace or flyback). It is the pulse developed during the *flyback interval* that is used by the other windings on the flyback transformer.

Keep in mind that voltages in solid-state circuits are much lower than in corresponding vacuum-tube circuits. However, the currents are much larger in solid-state. For example, the emitter-collector current in some horizontal output transistors is almost 1 A (800 to 900 mA is typical). It is not practical to measure such current, except possibly with a clamp-type probe. Also, resistances in solid-state coil windings are generally lower than in vacuum-tube circuits. Often, the horizontal deflection yoke winding is a fraction of an ohm.

6-2.2 Recommended Troubleshooting Approach

Many trouble symptoms caused by horizontal output circuits can also be caused by defects in other circuits. A dark screen (no raster) or insufficient width are two good examples.

If the low-voltage power supply is completely inoperative (or almost zero), the screen can be dark. If the low-voltage supply is producing a low output, the picture width can be decreased. Of course, in either case, the *sound is absent* or abnormal. On the other hand, if the horizontal oscillator and drive circuits are defective (no drive signal to the horizontal output tube), there will be no high-voltage or horizontal sweep, even though the sound will probably be normal. (In any service situation where there is a dark screen but with sound normal, never overlook the fact that the picture tube might be defective.)

The most logical troubleshooting approach is to analyze the symptoms and then isolate the trouble to the horizontal output circuits by an *input waveform check*. That is, if sound is normal (indicating a good low-voltage supply), measure the input waveform (from the horizontal drive circuit) at the horizontal output transistor base. Generally, this is on the order of 6 to 8 V and appears similar to that shown in Figs. 6-3 and 6-4. Check the waveform against the service literature.

If the input waveform is normal, then trouble is in the horizontal output circuit (unless the picture tube is bad). Of course, if the input signal is not normal, the next step is to check the *horizontal drive circuits*, as described in Sec. 6-3.

Dark screen. In the case of a completely dark screen, the next obvious check is to measure the high voltage at the picture tube second anode. Then measure the accelerating grid and focus voltages from the boost circuit. If the voltages are present and normal, the picture tube is defective. (Of course, if you haven't already thought of it, check that the picture tube filament is lighted.)

High voltage or boost voltage absent. If either the high voltage or the boost voltage is absent or abnormal, this will isolate the trouble to the corresponding circuit. If both voltages are absent (with the drive signal good), the problem is in the horizontal output transistor or related circuit parts (capacitors, flyback transformer, etc.).

High voltage only absent. If only the high voltage is absent, check for ac at the anode side of the high-voltage rectifier (unless this is not recommended in the service literature). When measuring the high voltage, always use a meter with a *high-voltage probe.* Observe all of the usual precautions when measuring high voltage. In addition, *do not make an arc test* of the high voltage with *any solid-state* set. The transient voltages can damage the transistors. In general, the author does not recommend an arc test, even with a vacuum-tube set.

Measuring voltages. Measure the boost voltage with a meter and *low-capacity* probe. With the exception of the high voltage, most of the voltages in a solid-state set can be measured using a low-capacity probe and a meter or oscilloscope, because the solid-state voltages are generally lower than in vacuum-tube circuits. Five hundred volts (peak-to-peak) is usually maximum for any solid-state circuit. Observe any and all precautions given in the service literature. Typically, there can be *warning notes* on the schematic such as "Do not measure high ac voltages."

Horizontal output transistor measurements. It is often helpful if you can measure the emitter-collector current of the horizontal output transistor. Usually this is not practical, so you must use dc voltage measurements to supplement the waveform checks. If the dc voltages and waveforms at the horizontal output transistor are correct, it is reasonable to assume that the circuit is good up to the flyback transformer.

Flyback and yoke checkers. Do not use horizontal system test sets (flyback checkers, yoke testers, etc.) designed for use with vacuum-tube TV on any solid-state circuit and vice versa. Solid-state flyback transformers have characteristics (Q, impedance, etc.) different from vacuum-tube transformers. Likewise, the drive voltages in solid-state circuits are much lower than vacuum-tube circuits (6 to 8 V compared with 90 to 100 V). If vacuum-tube-type horizontal checking equipment is used to test solid-state circuits or vice versa, either the TV set or the test set will probably be damaged. In any event the results will prove nothing. Except in rare cases, its possible to service the horizontal circuit of most sets using waveform and voltage measurements.

Common horizontal output troubles. The most common causes of trouble in the horizontal output and high-voltage circuits are (in order) capacitors, horizontal output transistor (since it operates at high current), diodes, and transformers.

Vacuum-tube versus solid-state horizontal circuits. Since solid-state horizontal circuits do not operate in the same way as the corresponding vacuum-tube circuits, the same symptoms do not always mean the same trouble. For example, the horizontal output transistor is in effect a *diode in parallel* with the damper diode. If the damper diode is open, the transistor can provide the same function and act as a damper. Of course, picture linearity will be very poor, especially on the *left side* of the screen. A shorted damper diode will usually make the entire circuit inoperative.

In many vacuum-tube circuits the picture tube boost voltages are supplied by the damper. If the boost voltage is present, the damper is good. In most solid-state circuits the boost is supplied by a separate diode, so the presence or absence of boost voltage has no bearing on the condition of the damper diode. Also, in most solid-state circuits, it is possible for a normal high voltage to be present even though there is an open in the horizontal deflection yoke. Of course, there will be no horizontal sweep (only a vertical trace on the picture tube screen), or the horizontal sweep will be impaired.

Voltage doublers. In some hybrid sets a voltage doubler circuit is used, as shown in Fig. 6-6. This is found mostly in large-screen (23-in. or larger)

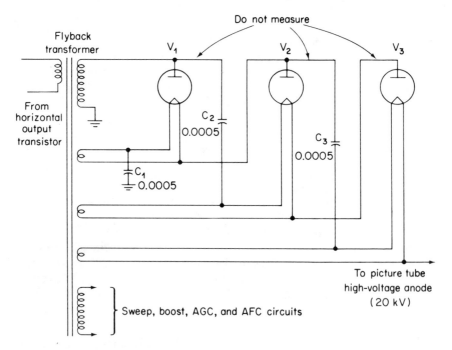

Fig. 6-6. Voltage doubling circuit found in high-voltage section of some hybrid sets

sets where the high voltage must be about 25 to 30 kV. Note that V_1 and V_3 conduct on positive peaks, whereas V_2 conducts on negative peaks. Capacitors C_1, C_2, and C_3 remain charged on both positive and negative peaks. This produces a dc output more than double that available across the transformer winding.

6-2.3 Typical Troubles

In the following paragraphs we will discuss symptoms that can be caused by defects in the high-voltage supply and horizontal output circuits.

Dark screen. If the picture screen is dark but sound is normal, it is likely that the horizontal output and high-voltage circuits are at fault. Of course, the picture tube could be defective, or the problem could be something simple such as a lack of filament voltage. Picture tube troubles are discussed in Sec. 6-8.

The most likely defects are open, shorted, or leaking capacitors; a leaking horizontal output transistor; open or shorted diodes; and a defective flyback transformer.

Before checking individual parts, there are some circuit tests that will help isolate the problem. These are as follows: Check the *drive voltage* waveform at the horizontal output transistor base, and the *sweep output* at the collector; check for *high voltage* at the picture tube second anode, observing precautions discussed previously; check for *boost voltage* and any auxiliary voltages (focus, accelerator grid, AFC or AGC pulses, etc.).

The first check to be made depends on whatever is the most convenient. For example, it is logical to check the horizontal output transistor waveforms before checking the high voltage. In some sets, the transistor may be at a very inaccessible location, so the best bet is to check the high voltage first.

If the drive waveform at the horizontal transistor base is absent or abnormal, the problem is ahead of the horizontal output circuits. If the base waveform is good but the collector waveform is absent or abnormal, the transistor is the first suspect. If the collector waveform is good but one or more of the output voltages is bad, check individual parts in the related circuit. Also check the dc voltages at each of the transistor elements.

Shorted capacitors (a common problem in the case of a dark screen) will show up when dc voltages are measured. Open or leaking capacitors are not located so easily. Generally, an open or leaking capacitor will produce an abnormal waveform. If a capacitor is suspected, try lifting one lead and checking the capacitor with a tester (for leakage, value, etc.), or try substitution if convenient. Again, it is not recommended that capacitors be checked by shunting with a good capacitor in solid-state sets unless the power is first turned off. The voltage surges can be damaging to transistors.

A leaking transistor will show up when waveforms are measured. The problem is confirmed further when the dc voltages are measured. Collector-base leakage in the horizontal output transistor is a common problem with solid-state sets. If leakage is bad enough to cause complete failure of the circuit, which results in a dark screen, the dc voltages at the transistor elements will be incorrect.

Keep in mind that the horizontal output transistor operates at or near zero bias or possibly with reverse bias. Thus, if there is any substantial forward bias, it is probably the result of leakage. Any collector-base leakage will forward-bias a transistor. In the case of a normally cut off horizontal output transistor, the undesired forward bias will attenuate the collector waveform in addition to producing incorrect dc voltages.

If the capacitors and transistor appear to be in order, check the diodes. If the sweep output waveform (transistor collector) is abnormal, check the damper diode. If the output voltages are abnormal (with a good sweep output), check the corresponding rectifier diodes.

A shorted damper diode is usually easy to pinpoint, since the transistor collector waveform and dc voltage will be abnormal. An open damper diode will usually not provide a dark screen (total failure of the circuit). If any of the other diodes are defective, this will show up as an absent or abnormal output voltage.

Note that in some all-solid-state circuits, such as shown in Fig. 6-4, the high-voltage rectifier is actually a group of several series-connected diodes. If any one of these diodes becomes shorted or develops excessive leakage, the remaining diodes can break down, because the normal voltage drop across the defective diode is placed on the remaining diodes, resulting in abnormally high peak voltage across the other diodes.

If the flyback transformer has an open winding or if a winding is shorted, the problem is usually self-evident. However, if there is only a partial short, leakage between windings, or a high-voltage arc, it may be difficult to check. Substitution is the only sure check, unless you have a flyback transformer tester suitable for the circuits (vacuum-tube or solid-state). Unfortunately, replacement of the flyback transformer is not an easy job. Therefore, do not try substitution except as a last resort (when capacitors, diodes, and transistor all prove to be good).

Picture overscan. Picture overscan or "blooming" in solid-state sets is the same as in vacuum-tube TVs. The picture becomes dim even with the brightness control full on, and there is an abnormal enlargement of the picture or raster. Usually the enlargement is uniform, but there may be some *defocusing.*

In some circuits, the brightness control operates in reverse. That is, rotating the control for an increase in brightness produces a decrease, after reaching a critical point of control. If picture overscan is not accompanied by insufficient width (a symptom of horizontal output failure), the probable cause is a failure in the high-voltage circuit only.

Any circuit problem that reduces (but does not completely eliminate) the high-voltage output to the picture tube second anode can cause overscan. The most likely defects are leaking high-voltage capacitors, defective high-voltage tube or rectifiers, leaking high-voltage leads, defects in high-voltage circuit protective resistances (if any), and shorted turns in the high-voltage winding of the flyback transformer.

The first obvious test is to measure the high voltage (observing all precautions). If the high voltage is normal (which is not likely), try a new picture tube. Also, there may be a *corona problem* (leakage from the high-voltage lead), especially in large-screen TVs, just as there is in vacuum-tube sets. The only practical cure is to replace the lead.

If the high voltage is low, check any high-voltage filter capacitors for leakage. In a hybrid set, try replacing the high-voltage tube. If the high-voltage circuit has protective resistors (R_3 and R_5, Fig. 6-3), as is the case in many hybrid circuits, check the resistance values. Resistor R_3 protects the high-voltage system in case of a short. Resistor R_5 sets the high-voltage rectifier tube filament to the correct value. If either resistor increases in value (say, due to overheating), the high-voltage output will be reduced.

It should be noted that many color TV pictures and rasters bloom when *both* brightness and contrast controls are turned full up. This is an inherent design problem (lack of high-voltage regulation).

Narrow picture. The most logical cause of a narrow picture that cannot be corrected by adjustment of the width controls is sufficient horizontal drive. This is usually accompanied by other symptoms, such as decreased brightness, picture distortion, and the like. Very often, a narrow picture (insufficient drive) is the result of a marginal breakdown rather than a complete breakdown. For example, if the horizontal output transistor has some collector-base leakage, the transistor will be forward-biased, and the sweep output will be decreased.

The most likely defects depend on symptoms that accompany the basic problem of a narrow picture. For example, if there is distortion on the left-hand side of the picture, look for an open or leaking damper diode. If there is right-hand distortion, check for a defective horizontal output transistor.

The first test is to measure the horizontal output transistor collector waveform as well as the sweep waveform to the horizontal yoke. These waveforms will rarely, if ever, be normal when the picture is narrow.

Note that in the circuit of Fig. 6-4 the horizontal transistor collector is connected directly to the horizontal yoke. In other circuits, such as shown in Fig. 6-3, the horizontal yoke is supplied by a winding on the flyback transformer.

The waveform measurements should be followed by voltage measurements at the horizontal output transistor elements. Abnormal voltage readings will show up such defects as leaking or shorted capacitors and diodes. As a final resort, look for a marginal breakdown in the flyback transformer, such as a partially shorted winding or some leakage between windings.

Do not overlook *insufficient horizontal drive*, especially when waveforms and voltages all appear normal but somewhat low. Check the drive waveform at the horizontal output transistor base. Look particularly for a drive waveform (at the base) that is normal but low in amplitude (below the usual 6 to 8 V). Compare the base and collector waveform amplitude and the horizontal yoke waveform amplitude very carefully against those shown in the service literature. In solid-state circuits, even a slight drop in waveform amplitude can indicate a large change in current (which can produce major problems in current-operated solid-state circuits).

Foldback or foldover. Horizontal foldback usually occurs only on one side of the picture screen. A portion of the picture is folded back on one edge of the display. In some rare cases, there will be a fold in the center.

Assuming that the horizontal drive signal is normal, the problem can be traced to the horizontal sweep components. The most likely defects are horizontal output transistor, damper diode, damper capacitors, horizontal yoke, or yoke drive winding on the flyback transformer. If foldback is on the right-hand side, look for a defective transistor or its related components. If foldback is on the left-hand side, check the damper diode and its related components.

The first test is to measure both the base (drive) and collector (output) of the horizontal output transistor. Then measure the horizontal yoke waveform (if it is different from the collector waveform). Invariably, the yoke waveform will be distorted.

Keep in mind that the horizontal sweep system is essentially a resonant circuit. The inductance of the flyback transformer combines with various capacitors in the circuit (such as C_1 in Fig. 6-3 and C_4 in Fig. 6-4) to resonate at about 50 kHz (typical). If the resonant frequency is not correct (generally low), the waveform will be distorted, resulting in foldback.

If foldback occurs after replacement of any part in the horizontal system, the problem is almost certainly one of an incorrect value: out-of-tolerance capacitor, flyback transformer with incorrect winding inductance (the dc resistance may be good), or a yoke with incorrect inductance. If foldback occurs before replacement of parts, look for parts that have gone out of tolerance. In solid-state horizontal sweep circuits, the maximum tolerance is usually 10%, with 5% producing better results.

In extreme cases, when all parts in the system appear to be good but foldback continues, check the duration of the drive pulse (*on time* compared to *off time*) against the service literature.

Nonlinear horizontal display. Any nonlinearity in the horizontal display is almost always found with at least one other problem (such as narrow picture, overscan or blooming, lack of brightness, foldback, etc.). Thus, the most likely causes of nonlinearity are the same as for the other symptoms. Also, the troubleshooting test sequence should be the same (waveform measurements followed by voltage measurements).

Horizontal nonlinearity should not be confused with the *keystone effect.* Horizontal keystoning is evident when the top and bottom widths of the picture display are not the same (picture is wider at the top than at the bottom or vice versa) and is *almost always* caused by a problem in the horizontal yoke. (One set of horizontal deflection coils has shorted turns and is unbalanced with the other set of coils.)

6-3. HORIZONTAL OSCILLATOR AND DRIVER

The horizontal oscillator and driver circuits provide the drive signal to the horizontal output and high-voltage supply circuits (described in Sec. 6-2). The horizontal oscillator and driver signals are at a frequency of 15,750 Hz and are synchronized with the picture transmission by means of sync signals (taken from the sync separator circuits, as described in Sec. 6-5).

6-3.1 Basic Circuit Theory

Most solid-state horizontal oscillator circuits include some form of AFC system to ensure that the horizontal sweep signals are synchronized (both for frequency and phase) with picture transmission, despite changes in line voltage and temperature or minor variations in circuit values. The AFC action is accomplished by comparison of the sync signals with horizontal sweep signals, both for frequency and phase.

Deviations of the horizontal sweep signals from the sync signals cause the horizontal oscillator to shift in frequency or phase as necessary, to offset the initial (undesired) deviation. For example, if the horizontal sweep increases in phase from the sync signals, the horizontal oscillator is shifted in phase by a corresponding amount but in the opposite direction (a decrease in phase to offset the undesired change).

There are three basic horizontal oscillator circuits used in solid-state TV. These include the balanced AFC (Fig. 6-7), the unbalanced AFC (Fig. 6-8), and the transistor AFC (Fig. 6-9). In all cases the circuit is composed of three sections: the horizontal AFC, the horizontal oscillator, and the horizontal driver. In a few cases the horizontal circuits are more elaborate.

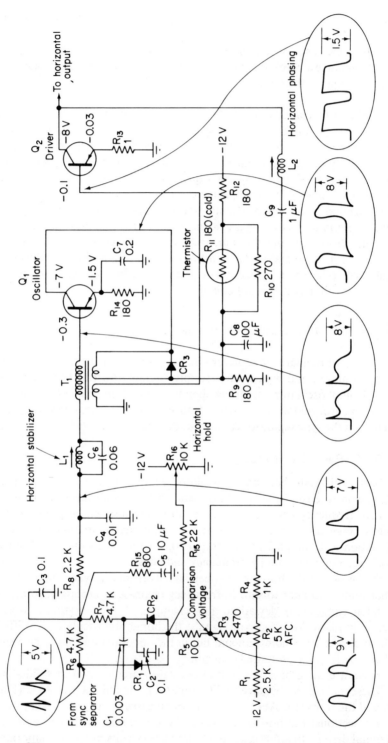

Fig. 6-7. Horizontal oscillator and driver with balanced AFC

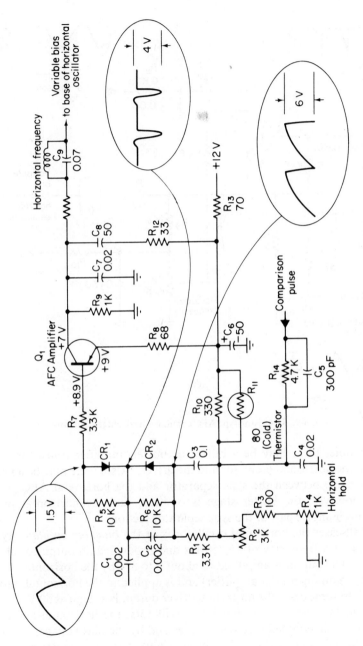

Fig. 6-8. Unbalanced AFC circuit for horizontal oscillator

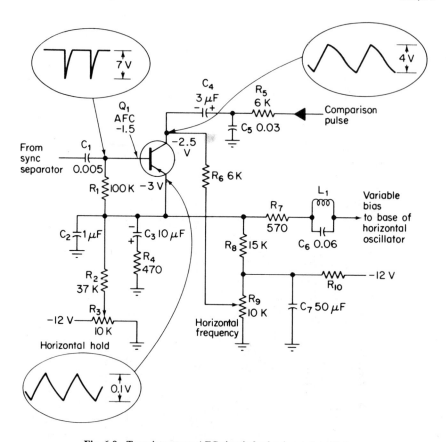

Fig. 6-9. Transistor-type AFC circuit for horizontal oscillator

For example, there may be a buffer or buffer/amplifier transistor stage between the horizontal oscillator and the driver. Also, there may be a phase inverter stage between the sync separator and the horizontal AFC. Generally, however, the buffer stage is omitted, and the horizontal phase inversion (if any) is part of the sync separator section.

In all cases the horizontal oscillator is of the *blocking oscillator* type, operating at a frequency of 15,750 Hz and producing an output of about 1 to 3 V. The output is amplified to about 6 to 8 V by the horizontal driver (a basic, common-emitter amplifier) and is applied to the horizontal output section. In some cases the horizontal driver output is as high as 50 V.

As in the case of most blocking oscillators, the horizontal oscillator frequency is determined by circuit values and by the bias voltage. In none of the typical circuits (Figs. 6-7 through 6-9) is the horizontal oscillator' triggered directly by the sync pulses. Instead, the sync pulses and comparison pulses produce a variable dc control voltage that is applied to the horizontal oscillator base. Any change in this control voltage shifts the horizontal oscillator frequency and phase. The control voltage is manually set by the horizontal adjustment controls (*horizontal frequency*, AFC, *horizontal hold*, etc., depending on the type of circuit) to an *average value*. Any deviation from the average value produces a corresponding change in the base bias, which, in turn, shifts the horizontal oscillator frequency as necessary.

In the balanced circuit of Fig. 6-7, the base bias for horizontal oscillator Q_1 is developed by the AFC diodes CR_1 and CR_2. The variable control voltage is filtered by an RC circuit with a long time constant. Any random noise pulses that may be mixed with the horizontal sync pulses are averaged out by the RC network, thus rejecting noise interference as much as possible.

Note that the sync pulses are applied to opposite ends of diodes CR_1 and CR_2, whereas the comparison pulses and the horizontal control voltages are applied to the diode common (CR_1 cathode, CR_2 anode). Diode conduction depends on the combined peak voltages of the sync pulses and the comparison pulses. The balanced AFC circuit requires a double-ended or push/pull input from the sync separator and is similar to a ratio detector used in FM detectors.

In the unbalanced circuit of Fig. 6-8, the input from the sync separator is single-ended, whereas the control voltage output is applied through an amplifier Q_1. The AFC diodes CR_1 and CR_2 receive both sync pulses and comparison pulses. Diode current depends on the instantaneous voltage of the two waveforms and thus changes from one instant to another. However, there is an average value of current in each diode over one complete cycle of operation. The amount of dc voltage produced is determined by the difference between these two average values.

The *relative phase* of the two waveforms across each diode changes if the horizontal oscillator drifts off frequency. Thus, the average value of rectified voltage also changes. One diode conducts more than the opposite diode, and the dc control voltage varies higher or lower, depending on the phase error (horizontal oscillator leading or lagging the sync pulses).

In the circuit of Fig. 6-9, transistor Q_1 is used in place of the AFC diodes. The Q_1 base is driven by the sync pulses, whereas the collector receives comparison pulses. The emitter waveform is a combination of both pulses. When there is a change in phase between the two sets of pulses, the dc emitter current changes. In turn, the dc control voltage applied to the horizontal oscillator changes in a direction that corrects the frequency and phase of oscillation.

6-3.2 Recommended Troubleshooting Approach

Failure of the horizontal oscillator and related circuits (AFC and horizontal drive) can produce symptoms similar to those produced by failure in other circuits. For example, if the horizontal oscillator stops oscillating, there will be no drive signal to the horizontal output circuit (Sec. 6-2). Thus, there will be no high voltage or boost voltage, and the picture tube screen will be dark. This same symptom can also be produced by failure of the horizontal output circuit (Sec. 6-2), the low-voltage power supply (Sec. 6-1), and the picture tube.

Also, operation of the horizontal oscillator circuits depends on signals from other circuits. For example, the horizontal AFC circuit must have sync signals from the sync separator (Sec. 6-5) and comparison signals from the horizontal output (Sec. 6-2) to operate properly.

Because of these conditions, the only practical troubleshooting approach for the horizontal oscillator circuits is to check waveforms at all outputs and inputs, followed by dc voltage measurements at all transistor elements.

If the driver output is absent or abnormal with good sync and comparison pulse inputs, the problem is definitely in the horizontal oscillator or driver sections. A waveform measurement at the input of the horizontal driver (transistor base) or horizontal oscillator output will localize the problem further.

If the driver output is present but the symptoms point to a horizontal circuit failure (horizontal pulling, jitter, distortion, etc.), the problem is most likely in the AFC circuits. Problems in the AFC section are usually isolated by means of voltage measurements followed by substitution of parts.

Note that the horizontal oscillator often appears to be reverse-biased, on the basis of dc voltage measurements. However, it is still possible for the oscillator to be forward-biased during the heavy collector current pulses (*on time*) and then drop back to reverse bias between pulses (*off* time). Therefore, the only test of the horizontal oscillator is the presence of correct waveforms at both the base and collector or the output winding of the horizontal oscillator transformer.

Also note that the horizontal stabilizer and horizontal phasing controls (if any) will have little effect on the dc voltages but will have considerable effect on waveforms. On the other hand, the potentiometer controls such as the horizontal frequency and horizontal hold will have considerable effect on the dc voltages. For example, in Fig. 6-7, the horizontal hold potentiometer will vary both the base and emitter voltages of the horizontal oscillator by more than 0.5 V. As a comparison, the presence or absence of sync pulses will mean less than about 0.3-V difference in dc voltages at the base and emitter of the horizontal oscillator.

The horizontal drive is usually zero-biased or slightly forward-biased. The horizontal oscillator output then drives the driver into heavy conduction, possibly to saturation. In circuits that use a transistor for the AFC (Figs. 6-8 and 6-9), the transistor is reverse-biased.

6-3.3   Typical Troubles

In the following paragraphs we shall discuss symptoms that could be caused by defects in the horizontal oscillator circuits.

Dark screen. If the picture screen is dark (no raster) but sound is normal, indicating that the low-voltage power supply is probably good, the first step is to measure the waveform at the horizontal drive output. If the waveform is normal, the problem is in the horizontal output and high-voltage circuit or the picture tube itself. If the waveform is not normal (weak, distorted, etc.), the problem is probably in the horizontal oscillator. The most likely defects are open, shorted, or leaking capacitors; leaking transistors; open or shorted diodes; and a defective blocking oscillator transformer.

Before checking individual parts, there are some circuit checks that will help isolate the problem: waveforms at sync input (from sync separator), comparison pulse input, horizontal oscillator base and collector (and/or output winding of blocking oscillator transformer), and voltages at all transistor elements.

If the sync pulses are absent or abnormal, the problem is in the sync separator rather than in the horizontal oscillator section. If the comparison pulses are not normal, with a good output from the horizontal drive, then the problem is in the horizontal output and high-voltage section (Sec. 6-2).

If the sync pulses and comparison pulses are both normal but the horizontal oscillator output is absent or abnormal, the problem could be in the AFC section or the horizontal oscillator. If the horizontal oscillator output is normal, the problem is in the horizontal drive circuit.

As usual, capacitors are the most likely cause of trouble in the horizontal oscillator circuits. For example, if C_8 in Fig. 6-7 is shorted, Q_1 will have no collector voltage and will stop oscillating. The same is true if there is a short in capacitor C_4. In this case, the dc control voltage from the AFC section will be shorted to ground, instead of biasing the horizontal oscillator base to the correct level.

If the TV set is old, look for worn potentiometer controls. For example, if R_2 in Fig. 6-7 is open, the bias on Q_1 will be abnormal. Q_1 may continue oscillating, but the frequency and amplitude will be off, probably both low. This will show up as an abnormal waveform and as incorrect voltages on the transistor elements.

If the capacitors and controls appear to be good, look for a defective diode. For example, if either CR_1 or CR_2 is defective, the control voltage on the base of Q_1 will be abnormal. Likewise, a shorted CR_3 will prevent the collector-to-base feedback necessary for Q_1 oscillation. If CR_3 is measured in circuit, it will appear shorted since the ohmmeter will measure the transformer winding resistance (typically only a few ohms).

Narrow picture. The most logical cause of a narrow picture that cannot by corrected by adjustment of the width controls is insufficient horizontal drive. If the problem is in the horizontal oscillator circuits, it will show up as a low output waveform from the horizontal driver and/or horizontal oscillator.

The most likely defects are an open emitter capacitor in the horizontal oscillator, collector-base leakage, shorted turns in the blocking oscillator transformer, and possibly off-value resistors. However, any number of defects could produce a low output.

The first step is to isolate the problem to the horizontal driver or the horizontal oscillator by waveform measurement. If the problem is in the horizontal oscillator, keep in mind that the trouble can be in the AFC section.

Horizontal pulling or improper phasing; loss of sync. Horizontal pulling is present when the picture pulls and appears in diagonal form. If the picture pulls completely into diagonal lines, this indicates a complete loss of sync. Note that the direction of the slant can provide a clue to the problem. If the lines slant to the right, the horizontal oscillator frequency is high. If the lines slant to the left, the frequency is low.

If the picture shifts to the right or left so that it is decentered, this indicates incorrect phasing. The horizontal oscillator is on frequency but not in phase with the sync signals.

It is possible that any of these problems can be the result of improper adjustment of controls. Thus, the first step is to adjust all of the horizontal controls. If this does not clear up the problem, check all of the waveforms and transistor voltages. Pay particular attention to the sync pulses (from the separator) and the comparison pulses (from the horizontal output circuits). If either of these is absent or abnormal, the AFC circuits will not operate properly.

For example, if the comparison pulses are absent, horizontal sync may be completely lost. However, the horizontal hold adjustment will become very critical, and phase shift will occur (the picture will be decentered to the left).

If the sync and comparison pulses are normal, the most likely defects are capacitors and/or diodes in the AFC section. However, there are many other defects that can cause these symptoms, so an analysis of the symptoms and measurements (waveform/voltage) must be made before checking individual parts.

Defects in the AFC diodes (CR_1 and CR_2 in Fig. 6-7) are quite common. Complete shorts or opens in these diodes will show up as severe distortion of waveforms or transistor voltages that are well out of tolerance. However, marginal defects in the AFC diodes can be difficult to locate. For example, a poor front-to-back ratio can result in a touchy horizontal hold.

If the picture pulls at the top, try adjustment of the horizontal stabilizer (phasing) control (L_1 in Fig. 6-7). If the adjustment has no effect on pulling, this points to a defect in the AFC circuits. Also try adjustment of the horizontal stabilizer control in the event of horizontal sync loss. (It is possible the oscillator is so far out of phase with the sync pulses that a proper sync is impossible.) If the horizontal hold control is very touchy and the images appear to overlap, it is possible that the stabilizing capacitor (C_6 in Fig. 6-7) is open. This is confirmed if the stabilizer control has little effect on the waveform at the base of the horizontal oscillator.

Horizontal distortion. There are many forms of horizontal distortion that can be caused by defects in the horizontal oscillator and driver circuit. The so-called "pie-crust" distortion is a typical example. With pie-crust distortion, the picture image appears to be made up of wavy lines, even though there is no pulling, loss of sync, or jitter. In solid-state TVs, any form of distortion is almost always the result of marginal performance in a particular component, rather than complete failure.

The most likely defects are in capacitors, particularly the filter capacitors that filter the variable dc control voltage from the AFC diodes to the horizontal oscillator. For example, a slight leakage in capacitors C_3, C_4, and C_5 in Fig. 6-7 can cause horizontal distortion. Transistor leakage is the next logical suspect. However, such leakage will usually cause other symptoms to appear (pulling, loss of sync, etc.).

6-4. VERTICAL SWEEP CIRCUITS

The vertical sweep circuits provide a vertical deflection voltage (vertical sweep) to the picture tube deflection yoke. The vertical circuits also supply a blanking pulse to the picture tube (usually through the video amplifier, as discussed in Sec. 6-8). This blanks the picture tube during retrace of the sweep.

The vertical sweep signals are at a frequency of 60 Hz and are synchronized with the picture transmission by means of sync signals (taken from the sync separator section, as described in Sec. 6-5).

6-4.1 Basic Circuit Theory

Figure 6-10 is the schematic diagram of a typical vertical sweep circuit. Note that there are three stages (oscillator, driver, and output). In some solid-state TV sets, the driver and output functions are combined into one stage.

The vertical oscillator is of the blocking oscillator type, operating at a frequency of 60 Hz and producing a pulse output of about 1 to 2 V. The oscillator pulse output is modified into a sawtooth sweep and is applied through the driver to the output stage. The final circuit output is a peaked

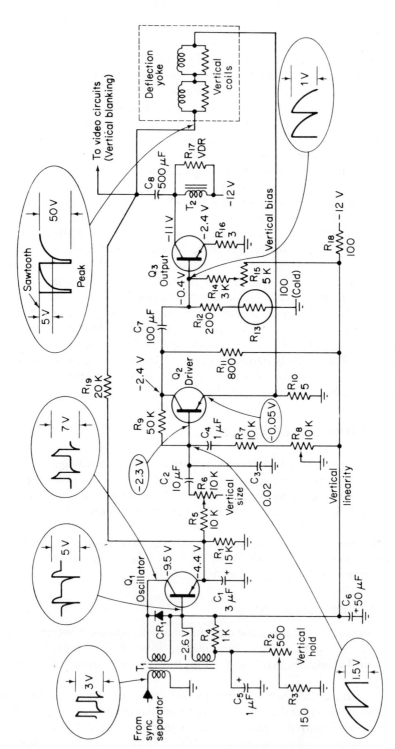

Fig. 6-10. Vertical sweep circuits

sawtooth waveform. The sawtooth portion (about 4 to 5 V) is applied to the picture tube deflection yoke, whereas the peaked portion (about 50 V) is used as the blanking pulse for the picture tube.

The vertical oscillator frequency is set by circuit values and by the bias voltage. However, the oscillator is locked in frequency by the sync pulses. As shown in Fig. 6-10, Q_1 is normally reverse-biased where dc voltages are concerned. Feedback is obtained through coupling of the collector and base windings of T_1. Whenever Q_1 conducts, the winding feedback makes the base more negative, increasing the collector conduction and charging C_1 to a high negative value. When the emitter of Q_1 goes sufficiently negative, Q_1 is cut off and remains so until C_1 discharges through R_1.

As with most solid-state blocking oscillators, Q_1 is on (conducting) for a very short period of time and is off (nonconducting) for a long period of time during each cycle. The point at which Q_1 starts to conduct (and thus the frequency of Q_1) is set by vertical hold control R_2, which controls the base bias. Actually, R_2 is set so that the frequency of Q_1 is just below 60 Hz. In practice, R_2 is adjusted until the vertical sweep is locked with the picture transmission. That is, the sync pulses trigger Q_1 into conduction *earlier* than Q_1 would otherwise start to conduct.

Diode CR_1 is often a source of trouble in the circuit. CR_1 acts as a blocking diode to prevent the Q_1 signal from going back into the sync separator. CR_1 also prevents an excessive peak voltage on the base of Q_1. If CR_1 opens, Q_1 can be destroyed. If CR_1 is shorted, Q_1 will not oscillate. If CR_1 is otherwise defective, the Q_1 pulses can feed back into the sync separator and disturb its operation.

In the circuit of Fig. 6-10, the Q_1 pulse output is converted to a sawtooth sweep by means of the emitter network (C_1R_1). The discharge of C_1 through R_1 produces the basic sweep waveform. The sweep is made linear through feedback of the output by R_{19}. The waveform is also shaped by the network $C_3C_4R_7R_8$. Vertical linearity control R_8 provides a manual control for the sawtooth waveform.

The amplitude of the sawtooth waveform to driver Q_2 (and thus the vertical size of the picture raster) is set by vertical size control R_6. Note that the emitter of Q_2 is connected to the deflection yoke. This provides negative feedback to ensure a more linear output.

The deflection yoke of a solid-state TV requires heavy current, even though the sweep voltage is low. For this reason, the output transistor Q_3 is of the power type (often mounted on a heat sink). Since there is always the danger of thermal runaway in power transistors, thermistor R_{13} is included in the base circuit of Q_3. If collector current increases, heating Q_3 and causing more current flow, the resistance of R_{13} will drop, lowering the emitter-base voltage differential. In turn, this will lower the emitter-base and emitter-collector current flows. Transistor Q_3 is further stabilized by unbypassed emitter resistor R_{16}. If it becomes necessary to replace transistor Q_3, it may be necessary to adjust vertical bias control R_{15}.

Most solid-state vertical output circuits have a bias control, since the characteristics of replacement transistors can be different from the original transistor. The peaked portion of the output waveform is obtained by inductive "kickback" of T_2. Typically, the peak will be about 50 V, even though the sawtooth portion is usually about 5 V.

Resistor R_{17} is a voltage-dependent resistor (VDR) found on many late-model sets. Resistor R_{17} keeps the amplitude of the output waveform within narrow limits. The resistance of a VDR is varied by the output voltage to the deflection yoke. If the output tends to increase, the resistance of R_{17} decreases, keeping the output constant. Resistor R_{17} should be checked first if the trouble symptom is a constantly varying vertical size or vertical height.

6-4.2   Recommended Troubleshooting Approach

Complete failure of the vertical sweep circuits is an easy symptom to recognize and is usually easy to locate. With complete failure, there will be no vertical sweep, and the picture display will be a horizontal line only. (With any type of TV set, vacuum-type or solid-state, *do not* operate the set when there is a complete failure in the vertical sweep. The bright horizontal line can burn out the picture tube screen.)

In the case of complete vertical sweep failure (all other functions normal), the logical first step is to check waveforms at the vertical oscillator output (driver input), driver output, and power output. If the vertical oscillator waveform is absent or abnormal, the problem is quickly localized to the vertical oscillator stage. (Note that in some vertical circuits, the oscillator will not go into oscillation unless there are sync pulses present. Therefore, it is always wise to check for sync pulses if the oscillator appears defective.)

If the oscillator waveform is normal but either the driver or power output waveforms are abnormal, this will localize the problem to these stages. If all of the waveforms appear normal but there is no vertical sweep, the deflection yoke is a logical suspect.

Marginal failure of the vertical sweep circuits is not so easy to recognize or locate. Such problems as loss of vertical sync (or very critical vertical sync), distortion (nonlinearity), line pairing or splitting (poor interlace), or lack of vertical height can be caused by any number of defects in the vertical sweep circuits.

In the case of marginal failure, the first logical step is to try correcting the problem by adjustment of the vertical controls. If the problem cannot be eliminated by adjustment, or if it is necessary to set a control to an extreme (one end or the other), then a marginal component failure can be suspected. Leaking capacitors, transistors, and diodes or worn potentiometers are likely causes. An open capacitor can also be a cause.

The next logical step in marginal failure is comparison of waveforms and/or voltages against those shown in the service literature. Usually, waveform measurement will be of the greatest value. One point to consider in making waveform measurements in solid-state vertical sweep circuits is possible *false distortion*. The sawtooth waveform developed by the vertical sweep circuits must be linear to get a linear vertical picture. A poor oscilloscope (one with narrow bandwidth or excessive input capacitance) can *indicate a nonlinear sweep* when a good sweep is present. Also, the oscilloscope can load the circuit and create nonlinearity.

6-4.3   Typical Troubles

In the following paragraphs we shall discuss symptoms that could be caused by defects in the vertical sweep circuits.

No vertical sweep. If the picture display consists of only a bright horizontal line, indicating that all circuits but the vertical sweep are probably normal, the first step is to measure waveforms in the vertical sweep circuits. Measure all of the waveforms (or their equivalents) shown in Fig. 6-10. Then measure the voltages at all of the transistor elements.

It is very unlikely that there will be a complete loss of vertical sweep without at least one of the waveforms and/or voltages being abnormal. A possible exception to this is where the vertical coils of the deflection yoke are open. Always check the yoke windings if the final output (peaked sawtooth) waveform appears normal but there is no vertical sweep.

Always compare the transistor voltages against the service literature when available. If literature is not available, use the following guideline: The vertical oscillator and vertical output transistors will be reverse-biased, whereas the driver is usually forward-biased. In two-stage vertical sweep circuits, the output transistor is zero-biased or slightly forward-biased, with the oscillator being reverse-biased.

Keep in mind that the reverse-bias condition of the vertical oscillator Q_1 is caused by the charge built up across the emitter capacitor C_1. If it were not for this charge buildup, Q_1 might be forward-biased (as it is during the brief *on time*). However, with Q_1 operating normally (oscillating), the average base-emitter differential will be such that Q_1 appears to be reverse-biased. Any solid-state blocking oscillator that shows a forward-biased condition is suspect.

If it becomes necessary to replace the vertical output transistor, be sure to follow the precautions for power transistors (proper heat sink, mounting, etc.). Also, if the power transistor is provided with a protective thermistor, check the thermistor whenever the power transistor is replaced. Generally, if the cold resistance of a thermistor is within tolerance, the thermistor is good. Sometimes, a thermistor can show a good cold resistance but still be defective when heating. As a precaution, check the base-emitter bias of the power transistor when power is first applied and as the circuit warms up. If the base-emitter bias voltage does not stabilize and the power transistor appears to overheat, remove the power and replace the thermistor.

Insufficient height. The lack of vertical height can be caused by improper adjustment of controls or by defective circuits. Therefore, the first step in troubleshooting this symptom is to adjust the vertical height control (sometimes called the vertical drive control or the vertical size control). There is no problem if height can be obtained with the vertical height control at midrange (or not more than about ¾-range). However, if the vertical height control must be full on, or near full on, to get the proper height, it is possible that the vertical output transistor is improperly biased. If the vertical output transistor is biased near the cutoff point, part of the sawtooth sweep will be clipped off, reducing the output. Thus, the next step in troubleshooting is to adjust the vertical output transistor bias.

Follow the service literature instructions if available. If not, adjust the bias control to produce the correct waveform amplitude at the vertical output transistor collector. In the circuit of Fig. 6-10, the sawtooth sweep is about 5 V, while the pulse peaks are about 50 V.

If there is no service literature, try the following procedure to set the vertical output bias. Set the vertical height control to midrange. Then adjust the bias control until the picture is at the desired height. If you get the correct height without distortion or other symptoms or vertical circuit failure, both the height and bias controls are properly set.

If it is impossible to get the proper vertical height by adjustment of the controls, the most likely causes are leaking capacitors, a worn vertical height control, and leaking transistors (particularly the output transistor).

Measure all waveforms and dc voltages in the vertical sweep circuits. Look particularly for low amplitude in sawtooth sweep at any point in the circuit. Aside from adjustments, the only logical causes for lack of vertical height are low-amplitude sweeps or a defective yoke. If the final output sweep is of the correct amplitude, check the yoke. If the final output sweep is low, work back to the vertical oscillator.

It is fairly certain that one or more of the waveforms and/or voltage measurements will be abnormal if there is insufficient vertical height. However, the abnormal indication may not pinpoint the defective component, particularly in the case of a marginal defect. The following are some typical examples using the circuit of Fig. 6-10.

If the vertical oscillator waveform is low, look for leakage in the bypass capacitor C_1 or the protective diode CR_1. If the drive waveform is low, look for leakage in waveshaping capacitor C_3. If the output waveform is low (with the yoke good), look for a defective VDR R_{17} (if used in your particular circuit) or a defective output transformer T_2. Usually, it will be necessary to check T_2 by substitution, since a marginal defect such as shorted turns or leakage between windings may not show up in resistance measurements.

Do not overlook the possibility of leakage in the output transistor Q_3. Any collector-base leakage will reduce the output. Keep in mind that it is possible for a transistor to be leaking and still show the correct base-emitter differential voltage. However, both voltages will be abnormally high in relation to ground or common, because collector-base leakage tends to forward-bias the transistor, resulting in more emitter current flow and a larger drop across the emitter resistor.

Vertical sync problems. The lack of vertical sync or a very critical vertical sync can be caused by improper adjustment of controls as well as by defective circuits. Thus, the first step in troubleshooting this symptom is to adjust the vertical hold (sometimes called vertical lock or vertical sync) control. There is no problem if vertical sync can be obtained with the vertical hold control at midrange. If the vertical hold control must be full on (or near full on) or it is necessary to readjust the control frequently, there is a problem in the vertical sync circuits.

If the problem cannot be cured by adjustment, the next step is to localize the fault. The same symptoms can be caused by a defect in the vertical portion of the sync separator (as discussed in Sec. 6-5).

First, check for the presence of proper sync pulses at the input to the vertical sweep circuits (at the primary of T_1 in Fig. 6-10). Then check the sawtooth sweep portion of the vertical oscillator (say, at the base of Q_1). With the picture rolling, the vertical sync pulse will appear to ride on the sawtooth, as shown in Fig. 6-11(a). Normally (picture not rolling), the sync pulse is hidden since it is close to the beginning (or end) of the sawtooth sweep.

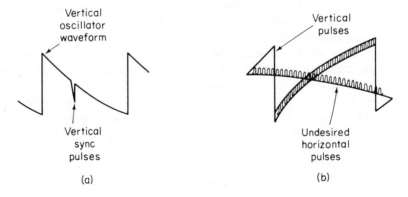

(a)   (b)

Fig. 6-11. Waveforms in vertical sweep circuits

If the input sync pulses are not normal, the problem is in the sync separator (Sec. 6-5). If the sync pulses do not appear in the vertical oscillator sawtooth waveform (with picture rolling), the problem is likely to be in the input transformer T₁ or related circuits.

With the picture still rolling, check the *rate of roll*. If the roll is steady, the vertical oscillator is probably good but the sync pulses are not of sufficient amplitude to lock the oscillator in sync. (Or the oscillator is biased such that the sync pulses cannot overcome the reverse bias.) If the roll rate is unsteady, with constant sync pulses of sufficient amplitude, the problem is in the vertical oscillator.

If the sync pulses are normal and it is impossible to get proper vertical sync, the most likely causes are leaking bypass capacitors (C_5 and C_6 in Fig. 6-10), a worn vertical hold control, a leaking or defective vertical oscillator transistor, or a defective vertical oscillator transformer (leakage between windings or some similar marginal defect).

If bypass capacitors are leaking, the vertical hold control will be partially shorted; it will not have the full resistance insofar as the circuit is concerned. The same condition will result if the vertical hold control has a worn resistance element. In the case of a very critical vertical sync (difficult to adjust or will not hold after adjustment), observe the amplitude of the vertical sync pulse riding on the sawtooth portion of the vertical oscillator waveform (Fig. 6-11). If the sync amplitude does not remain constant, look for problems in the sync separator (Sec. 6-5) rather than the vertical sweep circuit.

Vertical distortion (nonlinearity). There are several forms of vertical distortion. Some are easy to recognize, such as *keystoning*, where one side of the picture is much larger than the opposite side. Keystoning is generally caused by a defect in the deflection yoke or related circuits parts (such as a thermistor in the yoke circuit) but could be caused by a marginal defect in the vertical output transformer (T₂ in Fig. 6-10).

Other forms of vertical distortion are not so easy to recognize. In the extreme, there will be compression of the picture at the top with picture spreading at the bottom or vice versa. Often, the compression or spreading is slight. Use a test pattern transmission or a crosshatch pattern from a TV signal generator to check on linearity. If neither of these are available, a quick check of vertical linearity can be made as follows. Adjust the vertical hold control to produce slow rolling. Watch the blanking bar as it moves up or down on the picture screen. The blanking bar should remain constant in vertical height at the bottom, middle, and top of the screen.

Vertical distortion can be caused by improper adjustment of controls as well as by defective circuits. Thus, the first step in troubleshooting this symptom is to adjust the vertical linearity control. There is no problem if vertical linearity can be obtained with the control at midrange. If the vertical linearity control must be full on (or near full on), there is a problem in the vertical sweep circuits.

One problem here is the interaction of controls. For example, if the sawtooth sweep is low in amplitude (due to a defect not associated with nonlinearity), the vertical height control or the vertical output bias control may be advanced to get proper vertical height. These extremes in circuit resistance may make it impossible to get a linear picture no matter how the linearity control is adjusted.

If the problem cannot be cured by adjustment, the next step is to localize the fault. Start by measuring all waveforms and voltages in the vertical sweep circuits. Keep in mind that the vertical sweep will not be linear if the sawtooth waveform is not linear. Check waveform linearity, starting with the vertical output transistor and working back to the vertical oscillator.

Line splitting. The problem of line splitting (line pairing or poor interlace) is often associated with vertical sweep circuits. Although the trouble does appear in the vertical sweep circuits, the actual cause is usually in related circuits. For example, the trouble can be caused by an open capacitor in the integrator portion of the sync separator or by horizontal pulses leaking back into the vertical sweep.

If you suspect that the horizontal pulses are present in the vertical sweep, observe the vertical sweep output waveform with the oscilloscope retrace blanking function disabled. If there are any horizontal pulses mixed with the vertical sweep, they will appear as a pulse train on some portion of the vertical sweep output waveform. An example of this is shown in Fig. 6-11.

If the horizontal pulses are present, the most likely causes are to be found in the horizontal circuits, such as corona discharge from the high-voltage rectifier, leakage between leads in the two circuits (not too common in printed-circuit sets), and breakdown in the high-voltage sections (resulting in a spark discharge being picked up by the vertical circuits).

If there are no horizontal pulses in the vertical circuits but there is definite line splitting, the problem is in the integrator portion of the sync circuits (Sec. 6-5), with an open integrator capacitor the likely suspect. Generally, when line splitting is constant, the problem is in the integrator or is the result of leads being too close together. When splitting is intermittent, the cause is usually associated with the high-voltage section (spark discharge, radiation from high-voltage lead, etc.).

6-5. SYNC SEPARATOR CIRCUITS

The sync separator circuits function to remove the vertical and horizontal sync pulses from the video circuits and apply the pulses to the vertical and horizontal sweep circuits, respectively. The sync separator circuits also function as clippers and/or limiters to remove the video (picture) signal and any noise. Thus, the sweep circuits receive sync pulses only and are free of noise and signal (in a properly functioning set).

6-5.1 Basic Circuit Theory

Sync separator circuits are not standardized in solid-state TVs. For example, to get the desired clipper/emitter action, the input transistor of one sync separator is reverse-biased (class C). Only the peaks of the sync pulses turn the transistor on and appear at the collector. In another solid-state circuit, the input transistor is zero-biased (class B) or slightly forward-biased (class AB) so that the sync pulses drive the transistor into saturation at a level well above the signal and/or noise level. In other solid-state sync separator circuits, two stages are used, one with reverse bias and the other with zero bias (or slight forward bias). Thus, both clipping and limiting are obtained.

No matter what bias system is used, the vertical sync pulses are applied through a capacitor/resistor *low-pass filter* (often called the *vertical integrator*) to the input of the vertical sweep circuit (Sec. 6-4). The horizontal sync pulses are applied to the AFC section of the horizontal oscillator, where they are compared with the horizontal sweep as to frequency and phase (Sec. 6-3).

Figure 6-12 is the schematic diagram of a typical sync separator circuit. Note that the video input is negative-going and that the input transistor Q_1 is PNP. Thus, both the sync pulses and the signal/noise turn Q_1 on. However, with Q_1 biased near zero, the large sync pulses drive Q_1 into saturation at a level never reached by the signal and/or noise. The second sync separator transistor Q_2 is reverse-biased, so that the first portion (about 0.3 V typically) of the Q_1 output is clipped. This further removes any signal and/or noise.

The output of Q_2 is applied to the low-pass filter (vertical integrator) of C_6 and R_{11}. The 60-Hz vertical signals are applied to the vertical sweep circuit input (the primary of the vertical oscillator transformer) through series and shunt diodes CR_1 and CR_2. These diodes are arranged to pass the sync pulses to the vertical sweep input but prevent the vertical oscillator pulses from passing back into the sync circuits.

The output of Q_2 is also applied to Q_3, which acts as a phase splitter or phase inverter. The output from Q_3 is applied to the horizontal AFC

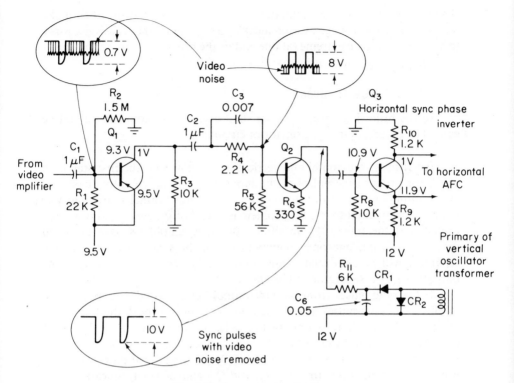

Fig. 6-12. Sync separator circuits

circuits. The phase inverter is required when the horizontal AFC circuits are of the balanced type (Fig. 6-7).

6-5.2 Recommended Troubleshooting Approach

Failure of the sync separator circuits can produce symptoms similar to those produced by failure in other circuits. For example, if the low-pass filter (vertical integrator) circuit fails, there will be a loss of vertical sync or poor vertical sync. This same symptom can be produced by failure of the vertical sweep circuits as well. Also, operation of the sync separator circuits depends on signals from other circuits. For example, if the video output is low, the sync separator output will be low or possibly absent.

Because of these conditions, the only practical troubleshooting approach for the sync separator circuits is to measure waveforms at all outputs and inputs, followed by dc voltage measurements at all transistor elements.

However, before going inside the set, the first logical step is to try correcting the problem by adjustment of horizontal and/or vertical controls.

If the problem cannot be eliminated by adjustment of controls, then make the waveform and voltage measurements. Keep in mind that if it is necessary to set a control to an extreme (at one end or the other) of if the control must be reset repeatedly, look for a marginal component failure.

6-5.3 Typical Troubles

In the following paragraphs we shall discuss symptoms that could be caused by defects in the sync separator circuits.

No sync. If horizontal and vertical sync are absent, the first step is to check for proper sync pulses at the input of the sync separator (from the video amplifier, Sec. 6-8). If these pulses are not normal (particularly if they are low in amplitude), the problem is ahead of the sync separator. For example, if the tuner, IF amplifiers, or video amplifiers have marginal defects (low gain, improper alignment, etc.), the sync pulses to the sync separator will be abnormal. Thus, both the horizontal and vertical sync outputs from the separator will be abnormal.

If the sync pulses from the video amplifier are good, the next step is to check the sync pulses at the *last point common to both* the horizontal and vertical sync. In the circuit of Fig. 6-12, such a point is the collector of Q_2. If the pulses at the Q_2 collector are not good with a good input from the video amplifier, the problem is in Q_1 or Q_2 (or the related parts). Further waveform measurements (between Q_1 and Q_2) and/or voltage measurements can be used to isolate the problem. The most likely causes of trouble in the circuits of Q_1 and Q_2 are open or leaking coupling capacitors, leaking transistors, and leaking bypass capacitors.

No vertical sync. If vertical sync is absent or critical (requires frequent adjustment and will not hold) but there is good horizontal sync, the problem is in the vertical integrator or the vertical sweep circuits. The first step is to measure the waveform at the vertical integrator output. In the circuit of Fig. 6-12, such a point is the primary of transformer T_1. If the pulses are good at this point, the problem is in the vertical sweep circuits (refer to Sec. 6-4). If the pulses are abnormal at the input to the vertical sweep, the problem is probably in the integrator.

When measuring waveforms of vertical and horizontal sync pulses, it is convenient to set the oscilloscope sweep to 30 Hz (for vertical) and 7875 Hz (for horizontal). This will display two cycles of the corresponding pulses. Some oscilloscopes designed specifically for TV service are provided with these sweep rates.

When measuring vertical pulses (sweep rate at 30 Hz) after the vertical integrator, there should be no horizontal pulses present (since the vertical integrator acts as a low-pass filter). However, when checking vertical pulses ahead of the integrator (say, at Q_1 or Q_2 in Fig. 6-12), the horizontal pulses may appear on the oscilloscope display. When the oscilloscope is set to measure horizontal pulses (set at 7875 Hz), the vertical sync pulses should not appear since they are so slow in relation to the horizontal sync pulse frequency.

When measuring ahead of the vertical integrator, which is the last point common to both horizontal and vertical sync, look for any noise or video signal that may have leaked through. If there is any noise or video at the output of Q_2, this is usually a sign of improper bias on Q_1 and/or Q_2. With improper bias, the undesired noise and video may not be removed by the normal clipping or limiting action. Of course, improper bias should show up as an abnormal bias.

When measuring the output of the vertical integrator, look for a *kickback* voltage from the vertical sweep circuit. Such kickback produces a display similar to that shown in Fig. 6-13. As discussed in Sec. 6-4, the

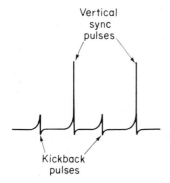

Fig. 6-13. Kickback pulses present in sync separator circuits

vertical oscillator uses feedback between secondary windings to sustain oscillation. The large surge of current through the secondary winding is reflected back to the primary and possibly into the sync separator circuits. The purpose of CR_1 and CR_2 is to prevent the kickback from entering the sync circuits. Note that not all solid-state circuits have diodes such as CR_1 and CR_2. However, there is usually some similar circuit.

No horizontal sync. If horizontal sync is absent or abnormal with good vertical sync, the problem is in the horizontal portion of the sync separator or the horizontal sweep circuits. The first step is to measure the waveform at the horizontal portion of the separator. In Fig. 6-12, such points can be the collector and emitter of Q_3. The waveforms should be identical in this case since Q_3 functions as a phase inverter and provides two pulses (180° out of phase) to a balance horizontal oscillator AFC circuit (Sec. 6-3). An unbalanced AFC requires only one horizontal pulse input from the sync separator.

If the pulses are good at the input of the AFC circuit, the problem is in the horizontal circuits (refer to Secs. 6-2 and 6-3). If the pulses are abnormal at the AFC input, the problem is in the horizontal output portion of the sync separator.

Keep in mind that horizontal sync troubles can be the result of poor high-frequency response, whereas vertical sync problems are generally caused by poor low-frequency response. Of course, this does not apply to sync problems caused by failure of a specific component.

Picture pulling. There are several forms of picture pulling common in solid-state sets. Often, the nature of the picture pulling symptoms can pinpoint the trouble. For example, when picture pulling appears to be steady and there is a bend in the image, this indicates that the horizontal AFC circuits are receiving distorted pulses. A likely cause is vertical kickback pulses entering the sync separator and distorting the horizontal output. If this condition is suspected, check for vertical kickback pulses in the sync separator and for distortion of the horizontal pulses with an oscilloscope.

If the picture pulling appears unsteady and particularly if the pulling tends to follow the camera signal, this indicates poor sync separation. That is, the video output is not being clipped and limited sufficiently to remove all camera signals from the circuits. If this condition is suspected, check transistor bias, particularly the bias of Q_1 and Q_2 in Fig. 6-12 (or their equivalents) in the circuit being serviced. This bias sets the signal levels at which clipping and limiting occur. In general, if bias is too high, the camera signal will be cut out but the sync signals will be attenuated. If bias is too low, the sync signals will be good, but some camera signal can also pass, resulting in picture pulling, distortion, and so on.

Once the picture pulling symptoms have been analyzed, the logical suspicion should be confirmed with waveform and voltage measurements. Start with a waveform measurement of the horizontal pulses to the horizontal AFC input. If those waveforms show a steady distortion, look for vertical kickback. If the waveforms show a fuzzy base line, look for poor sync separation and the presence of camera signal. For example, in Fig. 6-12 there may be a camera signal at the output of Q_1, but there should be none at the output of Q_2.

Do not confuse picture pulling with distortion of the raster. If the raster is bent, the problem is not in the sync separator circuits but in the horizontal or vertical sweep circuits. If the raster edges are sharp but the picture is pulling or bent, then the problem can be in either the sweep or separator circuits. Check the raster alone by switching the channel selector to an unused channel.

6-6. RF TUNER CIRCUITS

In tuner troubleshooting some technicians will only replace the tubes and transistors and possibly check alignment. They prefer to send defective tuners to specialized repair shops where mechanical parts can be replaced, adjusted, etc. For such technicians, repair is limited to application of a spray cleaner and lubricant to tuner contacts. While this may be the best way to go, it may be necessary to perform some further troubleshooting (at least to confirm that the problem is definitely in the tuner).

In any tuner, if the problem (weak picture, loss of picture, etc.) is observed on only one channel, the trouble is in the tuner. Troubleshooting should start with the circuits (coils and capacitors) that apply to the defective channel. If a problem appears on all channels, the trouble can be in the tuner or in the IF stages. Thus, the first step is to localize the problem.

6-6.1 Basic Circuit Theory

Figure 6-14 is the schematic diagram of a typical RF tuner. Note that the tuning coils are of the turret type, where a separate set of drum-mounted coils is used for each channel and the entire drum rotates when the channel is selected. Most transistor TV tuners are of the turret type, with a very small percentage of the switch type in use. (Switch-type tuners have series-connected coils mounted on wafer switches.)

All three stages of the tuner (RF amplifier, oscillator, and mixer) are connected in the common-emitter (CE) configuration. This is typical for transistor tuners, although a few tuners use the common-base (CB) configuration. The RF amplifier Q_{301} is neutralized by C_{308}. Neutralization is generally required for all solid-state RF amplifiers used in tuners. If the neutralization capacitor is open, the RF amplifier can break into oscillation.

The RF amplifier base is connected to the AGC network. Most solid-state TV AGC networks operate on a different basis from vacuum-tube systems. With solid-state, a strong signal increases the forward bias and drives the transistor into the saturation region of collector current, thus reducing gain. With vacuum-tube systems, strong signals increase the negative bias on the vacuum-tube grids to reduce gain. A few solid-state TV AGC systems also use increased negative bias to desensitize the circuits.

Most tuners (solid-state and vacuum-tube) are provided with at least one test point (often called the "looker" point). In the circuit of Fig. 6-14, the test point is at the collector of Q_{303}. This test point is used mostly to monitor the tuner output with an oscilloscope. The same point can also be used to inject a signal into the IF amplifiers.

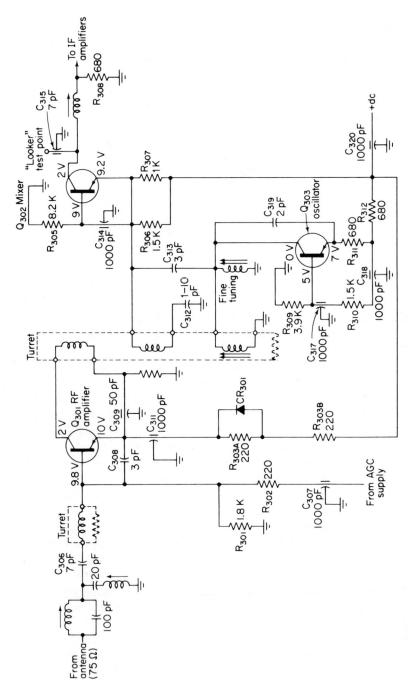

Fig. 6-14. RF tuner circuits

380

6-6.2   Recommended Troubleshooting Approach

It is often difficult to decide if the tuner or the IF amplifiers are at fault. The same symptoms can be produced by problems in either section of the set. For example, if the tuner amplifier gain is low, the same basic symptoms (weak picture and sound, poor contrast, etc.) will be produced as when the IF amplifier gain is low.

Some technicians prefer to make a quick check by shorting the RF amplifier AGC point (base of Q_{301}) to ground. This will remove the forward bias on the RF amplifier and usually cut the stage off. If this produces considerable change in the picture, the RF amplifier is probably good. If little change is noticed when the RF amplifier is cut off, the RF amplifier is probably at fault.

If adequate test equipment is available, the tuner should be checked thoroughly before replacing it or sending it to an outside shop. The best method is to apply a sweep frequency signal (with markers) to the antenna input and monitor the tuner output at the mixer test point with an oscilloscope. The basic procedures are described in Chapter 3. The same procedures can be used for tuner alignment. Always refer to the manufacturer's service literature for detailed alignment procedures. If the tuner output is good, as measured at the tuner looker point, the problem is likely to be in the IF stages.

If the tuner output is not good, the next step is isolation of the fault to one of the three tuner stages. One simple approach is to apply an RF signal (modulated by an AF tone) to various points in the tuner and observe the picture display. Tune the RF signal generator to the channel frequency. If the tuner is operating normally, a series of horizontal bars will appear on the picture tube screen. The number of bars depends on the frequency of the AF modulating tone.

If the picture display is normal with the RF signal injected at the collector of Q_{301} but not normal with a signal at the antenna, the problem is likely to be in the RF amplifier Q_{301}. Next, inject the signal at the base of Q_{301}. If the picture display is normal here, the problem is in the network between the antenna and the base of Q_{301}.

If it is not possible to get a signal through the tuner at any point, change the RF generator frequency to the frequency used by the IF amplifier. If the IF signal passes (produces a display on the picture tube), the most likely trouble is a defective oscillator Q_{303}.

6-6.3   Typical Troubles

In the following paragraphs we shall discuss symptoms that could be caused by defects in the RF tuner circuits.

No picture or sound. If a raster is present but there is no picture or sound, the first step is to inject an IF signal at the looker point or monitor the tuner output at the same point, whichever is most convenient. If the fault is definitely isolated to the tuner, the next step is to check the raster pattern for snow. If there is a complete absence of snow, the defect is likely to be in the mixer stage rather than in the RF amplifier. If the mixer stage and IF amplifiers are good, there will be some snow even if the RF amplifier is completely dead.

If it appears that either the RF amplifier or mixer is completely dead, measure voltages at the transistor elements and/or inject RF signals at both stages, as described in Sec. 6-6.2. Note that both the RF amplifier and mixer are forward-biased when operating normally. Keep in mind that the RF amplifier is usually forward-biased by the AGC network. If this forward bias is missing or abnormal, the problem can be in the AGC circuits rather than in the tuner.

A simple test is to apply the required forward-bias voltage to the AGC line where it enters the tuner. If this restores normal operation, check the AGC circuits for a defect. If the service literature is available, check for the correct amount of forward bias from the AGC circuit to the RF amplifier under no-signal conditions. Usually, the service literature will show both the no-signal and full-signal bias values. Use the no-signal bias value to make the test.

If an RF signal will not pass but an IF signal will pass as described in Sec. 6-6.2, the oscillator is suspect. As a quick test try injecting a signal at the oscillator frequency, with the set tuned to a normal, strong channel. Use an unmodulated RF signal generator. Preferably, the signal should be injected at the same point as is the tuner oscillator. In the circuit of Fig. 6-14, the oscillator signal is injected at the emitter of Q_{302} through C_{313}. If operation is restored when the external oscillator signal is injected, the tuner oscillator circuit is defective.

Poor picture or sound. The basic procedure for troubleshooting a poor picture and sound symptom (snow, weak sound, poor contrast, etc.) is the same as for no sound or picture. That is, the trouble must be isolated to the RF tuner and then to the particular stage in the tuner. However, it is necessary to have a greater knowledge of the circuit's capabilities to troubleshoot a poor performance symptom.

For example, if the tuner oscillator is completely dead, operation can be restored by injecting a signal at the oscillator frequency. This pinpoints trouble to the tuner oscillator. However, if the tuner oscillator is producing a weak signal, an injection test may prove confusing. Ideally, the service literature should be consulted to find the normal amplitude of the oscillator output.

Hum bars or hum distortion. In solid-state TV sets, the presence of hum bars on the picture, hum distortion, or poor sync accompanied by hum symptoms is generally the result of a failure in the power supply. In vacuum-tube sets, hum is often caused by cathode-heater leakage or some form of leakage in the tube elements. This is not the case in solid-state TVs. If there is any evidence of hum in the display, check the dc input to the RF tuner for the presence of 60- or 120-Hz hum. If 60- or 120-Hz signals are present on any of the dc lines, check the power supply as described in Sec. 6-1. Most solid-state sets now use full-wave rectifiers in the low-voltage power supply. Thus, hum will appear at 120-Hz signals on the dc line if the power supply filter or regulator fails.

Picture smearing and sound separated from picture. When the picture appears to be smeared, with trailing edges of images not sharply defined, the trouble is usually the result of poor response and is often accompanied by a separation of picture and sound. That is, when the fine tuning is adjusted for the best picture, the sound is poor, and vice versa. This symptom can be caused by poor alignment or problems in the AGC system.

Likewise, the problem can be in the IF stages. Thus, the first step is to localize the problem to the RF tuner with a response test as described in Sec. 6-6.2. Next, apply a fixed dc bias to the AGC line equal to the no-signal bias. (This is often known as *clamping* the AGC line.) Keep in mind that most solid-state TV AGC systems apply a forward bias to the RF tuner under both no-signal and full-signal conditions.

If the problem is eliminated with the proper forward bias applied, look for trouble in the AGC network, as described in Sec. 6-9. If the problem remains with the correct bias applied, check the response pattern of the tuner. If the response pattern is not good, try correcting the condition by alignment. Always follow the service literature instructions for alignment. If it is not spelled out by the service literature, clamping the AGC line with a forward bias is usually implied.

Many solid-state tuners are provided with a *delayed* AGC function. This is accomplished by diode CR_{301} in Fig. 6-14. CR_{301} requires about 0.5 V to conduct. If the signal increases to a point where the drop across $R_{303}A$ is more than 0.5 V, CR_{301} will conduct and short $R_{303}A$. This will place a large forward bias on Q_{301} and drive the transistor into saturation, thus reducing the gain.

Whenever tuner response is poor, with the AGC network and IF stages cleared, and the problem cannot be corrected by alignment, look for such defects as open neutralizing capacitors, open bypass capacitors, and loose or poorly grounded tuner shields. Any of these most likely defects can cause poor tuner operation but will not show up as incorrect dc voltages at the transistor elements.

Ghosts. If there are ghosts (double images or repeats) in the picture display, it is possible that the tuner is at fault. Of course, ghosts can be caused by propagation problems (where the TV transmission is reflected from a building or similar object, and the reflected signals arrive at a slightly different time than the nonreflected signals) or by antenna problems. Thus, the first step is to isolate the problem to the set rather than to the antenna or outside conditions. Generally, when ghosts are caused by reflections, the ghosts will change from channel to channel or will sometimes disappear on some channels.

A more positive test is to apply a signal from a pattern-type generator to the tuner input. If the ghosts disappear, they are the result of outside causes. If the ghosts remain with a signal applied to the tuner, they are definitely circuit ghosts and can be caused by problems in the tuner. Generally, circuit ghosts are the result of problems in the IF stages or video stages but can be in the tuner.

When there are circuit ghosts in the tuner, it is generally a case of very sharp response due to an abnormally high Q in one of the tuned circuits. This will show up as very sharp peaks in the tuner response curve. The most likely causes, other than poor alignment, are open capacitors, particularly bypass capacitors or the RF amplifier neutralization capacitor.

Picture pulling. As discussed, picture pulling is usually a problem in the sweep or sync separator circuits. However, the problem can also be caused by the tuner. As a first test, check the edges of the raster. If they are sharp and well defined, the sweep circuits are probably good. If an analyzer-type generator is available, inject an RF signal (complete with sync pulse) at the input of the IF stages (at the tuner looker point if convenient). If the pulling symptom disappears, the problem is in the tuner. If the pulling symptom remains, the problem is in the IF stages, video stages, or (more likely) the sync separator.

If the problem is definitely localized to the tuner, proceed with the localization and isolation steps discussed previously: Monitor the tuner response pattern at the looker point, measure the waveforms and voltages at each stage, try to correct the poor response pattern by alignment, and check the response pattern with the AGC line clamped and unclamped. One or more of these steps should pinpoint the problem.

Intermittent problems. When both the picture and sound are intermittent, the tuner is usually suspected. However, the actual cause can be in another circuit. For example, if the AGC circuit is defective and intermittently removes the forward bias to the RF amplifier transistor, the RF amplifier can be cut off, removing both sound and picture. At best, intermittent problems are difficult to locate. In the case of a tuner, the best bet is to *monitor* the circuits and watch for changes when the intermittent condition occurs.

As a minimum, monitor the tuner output at the looker point, the AGC line, and the dc voltage line into the tuner. If either the dc voltage or the AGC line voltage changes considerably when the intermittent condition occurs, the problem is likely not to be in the tuner but in the power supply or AGC circuits.

If neither the dc voltage nor the AGC line voltage changes but the tuner output display is affected during the intermittent condition, the problem is in the tuner. The most likely causes are cold solder joints, break in printed circuit wiring, and intermittent capacitors (particularly in the RF amplifier and mixer stages). In rare cases, the transistor will become intermittent. If all else fails to localize the trouble, try applications of heat and cold to the tuner transistors or simply try substitution of the transistors, whichever is more convenient.

UHF tuner problems. Figure 6-15 is the schematic of a solid-state UHF tuner. Note there is no RF amplifier as such, that a diode CR_1 is used as

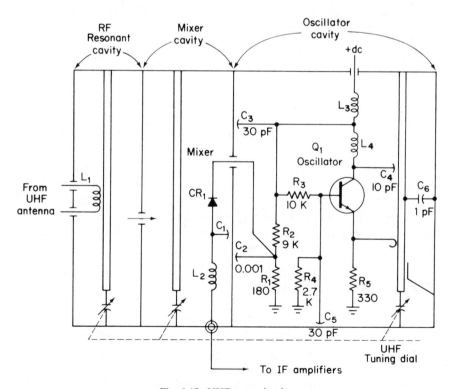

Fig. 6-15. UHF tuner circuits

the mixer, and that a transistor is used as the oscillator. Also note that the tank circuits are in the form of resonant cavities, tuned by variable capacitors connected to the UHF fine tuning knob. No AGC is provided for the UHF tuner.

All the problems that apply to a vacuum-tube UHF tuner apply to the solid-state version. For example, if the physical structure of the resonant cavities is changed, the resonant frequency will change. That is, if the tuner shields are removed, bent, or otherwise distorted or if shield screws are loose, the resonant frequency can change. A minor frequency change will make the UHF fine tuning knob markings incorrect. A major frequency change can make the UHF tuner totally inoperative.

Ideally, the UHF tuner should be checked by means of a UHF signal at the antenna input (preferably from a UHF sweep generator with markers) and an oscilloscope at the IF output. In this way, the tuner gain, bandwidth, and overall frequency response can be checked quickly and accurately. Unfortunately, many shops are not equipped with a UHF generator.

The first step in troubleshooting is to check operation of the set on a UHF channel with a known good signal. If the UHF reception is dead or very poor, with a known good signal and with good VHF reception, the problem is in the UHF tuner.

The next step is to measure all the voltages at the transistor elements. Note that Q_1 is forward-biased. Also note that the mixer diode CR_1 has a bias voltage placed on its cathode. If all voltages appear normal, the next step is to check the mixer diode CR_1 and oscillator transistor Q_1 by substitution. The front-to-back ratio of CR_1 can be checked if desired. However, if the diode appears to be defective, it must be substituted.

If none of these procedures localize or cure the problem look for open capacitors in the tuner. Usually, a shorted or leaking capacitor will produce an abnormal voltage at one or more of the circuit elements, where an open capacitor will leave the voltages normal but make the circuit operate improperly. There are exceptions of course. For example, if capacitor C_6 in Fig. 6-15 is shorted or leaking, the resonant cavity will be seriously detuned, possibly to the point where Q_1 might not oscillate. However, the circuit voltages will remain virtually unchanged from normal operation.

6-7. IF AND VIDEO DETECTOR CIRCUITS

The basic function of the IF and video detector circuits in solid-state sets are the same as for vacuum-tube sets. That is, the circuits amplify both picture and sound signals from the RF tuner-mixer circuit, demodulate both signals for application to the video amplifier and sound IF amplifiers, and trap (or reject) signals from adjacent channels. Thus, the basic troubleshooting methods for tube-type IF circuits can be applied to solid-state circuits.

6-7.1   Basic Circuit Theory

Figure 6-16 is the schematic diagram of a typical IF and video detector circuit. Three stages of amplification are used, with each stage in the common-emitter configuration. All stages are forward-biased. The first two stages Q_1 and Q_2 receive their forward bias from the AGC circuit. This same AGC line is connected to the RF amplifier in the tuner, as discussed in Sec. 6-6. On strong signals, the forward bias is increased, driving Q_1 and Q_2 into saturation and reducing gain. The third IF stage Q_3 does not have any AGC control. All three stages are neutralized to prevent oscillation in the IF circuits. If any of the neutralizing capacitors are open, the corresponding stage can break into oscillation.

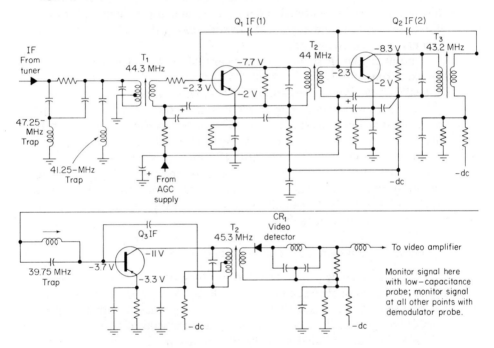

Fig. 6-16. IF and video detector circuits

Note that all of the functions provided by the circuit of Fig. 6-16 can be supplied by a single integrated circuit (IC). In such sets, the entire IC must be replaced as a package if trouble is isolated to the IF and video detector stages. In other IC sets, the video detector (CR_1) circuit components are separate from the IF stages. That is, all three IF stages (Q_1 through Q_3) are in an IC package (and must be so replaced), but the video detector stage is separate, and parts (CR_1, the related coils and capacitors, etc.) can be replaced.

In the circuit of Fig. 6-16 (which is typical for the majority of sets now in use), each stage is tuned at the input and output by corresponding transformers (T_1 through T_4). The stages are stagger-tuned; that is, each transformer is tuned to a different peak frequency. This gives the overall IF amplifier circuit a bandwidth of about 3.25 MHz.

Three traps are used. The 41.25- and 47.25-MHz traps are series resonant, whereas the 39.75-MHz trap is parallel resonant. As described in previous chapters, the traps are adjusted for a minimum output signal at the video detector when a signal of the corresponding frequency is injected at the IF input.

There are no test points as such provided on most solid-state IF circuits. However, input from the tuner is usually applied through a coaxial cable. This cable can be disconnected and used as a signal injection point for overall IF amplifier tests and alignment. In some cases, the looker point on the RF tuner can be used instead.

Output of the IF stages can be measured at the video detector CR_1. The video output is typically about 1 V for most solid-state sets. That is, the peaks of the sync pulses are about 1 V, with the picture signals about 0.25 to 0.5 V. Most solid-state IF circuits are mounted on printed-circuit boards so that the transistor base elements are accessible. Signals can be injected at any of the bases to trace performance through the IF stages.

In some solid-state sets, the entire IF function is performed by an IC. In such cases, the usual rules for IC troubleshooting apply. That is, a signal can be injected at the input, and the results are monitored at the output. Supply voltages to the IC can also be checked. If there is any defect in the IC, the entire IC must be replaced.

6-7.2 Recommended Troubleshooting Approach

The recommended approach for troubleshooting IF stages depends largely on the available test equipment. Ideally, an *analyzer-type generator* can be used. As discussed in Chapter 2, these generators duplicate the signals normally found at the mixer output and the video detector output (as well as several other signals). If the picture display is good with a signal injected at the video detector output (video amplifier input) but not good with a signal injected at the IF input (coaxial cable from the mixer), the problem is in the IF amplifiers. A possible exception is with defective AGC circuits. This problem can be eliminated by clamping the AGC line with the appropriate voltage. If the problem remains with the correct AGC bias voltage applied from a fixed external source, the problem is in the IF stages.

If an analyzer-type generator is not available, the next recommended test equipment setup is a *sweep generator* (with markers) and an oscilloscope. Refer to Chapters 2 and 3. (Many technicians prefer sweep/marker generator to the analyzer.) The sweep generator signal is injected at the IF input (cable from mixer), and the signal is monitored at various points throughout the IF stages with the oscilloscope. A *demodulator probe* is required for the oscilloscope if individual IF stages are to be monitored. A basic *low-capacitance* probe can be used if the output of the video detector is monitored. Keep in mind that signals through the IF stages of a solid-state TV are on the order of 1 V or less. Thus, the oscilloscope must have considerable vertical gain. Operate the horizontal sweep of the oscilloscope at 30 Hz so that two cycles of sync pulses are displayed.

If analyzer and sweep generators are not available, use an RF generator (modulated by an AF tone). Tune the RF generator to the approximate center frequency of the IF amplifiers and inject the signal at the IF amplifier input (cable from mixer) and at the bases of each transistor in turn. If the IF amplifier is operating normally, a series of horizontal bars will appear on the screen. (The number of bars depends on modulating frequency.)

6-7.3   Typical Troubles

In the following paragraphs we shall discuss symptoms that could be caused by defects in the IF amplifier stages.

No picture or sound. If a raster is present but there is no picture, no sound, or the sound is very weak and noisy, first inject an IF signal at the IF amplifier input. If operation is normal, the problem is in the tuner rather than in the IF stages. If the fault appears to be isolated to the IF stages, then clamp the AGC line and repeat the injection test. If the problem is cleared, look for trouble in the AGC circuits (as discussed in Sec. 6-9).

If the fault is definitely isolated to the IF stages, inject an IF signal at the base of each stage, as discussed in Sec. 6-7.2. Then measure the voltages at all transistor elements. This should isolate any problem serious enough to cause a no-picture/no-sound symptom.

Poor picture or sound. The basic procedure for troubleshooting a poor picture and sound symptom (snow, weak sound, poor contrast, etc.) is the same as for no sound or picture. That is, the trouble must be isolated to the IF circuits and then to a particular stage. This may not be so easy to isolate as when there is complete failure of a stage. For example, the service literature rarely gives the gain per stage for the IF circuits. However, there should be some gain for each stage. Typically, the *overall gain* for the IF stages should be on the order of 60 dB (from the tuner output to the detector input).

Generally, a poor picture and sound symptom is the result of low gain in one or more stages. Most of the troubles that cause low gain will show up as an abnormal voltage. For example, a leaking or shorted capacitor or an open transformer winding will produce at least an abnormal voltage at one or more transistor elements. Even if a transistor has low gain, the collector and/or emitter voltages will be abnormal. (Generally, the emitter will be low, and the collector will be high.)

Open capacitors can sometimes cause low gain, without substantially affecting voltages. For example, if a capacitor across an interstage transformer winding opens, the transformer resonant circuit will be affected. If the transformer is seriously detuned, gain will be reduced. If an emitter bypass capacitor opens, the ac gain of a stage will be reduced drastically, even though the transistor voltages remain substantially the same.

Hum bars or hum distortion. Unlike vacuum-tube sets where hum problems (hum distortion, hum bars in picture, and poor sync plus hum) can be caused by cathode-heater leakage in one of the tubes, hum in solid-state sets is generally the fault of the power supply filter. One possible exception is an open decoupling capacitor in one of the IF stages. Since the output of the IF circuit is fed into the video amplifier where vertical blanking pulses are applied (Sec. 6-8), these pulses can enter the IF stages if a decoupling capacitor is open. As is the case with any suspected open capacitor, try connecting a known good capacitor in parallel with the suspected capacitor.

Picture smearing, pulling, or overloading. Generally, these symptoms are associated with the video amplifier section rather than the IF, particularly when sound is good. However, if the IF stages are improperly aligned or if there is a defective part, making proper IF alignment impossible, the same symptoms will occur even though the sound channel may be good.

To eliminate doubt, inject a video frequency signal at the input of the video amplifier (out of video detector). If the picture problems are eliminated, the trouble is in the IF stages. Next, clamp the IF stage AGC line to see if the problem is eliminated. If the symptoms disappear with the correct bias applied, look for trouble in the AGC network, as discussed in Sec. 6-9.

If the problem remains with correct fixed bias, check the response pattern of the IF stages, as discussed in Chapter 3, following the manufacturer's service literature. If the response pattern is not good, try correcting the condition by alignment. If any particular stage cannot be aligned, look for such defects as open capacitors, poorly grounded IF coil shields, and open damping resistors.

Intermittent problems. When both the picture and sound are intermittent, either the tuner or IF amplifiers are usually suspected. However, the AGC circuit or the low-voltage power supply could be at fault. In any case the first step is to localize the problem to the IF stages by monitoring the tuner as described in Sec. 6-6.3. If the intermittent condition occurs but the tuner indications are good, the problem is in the IF stages.

Next, monitor the IF stages and watch for changes when the intermittent condition occurs. As a minimum, monitor the video detector output, the AGC line, and the dc voltage line of the IF circuits. In solid-state sets, the IF circuits are often mounted on a separate printed-circuit board with one dc voltage distribution point. The video detector output can be monitored by an oscilloscope with a low-capacitance probe. Any other point in the IF circuits (ahead of the video detector) will require a demodulator probe for monitoring.

If neither the dc voltage nor the AGC line voltage changes but the video detector output (or other point being monitored) is affected during the intermittent condition, the problem is in the IF amplifier stages. As in the case of any intermittent conditions, the most likely causes are cold solder joints, breaks in printed-circuit wiring, and intermittent capacitors. Intermittent transistors are rare. If all else has failed to localize the trouble, try applications of heat and cold to the IF transistors or try transistor substitution.

6-8. VIDEO AMPLIFIER AND PICTURE TUBE CIRCUITS

Video amplifier circuits have several functions, and there are many circuit configurations. Thus, it is very difficult to present a typical troubleshooting approach. However, most video amplifier circuits have three inputs and three outputs which can be monitored. If the inputs are normal but one or more of the outputs are abnormal, the problem can be localized to the video amplifier circuits.

6-8.1 Basic Circuit Theory

Figure 6-17 is the schematic diagram of a typical solid-state video amplifier circuit. Two stages, driver Q_1 and output Q_2, are used. Transistor Q_1 is connected as an emitter follower. Typically, the signal input and output at Q_1 are about 1 V. This signal consists of the sound, video, and sync pulses, as taken from the video detector.

The output of Q_1 is applied to the input of video output transistor Q_2, to the input of the sync separator circuits (Sec. 6-5), and to the input of the sound IF circuits (Sec. 6-10). In the circuit of Fig. 6-17, the output of Q_1 is applied to Q_2 through contrast control R_3. In some solid-state circuits, the contrast control is part of the video output transistor circuit (typically in the emitter of the output transistor).

The video output transistor Q_2 is almost always a power transistor and is mounted on a heat sink. Transistor Q_2 amplifies the 1-V output to a level of about 25 to 50 V depending on the picture tube screen size. Brightness control R_{12} sets the voltage level on the collector of Q_2. The output of Q_2 is applied through a 4.5-MHz sound trap to the cathode of the picture tube.

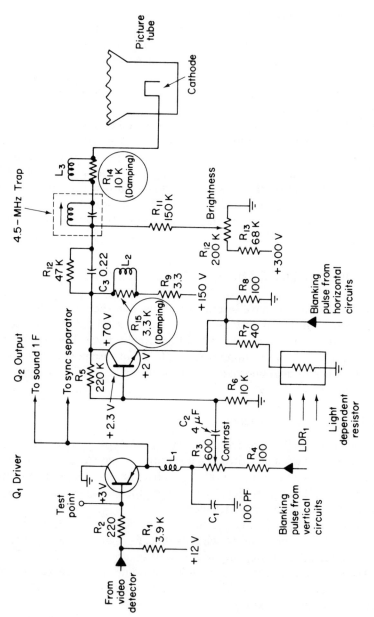

Fig. 6-17. Video amplifier and picture-tube (video) circuits

This trap, when properly adjusted, prevents any sound information from entering the video circuits. The Q_2 output should contain only the video information (picture signal) as well as the horizontal (15,750-Hz) and vertical (60-Hz) retrace blanking pulses. Since all of this information is applied to the cathode of the picture tube, the video (picture) signal is negative, whereas the blanking pulses are positive. (A positive pulse applied to the picture tube cathode decreases brightness, whereas a negative pulse increases brightness.)

In some solid-state picture tube circuits, the blanking pulses are applied to the first control grid of the picture tube (and in rare cases to the filament) rather than being mixed with the video pulses. In still other cases, the mixed video and blanking pulses are applied to the control grid. Some solid-state picture tube circuits do not have any horizontal blanking pulses.

Figure 6-18 shows some typical solid-state picture tube circuit configurations as well as typical voltages. Note that all of the voltages for operation of the picture tube (except for the filament) are supplied by the high-voltage portion of the horizontal output circuits (Sec. 6-2).

The vertical retrace blanking pulses from the vertical sweep circuits (Sec. 6-4) are applied to the emitter of Q_1. Horizontal blanking pulses are applied to the emitter of Q_2. Note that both Q_1 and Q_2 are forward-biased during normal operation.

The bias on Q_2 is partially determined by light-dependent resistor LDR_1, connected between the emitter and ground. As the ambient lighting around the set varies, the bias on Q_2 (and thus the gain of the video amplifier) varies. This variable gain feature provides the necessary changes in picture contrast to accommodate changing conditions of ambient light.

6-8.2 Recommended Troubleshooting Approach

The recommended approach for troubleshooting video amplifier stages depends largely on the available test equipment. Ideally, an analyzer-type generator can be used. These generators duplicate the signals normally found at the video detector output. If the picture display is not good with a signal injected at the video amplifier input (video detector output), the problem is in the video amplifier or possibly in the picture tube and its circuits.

If an analyzer-type generator is not available, the next recommended test setup is a sweep generator (with markers) and an oscilloscope. The sweep generator signal is injected at the tuner input or the IF amplifier input, and the signal is monitored at the video detector output. If the overall response through the RF (tuner) and IF stages is good but there are problems in the picture, the trouble is probably in the video amplifier.

As a quick check, the output of the video detector can be monitored with an oscilloscope (with the set tuned to an active TV channel). If there is

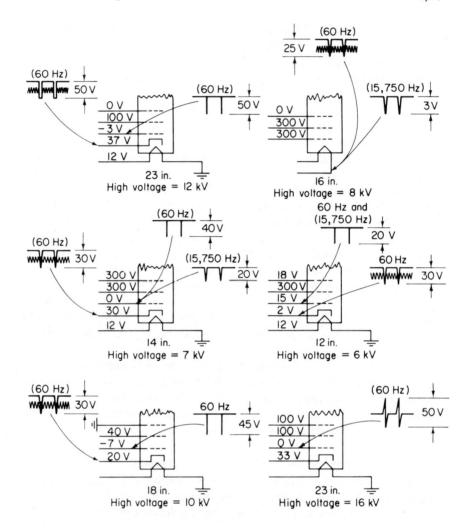

Fig. 6-18. Typical picture tube voltages and circuits

a signal of about 1 V at the video amplifier input, any picture problems are probably in the video amplifier circuits.

Some technicians prefer to make a square-wave test of the video amplifier circuits. A square wave of about 1 V is applied to the video amplifier input, and the resultant output is monitored on an oscilloscope at the picture tube cathode. This checks the overall response of the video amplifiers, including the effect of contrast and brightness controls. If square waves are passed without distortion, attenuation, ringing, and so on, the video amplifier circuits are operating properly.

6-8.3   Typical Troubles

In the following paragraphs we shall discuss symptoms that could be caused by defects in the video amplifier stages.

No raster. A no-raster symptom is usually associated with failure of the power supply or horizontal sweep circuits rather than failure in the video amplifier. However, complete cutoff of the picture tube can be produced by failure in the video amplifier circuits, even though the high-voltage and picture tube grid voltages are normal. For example, if brightness control R_{12} is open or making poor contact on the ground side, a high positive voltage is applied to the cathode of the picture tube. This will cut the picture tube off (no raster), regardless of signal conditions or other voltages.

If this condition is suspected, check all of the picture tube voltages against the service literature. Pay particular attention to the filament and cathode voltages. A low filament voltage can cause the filament to glow but will not produce enough emission for a picture. If any one of the voltages is not normal, trace the particular circuit and look for such defects as shorted or leaking capacitors, breaks in printed wiring, and worn brightness controls. If all of the voltages appear normal, try substitution of the picture tube.

No picture and no sound. With a raster present, a no-picture/no-sound symptom is normally associated with failure of the tuner or IF stages rather than the video amplifier. However, in the circuit of Fig. 6-17, failure of Q_1 could cut off both sound and picture, even though good signals are available at the video detector output.

The first step is to monitor the signal at the base and emitter of Q_1. Note that a test point is provided at the base of Q_1. If the fault is definitely localized to Q_1 (a good signal at the base and an absent or abnormal signal at emitter), the next step is to measure the voltages at all Q_1 elements. This should pinpoint the actual cause of trouble.

No picture and normal sound. With a raster present and good sound, a no-picture symptom is definitely in the video amplifier circuits. First monitor the signal at the base and collector of Q_2 and at the cathode of the picture tube. This should be followed by measurement of the Q_2 element voltages. As in the case of any amplifier, total failure of the circuit is usually easy to locate. Look for such defects as an open coupling capacitor and coils, cold solder joints, open or worn contrast controls, breaks in printed-circuit wiring, and the like.

Keep in mind that Q_2 is a power transistor and is subject to being burned out if its heat sink is defective (bad thermal conduction between heat sink and transistor). Also, other circuit defects can cause Q_2 to burn out. For example, if coupling capacitor C_2 is shorted or badly leaking, the forward bias on Q_2 could cause excessive current flow.

Contrast problems. Too little contrast (or weak picture) that cannot be corrected by adjustment of the contrast control is the result of poor gain or lack of signal strength. In the circuit of Fig. 6-17, the contrast control sets the signal level (or signal strength) applied to the video output transistor Q_2. The contrast control does not set the gain of either stage. This is typical for most solid-state video amplifier circuits, although there are a few solid-state circuits where the contrast control sets the stage gain (as is typical of most vacuum-tube sets).

Poor picture contrast can be the result of a weak (low-emission) CRT filament. In this case, the CRT must be replaced, or if it is a color CRT, its filament may be rejuvenated. The latter procedure is not always successful.

If the symptom is poor contrast, the first step is to monitor the input to the video amplifier circuits (video detector output). If the detector output (input to the video amplifier) is about 1 V but contrast is poor, the problem is in the video amplifier circuits (the video detector of a typical solid-state set rarely produces more than about 1 V).

The amount of gain provided by the video amplifier in normal operation depends, of course, on the circuit design. In the circuit of Fig. 6-17, output transistor Q_2 provides a gain of about 50. That is, the 1-V output from Q_1 is raised to about 50 V. In some solid-state video circuits (particularly those with three stages) there is an overall gain of about 100. That is, the video detector output is about 0.5 V, with 50 V applied to the picture tube. Keep in mind that these voltage and gain figures are typical for solid-state circuits. Always consult the service literature when available.

If the video amplifier gain is low or is suspected to be low, measure the transistor voltages, and look for the usual problems associated with low gain: excessive collector-to-base leakage, leaking capacitors, low power supply voltages, worn controls, and so on.

Too much contrast (that cannot be corrected by adjustment of the contrast control) is the result of too much gain or excessive signal strength. The first step is to monitor the video detector output. If the signal voltage is between 1 and 2 V, with excessive contrast, the problem is in the video amplifiers. If there is an excessive signal at the video detector output, look for trouble in the RF tuner or IF stages; if possible, consult the service literature for the correct video output signal level. It is also possible that the AGC circuits are failing to reduce excessive signals (refer to Sec. 6-10).

Excessive contrast is often associated with a defect in the contrast control circuit. Thus, the next step (after measuring the video detector output) is to monitor all voltages associated with the contrast control. If the voltages appear to be in order, monitor the signal voltages at the input and output of the video circuits.

Compare the signal voltage at the picture tube cathode against the service literature. If gain is too high, look for some circuit condition that is biasing the output transistor Q_2 to an incorrect level. Solid-state amplifier circuits are usually designed on the basis of minimum transistor gain. If a particular transistor has more gain than normal, it is possible that a slight change in bias can cause excessive gain. A bias change can result from aging of parts and from defects in circuits connected to the video amplifiers. However, this will show up as an abnormal voltage.

Another possible cause of excessive contrast is a defective heat sink for the video output transistor. If the heat sink is not making good contact with the transistor case (for good thermal conduction), the transistor temperature can rise and increase transistor gain. If the rise is not too great, the transistor will not burn out, particularly if the set is used for short periods of time. However, there will be excessive gain and contrast.

Sound in picture. A sound-in-picture symptom (picture appears to be modulated by the sound) can be caused by problems in the RF tuner or IF stages (including the AGC circuits that control these stages). Sound-in-picture is also often the result of poor alignment and improper fine tuning and adjustments.

First localize the trouble by monitoring the output of the video detector with an oscilloscope and low-capacitance probe. If the video display appears to be modulated by sound (video display is fuzzy and/or varies with sound), the problem is ahead of the video amplifier stages. Keep in mind that both sound and video are present at the base and emitter of Q_1 in normal operation. However, the sound should not modulate the video.

If the problem is definitely isolated to the video amplifier stages, first adjust the 4.5-MHz sound trap. In the circuit of Fig. 6-17, the sound trap is between the video output transistor and the picture tube cathode. Any sound present at the collector of Q_2 should be attenuated fully when the sound trap is properly tuned.

If the problem cannot be corrected by adjustment of the sound trap, look for defective components in the sound trap (open capacitor, shorted coil turns). Some solid-state video circuits do not have a sound trap, as such. Instead, the sound is removed through a transformer tuned to 4.5 MHz. The transformer secondary passes the sound signal to the sound IF stages. The transformer primary appears as a low-impedance short to 4.5-MHz signals in the video circuits, thus preventing the signals from passing to the picture tube.

Poor picture quality. When picture quality is poor (lack of detail, smearing, circuit ghosts, poor definition, etc.), the video amplifiers are usually suspected, particularly if the sound is good. However, these same symptoms can be caused by problems in the tuner and/or IF stages. If an analyzer-type generator is available, the first step is to inject a video signal into the input of the video amplifiers and watch the display on the picture tube. If the trouble symptom remains, the problem is in the video amplifier circuits.

If an analyzer generator is not available, the frequency response of the video amplifiers can be checked using square waves, as discussed in Chapter 3. In the case of video amplifier circuits, square waves of approximately 1 V are injected at the video amplifier input, and the resultant display is monitored by an oscilloscope connected to the picture tube cathode. A square-wave frequency of about 60 Hz is used initially. Then the frequency is increased to about 100 kHz.

If the square waves pass to the picture tube cathode without distortion, frequency response of the video amplifier is good. If there is any distortion, the waveshape will provide a clue as to frequency response. For example, if the leading edge (left-hand side) of the square wave is low or rounded, with the trailing edge square, this indicates poor high-frequency response or excessive low-frequency response. If the leading edge has greater amplitude than the trailing edge, low-frequency response is poor. If there is overshoot or oscillation on the leading edge, there is rising high-frequency response (or "ringing"). Keep in mind that symptoms such as smearing or poor definition are usually the result of poor high-frequency response, whereas circuit ghosts are caused by rising high-frequency response. (Of course, ghosts can be the result of antenna problems or broadcast transmission problems, as discussed.)

From a practical standpoint, any frequency response problems are the result of changes in tuned circuit. Look for such problems as leaking capacitors that can change the resonant frequency of peaking coils (L_2 and L_3 in Fig. 6-17), solder splashes that have shorted out damping resistors across peaking coils, or, in rare cases, shorted turns in a peaking coil.

The only frequency response problem unique to solid-state video amplifiers is excessive junction capacitance. Transistor junctions have a capacitance value that can change with age or (more likely) temperature extremes. Such a change can upset the resonant frequency (and thus the frequency response) of associated circuits. This problem is not too frequent in original equipment but can occur when video amplifier transistors are replaced. Always use the correct replacement (an identical type if at all possible.) Even with identical transistor replacements, it is possible that the junction capacitance can be incorrect (usually too much capacitance). Remember the old rule that a transistor (or vacuum tube) is good only if it operates properly in circuit.

Retrace lines in picture. In the circuit of Fig. 6-17, both vertical and horizontal blanking pulses are fed into the video amplifier to blank the retrace lines. As discussed, solid-state circuits do not provide blanking for the horizontal retrace. In other circuits, the blanking is applied to the picture tube elements via circuits other than the video amplifiers.

No matter what system is used, the obvious first step is to monitor the blanking pulses (with an oscilloscope and low-capacitance probe) at the point where they enter the video amplifier or at the picture tube element. If the blanking pulses are absent, trace the particular circuit back to the pulse source.

If the blanking pulses are low in amplitude, the typical symptom is the presence of retrace lines only when the brightness control is turned up. If this is the case, check the pulse amplitude against the service literature. Then trace the blanking pulses back to their source.

Intermittent problems. In a circuit such as Fig. 6-17, an intermittent condition in Q_1 or its related circuit elements will produce an intermittent sound and picture symptom. This same symptom can be produced by problems in the RF tuner and IF stages. Thus, the problem should be localized first to the video circuits by monitoring the tuner (Sec. 6-6) and the IF stages (Sec. 6-7). If the intermittent condition occurs but the tuner and IF indications are good, the problem is in the video amplifier circuits.

If the intermittent symptom is in the video only, the problem is then likely to be in the video amplifier circuits—probably in the video output transistor Q_2 circuits. This can be confirmed by monitoring the video signal at the base of Q_2 and the cathode of the picture tube. If the intermittent condition occurs without change in the video signals at the picture tube cathode and Q_2 base, the problem is in the picture tube. Monitor all of the picture tube voltages. Then try replacement of the picture tube. If the signal at the base of Q_2 remains unchanged but the picture tube cathode signal is intermittent, the problem is in the video output circuit.

6-9. AGC CIRCUITS

Most solid-state TV sets use a keyed, saturation-type AGC circuit. The RF tuner and IF stage transistors connected to the AGC line are forward-biased at all times. On strong signal, the AGC circuits increase the forward bias, driving the transistors into saturation, thus reducing gain. Under no-signal conditions, the forward bias remains fixed.

Although the AGC bias is a dc voltage, it is partially developed (or controlled) by bursts of IF signals. A portion of the IF signal is taken from the IF amplifiers and is pulsed, or keyed, at the horizontal sweep frequency rate of 15,750 MHz. The resultant keyed bursts of signal control the amount of dc voltage produced on the AGC line.

6-9.1   Basic Circuit Theory

Figure 6-19 is the schematic diagram of a typical solid-state AGC circuit. Transistor Q_1 is an IF amplifier, with its collector tuned to the IF center frequency of 42 MHz by transformer T_1. No dc voltage (as such) is supplied to the elements of Q_1. The keying pulses from the horizontal flyback transformer are applied to the collector through diode CR_1. This produces an average collector voltage of about 1 V. When Q_1 is keyed on, the bursts of IF signals pass through T_1 and are rectified by CR_2. A corresponding dc voltage is developed across C_4 and acts as a bias for AGC amplifier Q_2. Transistor Q_2 is connected as an emitter follower, with the AGC line being returned to the emitter. Variations in IF signal strength cause corresponding variations in Q_2 bias, Q_2 emitter voltage, and the AGC line voltage.

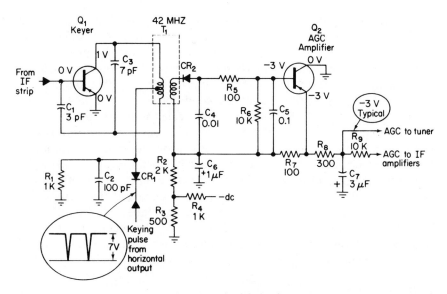

Fig. 6-19. Keyed AGC circuits

6-9.2   Recommended Troubleshooting Approach

If the AGC circuits are suspected of producing any problems (typically picture pulling, weak picture, brightness modulation, overloading, picture bending, picture washout, etc.), clamp the AGC line. That is, apply a fixed dc voltage to the AGC line equal to the normal AGC voltage. If the trouble symptom is removed with a fixed voltage applied, the trouble is likely to be in the AGC circuits.

The next step is to monitor the keying pulses to the AGC circuits and all of the voltages at the transistors. In the circuit of Fig. 6-19, the collector is the only element of Q_1 with a measurable voltage, and this voltage depends on the keying pulse.

A few solid-state AGC circuits are provided with an AGC control that sets the level of AGC action. This control is an internal screwdriver-type adjustment. Always try to correct AGC problems by adjustment before proceeding with troubleshooting when the AGC circuits are provided with a control.

6-9.3   Typical Troubles

In the following paragraphs we shall discuss symptoms that could be caused by defects in the AGC circuits.

No picture and no sound. When there is no picture or sound with a raster present, clamp the AGC line. If the picture and sound are restored, remove the clamp and measure the AGC line voltage. Check the keyed pulse. If the keying pulse is present but the AGC line voltage is abnormal (as it must be if the tuner and IF stages are completely cut off), look for shorted or open capacitors, defective diodes, breaks in printed-circuit wiring, cold solder joints, or similar conditions. If the keying pulse is absent, check the flyback transformer winding (Sec. 6-2).

Poor picture. When picture quality is poor (weak picture, overloaded picture, pulling, picture washout, or brightness modulation) and the AGC circuits are suspected, follow the same procedure as for a no-picture/no-sound symptom. That is, clamp the AGC line with the correct forward-bias voltage.

If the problem is cleared, indicating an AGC circuit defect, check the keying pulse and measure the AGC circuit voltages. This should pinpoint the circuit problem. A possible exception is open capacitors, which must be checked by substitution or by shunting with known good capacitors.

6-10. SOUND IF AND AUDIO CIRCUITS

The basic functions of sound IF and audio circuits in solid-state sets are the same as for vacuum-tube sets. The circuits amplify the 4.5-MHz sound carrier, demodulate the FM sound (remove the audio signals), and amplify the audio signals to a level suitable for reproduction on the loudspeaker. The audio portion of the circuits includes a volume control and usually a tone control.

6-10.1   Basic Circuit Theory

Figure 6-20 is the schematic diagram of typical sound IF and audio circuits. Two stages of IF amplification and two stages of audio amplification are used, with all stages in the common-emitter configuration. All stages are forward-biased. The final audio stage is push/pull and thus operates class B (with slight forward bias) or possibly class AB (with slightly increased forward bias). Unlike the video IF stages, the sound IF stages do not have any AGC circuits, and the forward bias is provided by fixed-resistance networks.

Both sound IF stages Q_1 and Q_2 are neutralized to prevent oscillation. Each stage is tuned at the output by corresponding transformers T_1 and T_2 to 4.5 MHz. The overall bandwidth of the sound IF amplifiers is typically 50 to 60 kHz in solid-state circuits.

There are no test points as such provided on most solid-state sound circuits. However, input from the video circuits is usually applied through a coaxial cable or shielded lead. This cable or lead can be disconnected, if it is required for testing, and can be used as a signal injection point for overall sound tests and alignment. The audio portion of the circuit can be tested by injecting an audio signal at the wiper arm of the volume control. The audio output voltage from the detector is typically less than 1 V.

Note that a ratio detector circuit is used. This is typical. However, some solid-state sets use a discriminator for the sound detector.

6-10.2   Recommended Troubleshooting Approach

The recommended approach for troubleshooting sound stages depends largely on the available test equipment. Ideally, an analyzer-type generator is used. As discussed in Chapter 2, these generators have a 4.5-MHz output that can be modulated (FM) by an audio tone (typically 1 kHz). This composite signal is injected to the sound IF input, and the resultant tone is monitored on the loudspeaker. The tone can be disabled so that the 4.5-MHz signal is used for alignment of the IF stages. Also, the audio tone can be injected (separately from the 4.5-MHz carrier) into the audio stages (usually at the volume control wiper arm). Thus, the audio stages can be checked separately from the IF stages.

If an analyzer-type generator is not available, the next recommended test setup is a sweep generator with markers and an oscilloscope. The sweep generator signal is injected at the sound IF input, and the signal is monitored at various points through the IF stages and at the detector output with an oscilloscope. A demodulator probe is required for the oscilloscope if individual IF stages are to be monitored. A basic low-capacitance probe can be used if the output of the sound detector is monitored. Keep in mind that signals through the sound IF stages of a solid-state TV are quite low (usually

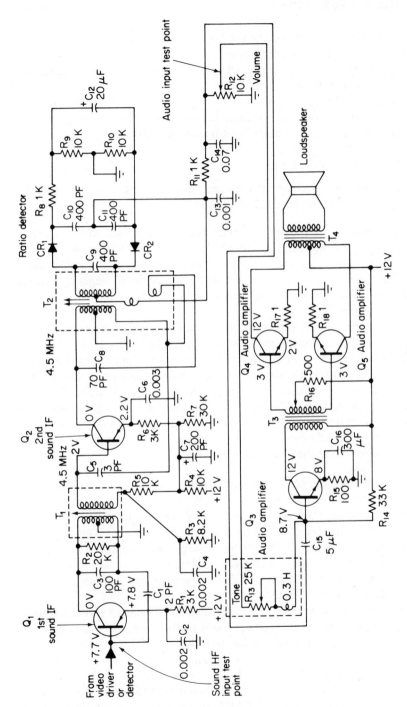

Fig. 6-20. Sound IF and audio circuits

on the order of 1 V). Thus, the oscilloscope must have sufficient vertical gain to display such signal amplitudes.

If analyzer and sweep generators are not available, use an RF generator to align the sound IF stages; then inject an audio generator signal to the audio stages (volume control wiper). If a problem such as distortion is definitely isolated to the audio section, a square-wave test can be made. The procedure is similar to that described for video amplifiers in Sec. 6-8.

6-10.3 Typical Troubles

In the following paragraphs we shall discuss symptoms that could be caused by defects in the sound IF and audio stages.

No sound. If the picture is normal but there is no sound first inject a 4.5-MHz signal at the sound IF input. Next, inject an audio signal (typically 1 kHz) at the audio amplifier input (volume control). If the audio signal passes but the IF signal is blocked, the problem is in the IF stages or possibly the detector. If the audio signal does not pass, the problem is likely to be in the audio section.

Next inject signals (audio or IF, as applicable) at the base of each transistor, and monitor the response on the loudspeaker. If this does not localize the problem immediately, measure the voltage at all transistor elements. This should isolate any problem serious enough to cause a no-sound symptom. A possible exception is an open bypass capacitor. If an emitter bypass capacitor should open, the stage gain (ac gain) will drop considerably, possibly without seriously affecting the transistor voltages. As usual, a suspected open capacitor can be checked by connecting a known good capacitor in parallel and reapplying power.

Poor or weak sound. The basic procedure for troubleshooting a poor sound symptom (weak sound, buzzing, distortion, etc.) is the same as for no sound. That is, the trouble must first be isolated to the IF stages, the audio stages, or the detector and then to a particular stage. This is done by signal injection. Next, the transistor voltages are measured. Generally, poor sound trouble is more difficult to isolate than a no-sound symptom. For example, the service literature rarely gives the gain per stage of either the sound IF stage or the audio stage.

As a rule of thumb, both the video detector (the input to the sound IF stages) and the audio detector (input to the audio stages) produce about 1 V in normal operation. This output can be as low as 0.5 V but is rarely greater than 2 V. In any event, if a 1-V audio signal is injected at the input of the audio stages (audio detector output or volume control) and the volume control is set to about midrange, there should be a loud tone heard on the loudspeaker. Likewise, a 1-V, 4.5-MHz signal (frequency-modulated by an audio tone) injected at the sound IF input should also produce a loud signal on the loudspeaker. With the modulated signal applied, it should be possible to measure about 1V at the audio detector output. Keep in mind that the ratio detector output is zero when the 4.5-MHz signal is not frequency-modulated.

If the trouble appears to be in the sound IF stages, try correcting the problem by alignment, as described in Chapter 3. The sound IF stages rarely go out of alignment during normal operation (except in the hands of a do-it-yourselfer). That is, circuit conditions usually do not change so much that the transformers must be drastically returned. However, aging of components may require periodic peaking of the alignment controls. When an IF transformer must be seriously retuned and it is confirmed that the stages have not been tampered with by unskilled hands, look for a defective component that is changing the resonant frequency of the transformer circuit. Open capacitors are likely suspects.

If the symptom is a weak signal and the IF stages are suspect, measure the gain of both stages (or all IF stages). Generally, most of the gain is produced by the last stage (the last stage Q_2 in a circuit similar to Fig. 6-20). Typically, the output from the second IF stage is about 3 to 4 V. If the gain of Q_2 is not at least 10 (usually higher), Q_2 is probably defective. Thus, if the base of Q_2 shows a 0.3- to 0.4-V signal, the output to the ratio detector should be 3 to 4 V.

If the sound is very low with a very weak background noise, the most likely suspects are leaking capacitors and/or leaking transistors. Of course, this condition should show up as abnormal voltage indications. Very weak sound can be caused by an open emitter (or base) bypass capacitor. This condition will reduce ac gain but leaves the dc voltages relatively unaffected.

If the sound is weak, with a loud buzz, the most likely suspect is the ratio detector. In effect, the ratio detector is not rejecting amplitude modulation. Since the sound signal may also contain some 60-Hz buzz from the vertical sync pulses (present at the video detector output), this signal may be amplitude-modulating the sound. Normally, the limiting action of Q_2 and the function of the ratio detector will prevent any audio from passing. If the ratio detector is defective, the most likely suspects are diodes CR_1 and CR_2 or stabilizing capacitor C_{12}.

As a general rule with audio circuits, if the signal is weak but the background noise is strong, look for an open condition in the signal path (open coupling capacitors, worn volume controls, etc.). If both the signal and background noise are weak, look for shorted or leaking components (capacitors, transistors, etc.).

When troubleshooting the audio section, *do not remove or disconnect the loudspeaker from the circuit.* If the final output stages of any solid-state audio circuit are operated without a load, the inverse voltage developed across the output transformer can cause the final transistors to break down. If the loudspeaker must be disconnected for any reason, connect a load resistance in its place across the transformer secondary winding. Use a resistance value equal to the loudspeaker impedance (typically, 3.2, 4, 8, or 16 Ω). The load resistance must be capable of handling the full output of the final stages (in watts). If a lower wattage resistor is used, it can burn out, removing the load and causing breakdown of the final transistors.

Index

A

B